AF691144

African Philosophy and Deep Ecology

This book investigates African philosophical contributions to the concept of deep ecology, which advocates rethinking human and non-human relationships within our ecosystems by promoting the inherent and earned worth of all beings.

With ecological crises impacting lives around the world, this book interrogates deep ecology thinking from African philosophical perspectives, highlighting the continent's important ontological, epistemological, ethical, aesthetic, and broad philosophical contributions. The book investigates issues such as the eco-phenomenology of human/non-human animal relations, Ubuntu and the environment, the superiorist fallacy, environmental belongingness, the impact of colonisation and modernity on non-human trauma, the politics of ecological narratives about African places, the question of moral status, African socialist perspectives, the question of degrowth, selective subordination, biodiversity loss, land ethics, the ontology of waste, and the concept of personhood in relation to global climate and ecological justice.

Providing a significant intervention in our understanding of the ecological crises and our duties toward ecosystems and the non-human other in the twenty-first century, this book is an important read for researchers, advocates, and other stakeholders working in the fields of environmental philosophy, climate change, indigenous studies, and African Studies.

Kenneth Uyi Abudu lectures at the Department of Philosophy, Ambrose Alli University, Ekpoma, Nigeria. He currently is a PhD student at the Department of Philosophy, University of Ibadan, Oyo State, Nigeria. His research interests include African Philosophy, with a keen interest in epistemology, disability, language, and the environment. He is the author of several essays including "Exploring Ignorance and Injustice in African Epistemology" (2023), "Africa and the Epistemic Normativity of Disability" (2024), and "An Afro-relational Account of the Non-human and Transhuman Entities" (2024).

Kevin Gary Behrens is Professor and Director of the Steve Biko Centre for Bioethics at the University of the Witwatersrand, South Africa. His research interests lie in the area of Applied Ethics, particularly in bioethics and environmental ethics. A major emphasis in his work is on applying African moral philosophical notions to ethical questions. He has published widely in international and national journals. He holds a rating of "Established Researcher" (C1) from the National Research Foundation.

Elvis Imafidon is Reader in African Philosophy and Director of the Centre for Global and Comparative Philosophy at SOAS University of London. He is also Research Associate at the African Centre for Epistemology and the Philosophy of Science (ACEPS). His research interests include African philosophy, philosophy of corporeality and disability, and philosophy of healthcare. He is the editor and author of several books and essays including *African Philosophy and the Otherness of Albinism: White Skin, Black Race* (2019), *Handbook of African Philosophy of Difference* (2020), and *Handbook of African Philosophy* (2023).

Routledge Studies in African Philosophy

Kimmerle's Intercultural Philosophy and Beyond
The Ongoing Quest for Epistemic Justice
Renate Schepen

African Epistemology
Essays on Being and Knowledge
Edited by Peter Aloysius Ikhane and Isaac E. Ukpokolo

African Ethics and Death
Moral Status and Human Dignity in Ubuntu Thinking
Motsamai Molefe and Elphus Muade

Intercultural Thinking in African Philosophy
A Critical Dialogue with Kant and Foucault
Marita Rainsborough

Endangered African Knowledges and the Challenge of Modernity
An Igbo Response
Donald Mark C. Ude

Ubuntu Ethics
Human Dignity, Moral Perfectionism, and Needs
Motsamai Molefe

Beyond Decolonial African Philosophy
Africanity, Afrotopia, and Transcolonial Perspectives
Edited by Joseph C. A. Agbakoba and Marita Rainsborough

Africa and the Metaphysical Empire
Frank Aragbonfoh Abumere

African Philosophy and Deep Ecology
Edited by Kenneth Uyi Abudu, Kevin Gary Behrens, and Elvis Imafidon

For more information about this series, please visit: https://www.routledge.com/Routledge-Studies-in-African-Philosophy/book-series/AFRPHIL

African Philosophy and Deep Ecology

Edited by Kenneth Uyi Abudu,
Kevin Gary Behrens, and Elvis Imafidon

Routledge
Taylor & Francis Group
LONDON AND NEW YORK

First published 2026
by Routledge
4 Park Square, Milton Park, Abingdon, Oxon OX14 4RN

and by Routledge
605 Third Avenue, New York, NY 10158

Routledge is an imprint of the Taylor & Francis Group, an informa business

© 2026 selection and editorial matter, Kenneth Uyi Abudu, Kevin Gary Behrens, and Elvis Imafidon; individual chapters, the contributors

The right of Kenneth Uyi Abudu, Kevin Gary Behrens, and Elvis Imafidon to be identified as the authors of the editorial material, and of the authors for their individual chapters, has been asserted in accordance with sections 77 and 78 of the Copyright, Designs and Patents Act 1988.

All rights reserved. No part of this book may be reprinted or reproduced or utilised in any form or by any electronic, mechanical, or other means, now known or hereafter invented, including photocopying and recording, or in any information storage or retrieval system, without permission in writing from the publishers.

Trademark notice: Product or corporate names may be trademarks or registered trademarks, and are used only for identification and explanation without intent to infringe.

British Library Cataloguing-in-Publication Data
A catalogue record for this book is available from the British Library

ISBN: 978-1-032-77023-9 (hbk)
ISBN: 978-1-032-77025-3 (pbk)
ISBN: 978-1-003-48091-4 (ebk)

DOI: 10.4324/9781003480914

Typeset in Times New Roman
by Deanta Global Publishing Services, Chennai, India

To Indigenous African communities, to whom we are indebted for the abundance of knowledge they reposit on how we could live better with the non-human.

Contents

Contributors

Kenneth Uyi Abudu lectures at the Department of Philosophy, Ambrose Alli University, Ekpoma, Nigeria. He currently is a PhD student at the Department of Philosophy, University of Ibadan, Oyo State, Nigeria. His research interests include African Philosophy, with a keen interest in epistemology, disability, language, and the environment. He is the author of several essays including "Exploring Ignorance and Injustice in African Epistemology" (2023), "Africa and the Epistemic Normativity of Disability" (2024), and "An Afro-relational Account of the Non-human and Transhuman Entities" (2024).

Kevin Gary Behrens is Professor and Director of the Steve Biko Centre for Bioethics at the University of the Witwatersrand, South Africa. His research interests lie in the area of applied ethics, particularly in bioethics and environmental ethics. A major emphasis of his work is on applying African moral philosophical notions to ethical questions. He has published widely in international and national journals. He holds a rating of "Established Researcher" (C1) from the National Research Foundation.

Mitchell Black is a PhD candidate in philosophy completing his thesis entitled *African Cosmic Complexity, Food Production, and the Situation of (Meta-) Polycrisis.* As a political agitator, conversationalist, and policy researcher, he explores the limits of Africa's contemporary political imagination, with a particular focus on envisioning alternative food system futures in Southern Africa. His work emphasises the importance of an agroecological food system as a foundation for the emergence of a beautiful, just, and flourishing future. His cosmocentric decoloniality seeks to deconstruct the Eurocentric character of the modern food system and develop strategies for its (re)organisation through the anamnestic (re)centering of African ways of knowing and being in relation to living complexity.

Ofuasia, Emmanuel (PhD) teaches in the Philosophy Department of Ladoke Akintola University of Technology (LAUTECH), Ogbomoso, Nigeria. He is also a research associate in University of Pretoria's Philosophy Department. His primary areas of research interest are: (Process) Metaphysics; African Philosophy (of Religion); Philosophical Logic, and Ifá. His publications in these areas have appeared as papers in journals and books in globally-renowned publishing firms like Springer, Brill, and Palgrave Macmillan. In addition to publishing, Emmanuel Ofuasia is also a recipient of local and international grants from organizations and institutions such as: Birmingham Centre for Philosophy of Religion in conjunction with the John Templeton Foundation (JTF), as well as Culture and Animal Foundation (CAF), New York, USA. In 2024, he published his first book – *Ìwà: The Process-Relational Dimension to African Metaphysics,* with Springer Publishing International. He is currently working on a book aimed at disclosing how panentheism serves as the most appropriate concept for talking about African traditional religions (ATRs).

Bjoern Freter is at Hagerstown Community College, Maryland, USA. He received his doctorate with his thesis "On Facticity and Existentiality" in 2014 from Free University, Berlin, Germany. Until 2023, he was Lecturer in World Philosophy at the School of Oriental and African Studies at the University of London, UK, and is now working as Visiting Assistant Professor at Gettysburg College, Pennsylvania, USA. He has published on a wide variety of topics, including pre-Socratic philosophy, Baroque and Classical German literature, as well as decolonization, white supremacy, and veganism. His current main research project aims at the desuperiorization of philosophy, i.e., developing a radically anti-oppressive moral philosophy, and at the Foundation of Superaltern Studies, a research area investigating the Western superiorist traditions and their self-representations.

Kai Horsthemke is Visiting Professor in the School of Education at the University of the Witwatersrand, South Africa, and at KU Eichstätt-Ingolstadt, Germany, and a fellow at the Oxford Centre for Animal Ethics, UK. He is the author of five monographs: *The Moral Status and Rights of Animals* (2010), *Animals and African Ethics* (2015), *Animal Rights Education* (2018), *Indigenous Knowledge: Philosophical and Educational Considerations* (2021), and *The Meaning of Death: A Philosophical Investigation* (2024). His edited book *Education, the Environment and Sustainability* was published in 2024.

Elvis Imafidon is Reader in African Philosophy and Director of the Centre for Global and Comparative Philosophy at SOAS University of London. He is also Research Associate at the African Centre for Epistemology and the Philosophy of Science (ACEPS). His research interests include African philosophy, philosophy of corporeality and disability, and philosophy of healthcare. He is the editor and author of several books and essays, including *African Philosophy and the Otherness of Albinism: White Skin, Black Race* (2019), *Handbook of African Philosophy of Difference* (2020), and *Handbook of African Philosophy* (2023).

Augustine E. Iyare lectures in the Department of Philosophy, Faculty of Arts, Ambrose Alli University, Ekpoma, Nigeria. He currently is a PhD student at the Department of Philosophy, University of Ibadan, Oyo State, Nigeria. His areas of interest include African philosophy and metaphysics. His recent publication is "Key Concerns in African Existentialism" (2023).

Bruce B. Janz is Professor of Humanities in the Department of Philosophy and co-director of the Centre for Humanities and Digital Research at the University of Central Florida, USA. His research interests include research on place and space, contemporary African philosophy, digital humanities, cultural philosophy, and contemporary European philosophy. His recent publication *includes African Philosophy and Enactivist Cognition: The Space of Thought* (Bloomsbury Academic, 2024).

Helen Kopnina is Assistant Professor of Sustainable Business at Newcastle Business School, Northumbria University, United Kingdom. Her research interests include four interrelated areas: sustainability, environment, education, and biological consequences. She is the author of over 200 peer-reviewed articles and (co) author and (co) editor of 17 books in these areas. Some of her publications include *Environmental Anthropology* (Routledge, 2015) and *Conservation: Integrating Social and Ecological Justice* (Springer, 2019), edited alongside Haydn G. Washington.

Olajide Akinleye-Martins, of School of Law, Northumbria University, Newcastle-Upon-Tyne, United Kingdom, is a Solicitor of the Senior Courts of England and Wales and is also qualified to practice as a Barrister and Solicitor of the Nigerian Supreme Court. She has practised extensively in commercial law in Nigeria and the United Kingdom and has worked as in-house counsel in construction and manufacturing organisations. In addition, she is a qualified Chartered Accountant, banking, and tax professional with requisite experience in financial audit and taxation. Olajide's research interests are in the areas of international investment arbitration and international environmental law. She obtained her LLB from the University of Ibadan, Nigeria, and her LLM from the University of Birmingham, United Kingdom. She is currently undertaking her PhD (Law) at Northumbria University, Newcastle, United Kingdom, with her research study on investment arbitral tribunals and environmental policies of host states within the African fossil fuel industry.

Dennis Masaka is a lecturer in the Department of Ethics, Philosophy, Religion, and Theology at Great Zimbabwe University. He is a Research Fellow in the Department of Philosophy and Classics at the University of the Free State, South Africa. He is currently a visiting fellow at the Uehiro Oxford Institute, part of the Africa Oxford Initiative (AfOx) Visiting Fellowship programme at the University of Oxford, working under the guidance of Dr Alberto Giubilini. Dennis' research focuses on epistemic (in)justice, communitarian ethics, and the socio-political dynamics surrounding land redistribution in Zimbabwe. He has published a number of journal articles, book chapters and an edited volume on these research themes.

Thaddeus Metz is Professor of Philosophy at the University of Pretoria, South Africa. Metz has published more than 300 scholarly books, chapters, and articles. Many of them take an analytic approach to the meaning of life, philosophy of religion, African morality, human rights, biomedical ethics, the proper function of a legal system, the role of a university, and a range of related topics in value theory and moral-political philosophy. Some of Prof Metz's recent books are *A Relational Moral Theory: African Ethics in and Beyond the Continent* (Oxford University Press, 2022) and *God, Soul and the Meaning of Life* (Cambridge University Press, 2019). Recent journal publications include articles in *Mind, The Monist, Royal Institute of Philosophy Supplement, International Journal for Philosophy of Religion, International Journal of Philosophy and Theology, Ethik in der Medizin*, and *Public Health Ethics* (all published in 2021).

Naziru Zakari Muhammad is Conservation biologist. He studied BSc (Zoology) at the University of Jos, Nigeria, and earned a PG Diploma in Wildlife Management at Mweka College, Tanzania. He is one of the wildlife officers in Nigeria trained to sustain the legacy of the Global Environment Facility Project Nigeria (2004–2009) on Protected Areas and Biodiversity Management. He presently works in Yankari Game Reserve, Bauchi, Nigeria, a premier wildlife destination in West Africa. His research interests include areas relevant to conservation, such as community conservation, studying ecosystems, trophic levels, and the impact on the biodiversity of an area due to factors such as climate, invasive species, and anthropogenic causes.

Juan Oeschger is a PhD candidate and Assistant Lecturer in the Philosophy Department of the University of Pretoria. Conducting his research under the guidance and supervision of Professor Jonathan O. Chimakonam, he focuses on Deep Ecology, African Relational Philosophy, and Philosophy of Education.

Abidemi Israel Ogunyomi is Lecturer at the Department of Philosophy and Religious Studies, Augustine University, Ilara-Epe, Lagos, Nigeria. He currently is a PhD student at the Department of Philosophy, University of Ibadan, Oyo State, Nigeria. His research interests include environmental philosophy, metaphysics, existentialism, and African philosophy. He has published in these areas. Some of his recent publications include: "Multiculturalism, Identity and Language: Some Critical Remarks on Molefi Asante's Idea of Afrocentrism" (2024) and "An Existential Interpretation of Evil: A Critique of Ebun Oduwole and Kazeem Fayemi on the Philosophical Problem of Evil in Yoruba Thought" (2024).

Angela Roothaan is Assistant Professor of Philosophy at the Faculty of Humanities at Vrije Universiteit Amsterdam, the Netherlands. Her research focuses on African intercultural philosophy, philosophy of values, and spirit ontologies. She has published books (in Dutch) on Spinoza, nature in ethics, truth, spirituality, and ghosts/spirits in modern culture. She is the author of *Indigenous, Modern and Postcolonial Relations to Nature: Negotiating the Environment* (Routledge, 2020). Together with Bolaji Bateye, Mahmoud Maseli, and Louise F. Muller, she edited the volume *Beauty in African Thought: Critical Perspectives on the Western Idea of Development* (Lexington Books, 2023).

Mike Russell is Senior Lecturer and Researcher at the Faculty of Business and Economics, Amsterdam University of Applied Sciences. His research interests explore the applications of emerging technologies and their implications for sustainability, transparency, trust, and philanthropy.

Aïda C. Terblanché-Greeff was previously appointed as an extraordinary researcher at the Social Transformation research unit and is currently Senior Lecturer at the School of Philosophy, Faculty of Humanities, at North-West University (South Africa). She holds degrees in various academic fields obtained from this institution. More specifically, she has a BA in Humanities, a BA (Hons) in Psychology, and an MA in Development and Management (specialising in disaster studies). She also has a PhD in Philosophy obtained from the University of Johannesburg (South Africa). In general, her research is interdisciplinary and empirically engaged. Her research interests include African philosophy, social philosophy, cross-cultural studies (viz., psychology and philosophy), and environmental ethics. She has published academic articles and book chapters with reputable national and international journals and publication houses.

Acknowledgements

I wish to express my deepest gratitude to the individuals whose invaluable support, both personal and professional, made the successful completion of this project possible. I am particularly thankful to my esteemed co-editors, Prof. Kevin Behrens of the Steve Biko Centre for Bioethics, University of Witwatersrand, South Africa, and Dr. Elvis Imafidon of the School of Oriental and African Studies (SOAS), University of London. Their inspiring words, expert insights, constructive feedback, and unwavering editorial commitment were instrumental in transforming this vision into reality. Despite their demanding schedules, they graciously agreed to serve as co-editors and contributed chapters of exceptional quality, for which I am deeply appreciative. This project would not have reached fruition without their dedication.

I am equally grateful to the contributors for their patience and steadfast determination, which ensured the successful completion of this book. My sincere thanks also extend to the editorial team at Routledge, including Helena Hurd and Katerina Lade, for their meticulous and tireless efforts in bringing this project to publication.

Finally, I wish to acknowledge the guidance and sustenance of the nonpareil being, whose presence illuminates my path and inspires my journey.

Kenneth Uyi Abudu, Ekpoma, Nigeria

Introduction

Kenneth Uyi Abudu, Kevin Behrens, and Elvis Imafidon

The human–non-human relationship takes a central place in contemporary environmental philosophy. This branch of philosophy examines our interactions with the natural environment, reevaluating our understanding of nature, its value, and its entitlements. Environmental philosophy also explores the metaphysical assumptions underlying our perceptions of nature, the epistemologies shaping our understanding of the natural world, and the aesthetic and moral values attributed to the environment in diverse cultural contexts (Mathews, 2014, p. 543). Since its inception in the 1970s, debates in environmental philosophy have evolved and become more intense largely due to a glaring deterioration in the relationship between the human and the non-human and the unavoidable consequences that have followed, such as the numerous environmental crises of global scale, climate change, biodiversity loss, air pollution, and oil spills. These crises have exposed the exploitative tendencies of humans toward the non-human world and underscored the dominance of anthropocentrism in environmental thought.

Anthropocentrism, often referred to as the conventional and classical worldview in environmental philosophy, assigns in its numerous forms intrinsic value primarily to humans while viewing the worth of non-humans as derivative of their utility to humanity (Horsthemke, 2015, p. 5). This human-centred perspective is characterised by two core theses: first, that humans are fundamentally distinct from the natural environment, and second, that nature exists primarily for human exploitation and control. Consequently, the natural world is often regarded as having primarily an instrumental value. Edwin Etieyibo identifies five key themes of this worldview: (1) Nature is hierarchically ordered, with humans at the apex; (2) There is an ontological divide between humans and non-human nature; (3) Nature operates as a deterministic system of inanimate particles; (4) Only humans are intrinsically valuable; and (5) The moral community is exclusive to human beings (Etieyibo, 2017, p. 149).Anthropocentrism as thought and practice is further fuelled by the prioritisation of science, technology, and materiality in our age, which has compounded the exploitative tendencies of humans towards the non-human in the natural environment and which has continued to fuel the culture of extinction across the world, a process made possible by the use of science and technology – an idea that is considered dominant in the Euro-American world. This argument is clearly emphasised by Miller G. Tyler that:

DOI: 10.4324/9781003480914-1

> The American attitude (and presumably that of most industrialized nations) towards nature can be expressed as eight basic beliefs [four of which are reproduced here]. 1. Humans are the source of all value. 2. Nature exists only for our use. 3. Our primary purpose is to produce and consume. Success is based on material wealth. 4. Production and consumption must rise endlessly because we have a right to an ever increasing material level of living (1979, p. 489).
>
> (Tyler, 1979, p. 489)

Arguing in line with Tyler's assertion , Bill Devall submits that science and technology are fused together in such a way that they create methods for manipulating natural phenomena, including weather manipulation. The goal of change, or "planned obsolescence," is self-serving and as such, the present is valued more than future generations, and the new is cherished more than the ancient. Thus, a greater level of material human life and the fulfilment of one's own desires are the objectives of human beings, as evidenced by the ownership of goods such as homes, cars, recreational vehicles, to mention a few (Devall, 1980, p. 301). True to Devall's submission, humans consider the non-human as a means to ends, with the resultant effect being some of the environmental crises mentioned earlier. Think, for example, about the plight of the Amazon Rainforest in Brazil, a rainforest considered to host more than fifty percent of the world's largest tropical forest.For the past decade humans' exploitative activities such as deforestation, burning of fossil fuel, overexploitation of resources, and urbanization, have been responsible for the decline in the Amazon forest, thereby leading to unwanted environmental consequences for not only the indigenous people living around this forest, but also humanity at large. Thus, these exploitative activities of humans and the continued decline of the forest are responsible for over seven percent of the global warming in the world today. The humans' exploitative activities in this Rainforest show clearly how over time, environmental degradation continues to gain momentum against the practice of conservation. Another example that comes to mind here is the Niger Delta region of Nigeria, where environmental degradation caused by the techno-industrial culture of exploiting crude oil has affected the indigenous people, thereby leading to environmental crises such as flood, oil spillage, and natural gas flaring. It is instructive to note here that these environmental problems are not peculiar to only the Amazon Rainforest and the Niger Delta region of Nigeria. Hence, these examples do not provide an exhaustive explanation for the environmental crisis being witnessed across the world.

The above thus explains the upsurge of interest in the philosophy of deep ecology, both in research and in practice, especially in the quest by philosophers to address the fractured human–non-human relationship. The term "Deep Ecology" was introduced by Naess (1973) in an article entitled "The shallow and deep, long-range movement: A summary." Here, Naess distinguished between the shallow ecology movement and the deep ecology movement. In the shallow ecology movement, the focus is primarily on the fight against pollution and resource depletion, with its central objective as the prioritization of the health and affluence of people in developed countries. For the deep ecology movement, there are two main defining characteristics: one is a rejection of the human-in-environment image in favour of the relational, total-field image. Here, organisms are seen to be in a web of relationships, and by extension, they are considered to have intrinsic value. The second is biospherical egalitarianism – in principle. This suggests that all organisms have equal rights to blossom, and the quality of a life is contingent upon the pleasure that is received from the close partnership with other forms of life (Naess, 1973). Thus, beyond the need to recognize the non-human and other forms of life as having intrinsic values, of particular interest to deep ecologists is to challenge the anthropocentric paradigm that is characterized by the culture of dominance through science and technology. Deep ecology is then seen as a normative worldview as well as a call to ecological consciousness. Thus, we might characterize deep ecology as a type of ecosophy (Naess, 1989; Guattari, 2000) or the quest for ecological wisdom that examines how philosophy understood here in the inclusive and global sense intersect with issues and debates in ecology, environment, related studies and policies, and ethical and epistemological concerns. It is instructive to point out here that for Arne Naess, deep ecology is a practical philosophy rather than a theoretical one aimed at providing meaningful policies, normative goals, and value principles that will assist us in confronting the environmental challenges that we face all over the world today (Naess, 1973). To further develop deep ecology as a philosophy, in 1984, George Sessions, an American philosopher, in collaboration with Arne Naess, developed what is today known as the deep ecology principles. These principles are also known as the practical guides to any form of philosophy that can be considered as focusing on deep ecology:

1. The well-being and flourishing of human and non-human life on Earth have a value in themselves. These values are independent of the usefulness of the non-human world for human purposes.
2. The richness and diversity of life forms contribute to the realization of these values and are also values in themselves.
3. Humans have no right to reduce this richness and diversity except to satisfy vital needs.
4. The flourishing of human life and cultures is compatible with a substantial decrease of the human population. The flourishing of non-human life requires such a decrease.
5. Present human interference with the non-human world is excessive, and the situation is rapidly worsening.
6. Policies must therefore be changed. These policies affect basic economic, technological, land ideological structures; the resulting state of affairs will be deeply different from the present.
7. The ideological change is mainly that of appreciating life quality (dwelling in situations of inherent value) rather than adhering to an increasing higher standard of living. There will be a profound difference between big and great.

Each of these statements calls for philosophical scrutiny and there have been many such scrutiny that chapters in this book would in one way or the other lend their voice to. But they collectively provide a sufficient basis to enter into conversations about a deep ecology philosophy. What this echo is that deep ecology strongly condemns and rejects anthropocentrism and urges its replacement with forms of ecocentrism; identification with all forms of life; the need to care for the environment; the rejection of instrumental rationality in our relationship with the non-human and an embrace of intrinsic relationality (Katz *et al.*, 2000). But these principles and basic assumptions of deep ecology will be paradoxical if they fail to draw the needed ecosophy or ecological wisdom and philosophy from all forms of life. It would be a performative contradiction if the philosophical discourse of the ecosystem relies heavily on Western-centric thought or philosophies from the Global North without actively exploring Global Southern or non-Western perspectives and how traditions and peoples in non-Western places theorised and practised deep ecology and continue to do so. It is therefore the African voice on deep ecological matters in terms of theory and practice that this book aims to explore, for the central assumptions and arguments of deep ecology might be conceptually novel but are not in thought and practice , especially when examined from non-Western intellectual traditions such as the robust and rich traditions of Africa and Asia. In these non-Western intellectual traditions, there have always been thoughts and practices on how to commune with the environment – ways that show reciprocity, relationality, and solidarity in addressing the parasitic relationship of the human towards the non-human. This book therefore sheds light on the thoughts and practices of an underexplored tradition in the deep ecology discourse–the African philosophical perspectives on deep ecology.

African philosophy in academic settings has made significant progress and strides in its contribution to global philosophy in the last half a century or so. The began with a distracting inquiry into its existential status resulting in the early stages in its denial by not only Eurocentric philosophers such as Lucien Levy-Bruhl but also philosophers of African traditions who were trained in the Western paradigm of philosophy, such as Kwasi Wiredu, Kwame Anthony Appiah, Peter Bodurin, Pauline Hountondji, among others. But the question of the existence or non-existence of African philosophy has quickly been replaced with doing African philosophy in actuality and theorising how African philosophy answers some of the most deepest questions in philosophical discourses and how it can be applied to some of the most fundamental issues that confront humans and non-humans today.

In the discussion of African philosophy, there is a preoccupation with Africa broadly construed. The term "Africa" in relation to African philosophy is used to denote several interrelated intentions. Firstly, the geography of reason is brought to bear. This seems limited, though, as it leaves out a lot of linkages when considering African thought in relation, for example, to diasporic interpretations and experiences. In the second sense therefore, it would denote more broadly experiences, ideas, and revelations that may be deeply rooted in the African locale but go beyond it. Thus, Africans and members of the African Diaspora are always in one way or the other in focus in the discourse of African philosophy. This would include shared ontologies, epistemologies, ethics and aesthetics, memories, experiences, and meaning making that emerges from African and African diasporic places. African philosophy is therefore, the African formulative and critical responses to human philosophical questions such as the nature of reality, being, knowledge, and morality, the essence of beauty and of existence, and the nature of the human mind and of personal identity, responses that are embedded in the interconnected indigenous and postcolonial repositories of knowledge in African and African Diasporic spaces (Ukpokolo, 2023, p. 5). Segun Gbadegesin (1991) provides four accounts of African philosophy that captures this:

1. For one group, African philosophy is the philosophical thought of Africans as could be examined from their various worldviews, myths, proverbs, etc. In this sense, it is the philosophy indigenous to Africans, and untainted by foreign ideas.
2. The second group understands African philosophy as the philosophical reflection on, and analysis of, African conceptual systems and social realities as undertaken by contemporary professional philosophers. This view however limits the meaning of African philosophy to the contemporary work of professional African philosophers.
3. The third group understands African philosophy as the combination of these two approaches, without suppressing or looking down on any. This would involve sifting philosophical thoughts of Africans extracted from their various worldviews, myths, proverbs, etc., and reflecting on them by professionally trained African philosophers.
4. The fourth group argues that African philosophy is not any of the above; however, its proponents (often attributed to the Beninese philosopher Paulin J. Hountondji regard African philosophy as any collection of texts produced by Africans and specifically described by their authors as Philosophy. This obviously in many ways resembles (2) above and again provides a very narrow conception of African philosophy that ignores a robust heritage of non-textual philosophical thought.

The third of these four accounts brings us closer to how we think about African philosophy in the pages of this book as the philosophical systems of indigenous African peoples (either in African or African-diasporic locales), existing in textual or non-textual forms that consist of their ontology, epistemology, ethics, aesthetics, political philosophy and other philosophical thoughts as well as the critical reflection on and interrogation of such in the many forms in which it happens. Many of the chapters in this book draw heavily from such philosophical thought in conceptualising the ecological wisdom that emerges from African heritages. Such ecological wisdom which or ecosophy include African communitarianism, African theories of personhood and identity, African relational ontology and ethics, and African philosophies of interconnectedness such as Ubuntu and Ukama. Thus, this book aims to bring together novel ideas on a discourse in African philosophy that is yet to be substantially explored in contemporary literary discourse of deep ecology.

The essays in this book show that the rich and robust African philosophical discourses on the ontology and normativity of community, personhood, duties and rights, Ubuntu and relationality, being, beauty, spirituality, relational environmentalism, moral status, and other related discourses can contribute to the deep ecology discourse, These rich and interesting African philosophical perspectives are still very much underexplored in relation to global discourses in deep ecology. It is the goal of this volume to fill this gap in literature. Thus, far from focusing solely on the experience of ecological crises in Africa often found in the discourse of ecology in relation to Africa, the exploration of what African philosophical theories, concepts and ideas about being and deep ecology could bring to the table in understanding these and global ecological crises and addressing the shifting environmental demands is addressed in this book. Chapters within the book, therefore, analyses from African philosophical perspectives themes and issues such as the ontology of waste, eco-phenomenology and African lived-ecology of the non-human animals' relations, Ubuntu and the environment, the superiorist fallacy in ecological discourses, deep ecology and foreign policy initiative in Africa, the impact of colonization and modernity on collective non-human trauma, the politics of ecological narrative about African places, and the questions and debates on moral status, degrowth, selective subordination, biodiversity loss and land ethics. .

The book consists of fifteen chapters each of which examines several related aspects of African philosophical frameworks including ontologies, epistemologies, and ethics in the theorisation of deep ecology and how they speak to specific themes in the discourse of deep ecology. In the opening chapter, "From Intrinsic Value to Ontological Reality: The Philosophy of Life Strength or Vital Force," Angela Roothan discusses Bantu relational ontology as a fundamental foundation for deep ecology. This ontology challenges anthropocentrism, affirms the intrinsic value of non-human entities, and advocates for environmental activism aimed at preserving the ecosphere for purposes beyond human utility. Framing her argument as a response to Arne Naess's deep ecology, Roothan seeks to enrich Naess's ecosophy through insights drawn from Placide Tempels's seminal work, Bantu Philosophy. Roothan further engages with Patrick Monga-Kasimb's recent critique of the hierarchical dimensions of Bantu relational ontology, arguing instead that its dynamic, non-hierarchical structure subverts traditional anthropocentric and theocentric frameworks. By decentring both God and humans from their presumed positions of supremacy in the moral and ecological order, she positions Bantu ontology as a robust conceptual foundation for addressing ecological challenges – an area deeply resonant with the concerns of deep ecology. In Chapter 2, "The Environment in Yoruba Collaborative Ontology," Abidemi Israel Ogunyomi offers a perspective on deep ecology rooted in Yoruba indigenous collaborative ontology. Ogunyomi argues that within Yoruba thought, the environment comprises an intricate web of relationships encompassing both human and non-human entities. Yoruba collaborative ontology emphasises the cooperation, interdependence and interconnectedness of all entities within the environment. Central to this discourse are Akinsola Akiwowo's socio-anthropological theories of Alájọgbé (co-tenants/cohabitants) and Alájọbí (those bound within the same web of birth).

These theories articulate the ideal relationships between human and non-human entities, underscoring the mutual responsibilities within this interconnected web. Ogunyomi further illustrates these ideas by analysing verses from the Ifa corpus, which encapsulate Yoruba indigenous understandings of ecological balance and their relevance to deep ecology.

Chapters 3 and 4 explore the African philosophical concept of Ubuntu and its connections to deep ecology, particularly its implications for human and non-human relationship. In Chapter 3, titled "Some Critical Thoughts on Mogobe Ramose's Philosophy of Ubu-ntu and a Proposal for its Desuperiorisation," Bjorn Freter critically examines Mogobe Ramose's interpretation of Ubuntu. Freter focuses on the ontic reading of the maxim, *Umu-ntungumuntungabantu* ("A human being is a human being through other human beings"), which he argues reserves the notion of "human being" exclusively for humans, thereby marginalizing non-human entities. Freter contends that this anthropocentric interpretation fosters speciesism, rooted in a "superiorist fallacy," wherein the value or moral worth of non-human entities is measured against that of humans in a subordinate manner. Such a perspective, he asserts, perpetuates anthropocentrism with long-term implications for human-non-human relationships. To address this limitation, Freter advocates for a recalibration of the Ubuntu maxim to include non-human entities. He proposes the isiXhosa phrasing: *umntungumntungabantukwaye abo bangengobantu (umzekelo: invelo, umhlaba, konkeokunempilo*, which translates to "A human being is a human being through other human beings and through those who are not human beings (for instance, nature, the earth, and all that has life)." This recalibration broadens the scope of Ubuntu to recognize the intrinsic worth of non-human nature, aligning it with the principles of deep ecology. In Chapter 4, titled "Interrogating Deep Ecology within the Framework of Ubuntu Philosophy," Dennis Masaka examines Ubuntu's potential to foster recognition of the intrinsic worth of the environment. Masaka critiques the notion that Ubuntu's normative stance is based solely on ontological properties. Instead, Ubuntu affirms the ontological and normative necessity of all entities in the environment. Masaka introduces "qualified deep ecology," a variant grounded in Ubuntu, which asserts the environment's intrinsic rather than instrumental value. This perspective directly challenges the anthropocentric worldview that has long dominated human–non-human relationships. In Chapter 5 titled "An African Theory of Moral Status: A Relational Alternative to Individualism and Holism," Thaddeus Metz presents a relational theory of moral status common in sub-Saharan African communities. Contrasting it with Western theories such as individualism and utilitarianism, Metz's "modal relationalism" posits that a being's moral status correlates with its relational or communal capacity. This framework extends moral consideration to humans, animals, plants, and other natural entities. Metz argues that this African relational approach enriches deep ecology by promoting the intrinsic worth of non-human entities, offering a stark contrast to the instrumental valuation found in shallow ecology. Kai Horsthemke, in Chapter 6 titled "African Ecological Ethics and Moral Status of Non-human Nature," explores African ecological ethics as an underexplored aspect of environmental philosophy. He critically examines the moral standing of non-human nature within African philosophical traditions, often critiqued for shallow environmentalism. Horsthemke explores how these traditions can contribute to a more egalitarian environmental ethic. By analysing emergent African ecological ethics, he highlights its potential as a foundation for a profound, less anthropocentric human and non-human relationship and ecological stability.

Chapter 7, "African Relational Environmentalism and Moral Considerability" by Kevin Behrens, critiques the assumption that African worldviews are inherently anthropocentric. Behrens argues that moral considerability in African thought is based on relationality – the interconnectedness of humans, non-human entities, and nature. Behrens asserts that this relationality challenges the instrumental valuation of nature, prevalent in Western thought, by emphasizing that all entities possess life force. He acknowledges criticisms regarding the inability of non-human entities to pursue their interests but affirms that the African relational approach still aligns with deep ecology's biocentric goals. In Chapter 8, "Deep Ecology, Ontic Relationality and Positionality: Exploring the Ontology of Anthropogenic Waste," Elvis Imafidon examines the ontological roots of the global anthropogenic waste crisis. He attributes this to the exploitation and objectification of non-human entities, leading to ontological disequilibrium. Imafidon proposes two features of sub-Sahara African ontologies, being-in-place and being-together, as important to counter human exploitative practices that produce waste. These frameworks highlight moral duties toward non-human entities and acknowledge their ontological significance, offering a pathway to re-actualising ecological balance and challenging the instrumentalization of non-human beings. Bruce Janz chapter titled "Peripherality, Non-philosophy and Ecology in African Philosophy" (Chapter 9), underscores the growing corpus of African ecological philosophy. Janz draws parallels between deep ecology and African thought systems such as Ubuntu, advocating for a non-human-centric ethic. He suggests a shift in ethical paradigms, proposing that deep ecology upends traditional ethics by embedding itself in our ontological understanding of reality. Janz emphasizes the urgency of rethinking human engagement with nature beyond mere ethical decision-making.

In Chapter 10, "Selective Subordination for Harmony and Holistic Balance from an African Perspective," Emmanuel Ofuasia critiques egalitarianism and its limitations in addressing the human–non-human relationship. Drawing from Yoruba Ifa corpus, Ofuasia introduces "selective subordination for harmony," a framework grounded in natural hierarchy and spiritual balance. He explores the role of animal sacrifices in maintaining ecological and spiritual harmony, demonstrating how this practice underscores the interconnectedness of the physical and metaphysical environments. In Chapter 11 titled "Biodiversity Conservation in Nigeria: Contrasting an Anthropocentric and a Deep Ecology Perspective," by Olajide Akinleye-Martins, Naziru Zakari Muhammad, Helen Kopnina, and Mike Russell, investigates biodiversity conservation in Nigerian indigenous communities. They highlight the detrimental effects of anthropocentrism on biodiversity and explore intrinsic motivations rooted in indigenous ontologies. Focusing on the Yankari Game Reserve, the authors contrast the economic motivations tied to Western capitalist progress with the intrinsic values emphasized by local conservation efforts, advocating for a paradigm shift towards deep ecological principles. Kenneth Uyi Abudu, examines in Chapter 12 titled "Ecophenomenology and Deep Ecology: Toward an African Lived-ecology of the Human–Non-human Relationship," the convergence of ecophenomenology and deep ecology. He highlights African conservation practices such as totems, taboos, and sacred forests, emphasizing experiential and phenomenological relationships with the non-human. Abudu's concept of African lived-ecology rejects ontological dichotomies between humans and non-humans, advocating for symbiotic conservation practices to promote the flourishing of all entities.

In Chapter 13 titled "Deep Ecology, Irreducibility, and African Relational Thinking," Mitchell-Ron Black and Juan Oeschger explore the intersections of African Relational Thinking (ART), deep ecology, and Critical Complexity Theory (CCT). They critique neo-colonial ecological governance in Africa, focusing on coal extraction in South Africa. By employing deep ecology, they propose solutions to mediate tensions between local mining proponents and climate justice movements, advocating for ecological frameworks that resist neo-colonial dominance and foster relational ecological governance. Chapter 14 titled "Integrating Deep Ecology, Degrowth, and Ubuntu: Foundations for an Ecosophy" by Aïda C. Terblanché-Greeff (Botha), introduces "Ubuntu Degrowth," as an ecological wisdom that integrates deep ecology, degrowth, and Ubuntu. Botha critiques anthropocentrism and capitalism, proposing a harmonious relationship between humans and nature based on intrinsic value and ecological stability. This synthesis offers a robust philosophical foundation for addressing ecological challenges within African and global contexts.. The final chapter titled "The Non-Human in the Context of Colonisation, Modernity and Trauma" by Augustine E. Iyare, sharply zooms the focal lens of deep ecological discourses in Africa to the non-human by exploring the anthropocentric legacy of colonisation and modernity, particularly their exploitative impact on non-human entities. Iyare examines biospherical trauma in regions like Nigeria's Niger Delta and South Africa's platinum mines. He advocates for deep ecology as a framework to foster equitable and sustainable resource utilization, emphasizing the importance of intergenerational justice and relational ethics in addressing ecological exploitation.

We hope that the chapters in this volume will actively encourage the reader to reevaluate and give serious consideration to the fractured human and non-human relationships, both in the African context and across the globe and ignite further conversations that take seriously global perspectives.. This is especially important as humanity faces a plethora of bold, difficult, and complex environmental crises that are fundamentally emerging from how humans relate with the non-human. The exploration of deep ecology from the African perspective provided here does not pretend to provide an exhaustive exploration of both the theoretical and practical frameworks and issues that could emerge from such a robust and multifaceted tradition of human thought. There is still much left to be explored, and the essays in this book provide a rich and solid position to begin from.

References

Devall, B. The Deep Ecology Movement. *Natural Resources Journal*, 20, 1980: 300–322.

Etieyibo, E. Anthropocentrism, African Metaphysical Worldview, and Animal Practices: A Reply to Kai Horsthemke. *Journal of Animal Ethics*, 7(2), 2017: 145–162.

Gbadegesin, S. *African Philosophy: Traditional Yoruba Philosophy and Contemporary Realities*. New York: Peter Lang, 1991.

Guattari, F. *The Three Ecologies*.London; and New Brunswick: The Athlone Press, 2000.

Horsthemke, K. *Animals and African Ethics*. Hampshire: Palgrave Macmillan, 2015.

Kratz, E., Light, Andrew, and Rothengerg, D. Introduction: Deep Ecology as Philosophy. In E. Kratz, A. Light, and D. Rothenberg (Eds.), *Beneath the Surface: Critical Essays in the Philosophy of Deep Ecology*. Cambridge, MA and London, England: The MIT Press, 2000: x–xxiv.

Mathews, F. Environmental Philosophy. In O. Graham and N.N. Trakakis (Eds.) *History of Philosophy in Australia and New Zealand*. Dordrecht Heidelberg; New York; and London: Springer,2014: 543 - 592.

Næss, A. *Ecology, Community and Lifestyle: Outline of an Ecosophy*. Cambridge: Cambridge University Press, 1989.

Naess, A. The Shallow and the Deep, Long-Range Ecology Movement: A Summary. *Inquiry*, 6, 1973: 95–100.

Tyler, G.M. *Living in the Environment*, 3rd edition. Belmont, CA: Wadsworth, 1979 .

Ukpokolo, I.E. The Meaning of African Philosophy. In E. Imafidon, M. Tshivase, and B. Freter (Eds.), *Handbook of African Philosophy*. Switzerland: Springer Nature, 2023: 3–10.

1 From Intrinsic Value to Ontological Reality – the Philosophy of 'Life Energy' or 'Vital Force' in *Bantu Philosophy*

Angela Roothaan

Deep Ecology and Bantu Philosophy

Deep Ecology has added significantly to our normative thought on human-nature relations by centering the intrinsic value of more-than-human beings. One of the founding publications establishing deep ecology was the 1973 paper by Arne Naess, titled 'The shallow and the deep, long-range ecology movement: a summary' (Naess 1973). With his paper, Naess was in tune with the early 1970s concern about the effects of a growth economy, the steep increase of human populations worldwide, and the decline of the natural environment under the pressures of both. In the article, Naess proposes to decenter the human species when thinking about ecology and critiques the injustice of unequal wealth, which tends to make ecology only about conserving an ecosphere that is livable for the rich part of humanity. He argues for an ecosophy, which to him is a philosophy of ecology that is not primarily focused on understanding and explaining but that is normative and prescribes what we ought to do to return from an anthropocentric and detached form of philosophical ethics that benefits the rich only – and replace it with an activism concerned with normative theory based on the value of all beings in their interconnectedness.

The work of Naess reflects a general orientation in the deep ecology movement, which was always characterised by a close connection between activism and scholarly work (see Drengson et al. 2011). Deep ecology has a holistic outlook and tries to refocus (after an era characterised by a mostly anthropocentric outlook in philosophy, science, and society) on the deep connectedness of the human being with all other living beings and the earth as the host to all. It opened ethicists' minds to the idea that nature is not only valuable in relation to us, the human species, but that all beings in nature have intrinsic value and therefore are deserving of conditions to live and flourish. Deep ecology is thus relational in a double sense – it centres the relatedness of all humans as having an equal right to flourish (countering a classist division of sources for wellbeing), and it centres the relatedness of all natural beings as the basis of ethics (instead of the relatedness only of those bound by a social contract). The key argument of deep ecologists for a post-anthropocentric ethic is that all natural beings have intrinsic value. Even though value can be understood as having an ontological status,[1] in contemporary value theory, this status is understood in relation to human agency only, and not in the

DOI: 10.4324/9781003480914-2

sense of a more-than-human ontology. Charles Taylor, for example, investigates what he calls 'moral ontology', stating that values ('hypergoods') have ontological status only insofar as they determine human actions towards the good.[2] Similarly, when deep ecologists extend the realm of value beyond the human and include more-than-human life, this inclusion concerns the (good) human will and agency. The core issue for deep ecology thus remains bound to human morality: how can we, humans, change our orientation toward the good in such a way that it includes life beyond the human as intrinsically valuable.

In what follows, I will argue that African philosophy, insofar as it is based upon African wisdom traditions structuring the world that humans live in, moves beyond the ethical viewpoint described above. The wisdom that this philosophy aims to articulate involves an ontological understanding beyond what pertains only to (human) morality. When the more-than-human is included in understanding what is, and what is good, humans may experience orientations that transcend what is good according to us humans. To make this concrete: the practice of making sacrifices to non-human actors is not understandable from a purely human ethics of responsibility, but can be made understandable from a more-than-human ontology. Such African 'gnosis', as Valentin Mudimbe has called it,[3] includes the material and the spiritual being of the more-than-human as a basis for moral reflection, and thus is wider or deeper than the ethic that remains within the boundaries of the human will and human agency. Instead, it envelopes all normative thought in a comprehensive understanding of more-than-human reality as relational.

Inspired by the reflection above, I will explore in this chapter an alternative grounding for deep ecology on the basis of an African relational ontology. To this effect, I will focus on the work that first presented such an ontology to a European philosophical readership, the 1945/1946 Bantu Philosophy written by the Belgian missionary Placide Tempels. Struggling with the Aristotelian terminology of essence and existence familiar to him through his priestly studies, Tempels aimed to introduce European readers to a philosophy expressed in Bantu languages, especially Kiluba.[4]Bantu Philosophy, I will argue, has not been well understood as a critical ontology of life energy or life force, due mostly to the problematic translations in existence.[5]Even though Bantu Philosophy has been severely criticised for being a settler colonial manual (Serequeberhan 2012 [1994]),[6] just a type of *ethno-philosophy* (Hountondji 1996 [1976]), having led to the 'bantustanization' of African Philosophy (Matolino 2022) and being a failed Western appropriation of African indigenous philosophy (Mungwini 2022), Bantu Philosophy was also taken up by many authors as an important source to be critically re-read (cf. e.g., Bujo 2016; Diagne 2013; Katchekpele 2019). For our purpose here, it is important to note that the concept of 'vital force' coined by Tempels has been positively taken up and discussed in many studies in African Environmental Philosophy (cf. e.g., Eze 2017; Ibanga 2023; Mosima & Shang 2023). With this chapter, I want to contribute to the further elaboration of such work by providing a critical reading of Bantu Philosophy *based* on tarchival and textual study.

In the following, I will show how a careful translation of the Dutch original text of Bantu Philosophy, paired with a thorough study of the archival material on the work, its author, and his correspondents, can bring a deeper understanding of the work. My focus will be here on its critical appreciation of the indigenous 'Bantu' relational understanding of nature and of humanity's place in it. In it may lie a thorough basis for an ontological ecological philosophy or ecosophy that moves beyond understanding normative thinking as the moral responsibility of humans for ecology, and develops into a more-than-human relational ontology. To gain a better articulation of the indigenous understanding of more-than-human nature based on the basis of the research into Bantu Philosophy I will do three things. I will (1) clarify the structure of the book, and repair some of what was lost in the existing translations, (2) dig into the ontological debates that the book deals with and shows a way out of, and (3) further clarify the key concept of life energy, or as it traditionally has been translated, 'vital force' as the core of an African relational ontology. Through these three steps, I will show the intricate 'web'-like structure of reality that is implied in the concept of life energy, and argue from there for a new look at an indigenous ontological understanding of nature as a basis for actively practising ecological being in relation.

The Critical Intent and Structure of Bantu Philosophy

To understand the argumentative structure and philosophical approach of Bantu Philosophy, we have to restore the work to its original form as it appears in the Dutch edition of 1946. I will argue that, contrary to common characterizations of the work as a form of ethnophilosophy (Hountondji 1996 [1976], chapter 1), in this original version it demonstrates an explicit critical philosophical purpose, and not just a descriptive, ethnological one. To do so, I will discuss in this section some structural changes made in the current English translation by Colin King (1959), and by repairing them, I will prepare the ground for a more appropriate discussion of the relational ontology proposed in the book. Before we can discuss the structure of the book as it was designed, a few words must be said of its complex publication history and Tempels' intentions in writing it and about its author.

Placide Tempels arrived in the Belgian-dominated Congo in 1933 as a young missionary without any previous experience with life in Africa. At that time, the country that is now the Democratic Republic of Congo had been (since 1908) under the governance of the Belgium government.[7] For about ten years, Tempels tried to carry out missionary work as it was assigned to him, mostly travelling on foot or on motorbike to rural villages, where Christian conversion and integration of the Congolese into a European-style schooling system was to be attained. During that period, however, Tempels began to doubt the mono-directional conversion method that focused on deconversion from original African religion and replacing its beliefs and rituals with Christian ones. Upon learning the indigenous languages and asking people about their traditional beliefs, he became convinced that the religious replacement strategy was wrong and that another approach had to be designed.[8] Missionaries, Tempels thought, should try to understand the structure and core of indigenous thought systems and practices to see where they could connect to a Christian core that , in turn, should be freed from overly European cultural additions. To work towards such an Africanized Christianity, he started to work on a 'Bantu Catechism' to help the priesthood in the Congo work *with* their parishioners, instead of transforming them into Europeanized Christians.

In that phase of his life, he met the Belgian magistrate Émile Possoz, who had come to similar conclusions as Tempels in his juridical work in the Congo. Possoz studied (and published on) indigenous law and critiqued the colonial administration for halting the true progress of Congolese society by not working with what was good in the existing indigenous legal institutions, but by trying to thoughtlessly replace them with a European system.[9] Possoz convinced Tempels that he should write a Bantu philosophy first, to give grounding to his missionary writings, and the rest is history. From that moment, the two men remained correspondents for the rest of their lives, which has provided posterity with the most helpful archive on the publication and reception history of Bantu Philosophy. After making a first brief concept of the work, Tempels started to publish its chapters, in Dutch, in two Congolese colonial research journals.[10] Possoz (who was bilingual) started to translate them into French, and the work was offered to *Lovania*, the publishing house that was connected to the newly founded Congolese branch of Leuven University in Belgium. Lovania's editors, however, doubted the quality of the Possoz translation, for being complex and too radical. Another connection of Tempels, Father Rubbens, was asked to make a more palatable translation, for which Possoz could only write a foreword (cf. Smet 2001). This became the first book publication of Bantu Philosophy, published in 1945.

Tempels nor Possoz were happy, however, with Rubbens' work, and at the instigation of Possoz, Tempels created a Dutch version of the book, using and adapting the original articles, and correcting and improving the changes and mistranslations made by Rubben. Under Rubbens' hands, the Bantu appeared to be inferior to the white colonials, contrary to Tempels' original intentions.[11] The new Dutch version, published in Antwerp in 1946, has the words 'original text' on its front page and is therefore taken by me to be the authoritative version of the book. When the influential Paris-based publication house Présence Africaine wanted to republish the work in French, Tempels tried to correct the Lovania version again to fit his new Dutch version, but many of his corrections were ignored in the final print that appeared in 1948.[12] The publication history became a frustrating adventure for its author, who also suffered from opposition by influential persons in the colonial branch of the Catholic Church in the Congo, as well as restricting measures on his daily life brought about by his own Franciscan order.[13] When finally in 1959 the first English translation appeared, after a not-so-well-known German one in 1956, translation issues persisted, possibly even worsened, as its translator, pastor Colin King, adapted the text freely, consistently erasing the critical philosophical structure of the work.

This structure is very clear in the Dutch original: from the seven chapters, the middle five form the core of the book. The first and the last chapter are meant to clarify to its potential readership of colonial intellectuals why they should read the book, and what are its most important contributions. These chapters address primarily non-African readers and start out from the presuppositions and prejudices of those readers. They convey the importance of Bantu Philosophy for missionaries, ethnological researchers, and colonial administrators, expressing hopes that they do their work more in dialogue with Africans and from an understanding of, and respect for, the indigenous traditions of thought and life practices. The remaining five chapters can again be divided into two groups. The first group, consisting of Chapters 2 and 3, contains the foundational or theoretical part of the philosophical work – proposing a Bantu ontology and epistemology, as a hypothesis to be strengthened or corrected by future findings. The second group, Chapters 4–6, contains the empirical part of the work – the investigation of Bantu psychology, ethics, and a chapter called 'restoration of life', describing restorative legal, social, and spiritual practices in case of transgressions of societal, juridical, and moral/spiritual norms.

The relation between the two types of chapters, the theoretical and the empirical ones, is twofold. The theoretical part is meant to explain the very norms and institutions described in the empirical part. If it does its work well, it makes sense of those practices, helping to understand them as rational and logically consistent in their relatedness. The empirical part is meant to corroborate the hypothetical theoretical structure – to show whether the Bantu ontology and epistemology, as laid out, make sense and if they really touch the deep structure of everyday practical norms and institutions in Bantu societies. With this design, Bantu Philosophy explicitly was meant to have a critical philosophical function: the work should distinguish between practices and institutions that are right or wrong, seen from the core structure of Bantu ontology and epistemology. It should show when such practices deviate from the deep structure of Bantu philosophy by understanding it incorrectly or not following the ontological laws inherent in it. This aspect makes Bantu Philosophy a critical ontological philosophy.

That the structure of the book is rarely discussed in the literature on Bantu Philosophy is mostly due to the mentioned disfigurements in the translations. Very clear is this exactly concerning the described critical function of the work, as King erased in his translation all textual passages where Tempels compares Western and African ontologies and epistemologies and stresses them to be different but equal. I will just give one example of such an erased passage, on page 24 of the original Dutch edition, that is, in the ontological Chapter 2. Tempels writes there:

> The insight of the indigenous in the nature of things, as well as the many distinctions made by learned scholars, are *both* intellectual knowledge of essential being. Both forms of understanding don't differ in their being knowledge. Both are knowledge of essential beings, metaphysics, and the thought system they built on a certain understanding of essential being, is philosophy.[14]

It is clear that Tempels tries to counter the common prejudices among colonials of his time that 'primitive' people (he uses the term but denies its denigrating meaning in the same sentence) are incapable of rational thought and of philosophy, especially because (in the Dutch original version) he repeats the message several times over in this chapter.[15] It is no surprise that the translation that has all sentences of such intent erased led to interpretations of the book as either a philosophical appropriation of African thought for colonial purposes or a non-critical ethnological description of an exotic (from a European viewpoint) philosophy. Only after repairing the text thus can we again see the critical ontological structure of Bantu Philosophy. From this basis, we can further interpret the specifics of its ontology and how it can work for an African ecological ontology.

Being and/or Essence; Force, Strength, or Energy

To get a clear view of the way Tempels positioned his understanding of African ontology, we have to take note of a few key issues. First, and always when we use his text, we have to note the issues inherent in it. Tempels interpreted Kiluba concepts in his own Flemish-Dutch, using professional philosophical Latinized terminology from his studies in Aristotelian, (neo-)Thomist, and other European philosophical systems of thought. From the correspondence with Possoz, it is clear that Tempels was well aware of the translation issues inherent in his work. He wrote, for instance, in a letter from 1947, reacting to a reviewer who critiqued his work for conflating terminology from different Bantu languages into general 'Bantu' concepts, that he sees language as a source for philosophical thought which as such transcends bare language itself:

> I say, that I only began to understand the language through the philosophy, and that I now see that the dictionaries should be refined towards that philosophy. Also that famous dictionary of Jousse and P. Van den Bulcke. I would like to see how they will have translated and circumscribed therein: *tata* (father), *mwana* (?), force, etc. etc.[16]

And only a few weeks later, he stresses in another letter to Possoz, where he explains a key concept in Bantu Philosophy (the dynamic), how difficult it is to capture the Bantu concepts in his own language, Flemish-Dutch, and warns: 'Who understands that terminology in a Western philosophical sense, is for certain in deep trouble [...]'.[17] Consistent with his awareness of the sensitive matter of translating Bantu concepts, he always gives the original words in his work, leaving them untranslated and then explaining them through European concepts. He leaves them as marks, so to speak, for the European reader, indicating that they express and hide a world of thought and practice that is foreign to her/him and can only be approached by careful and consistent attempts to learn the language and participate in that world. All the same, Tempels is writing primarily for a non-African audience, as he wants to point European readers the way to start to take Bantu reality seriously and into account. By keeping the Kiluba concepts in their original form, Tempels gave them a place in global philosophical language, that was and still is strongly dominated by Greco-Latinized European conceptual frameworks.

Now, when we focus on the most famous concept from *Bantu Philosophy,* vital force, discussions on its meaning, appropriateness, and effect on African Philosophy as a field also abound (cf. Mosima 2016, chapters 1 and 2; Roothaan 2022). In the early days, Tempels' most acute readers resembled himself in the position they held in the world. Mostly, they were catholic missionaries with an ethnological and linguistic interest in African cultures and/or philosophers. In their critical reviews, they tended to focus on questions such as whether Bantu Philosophy's descriptions of an African ontology depended too much, or too little, on Thomas Aquinas and Aristotle. They doubted whether Tempels used his classic ontological terminology well, and if the whole enterprise was not venturing too far from the metaphysical underpinnings of catholic theological orthodoxy. Studying the discussions around metaphysics from our present context may serve another purpose. When we better understand if and where Tempels ventured away from the catholic Aristotelian categories, it can give (1) more insight into how he understood Bantu ontology and (2) if such an ontology can help to ground a deep ecological understanding that is relational and holistic.

When we look at chapter 2 of Bantu Philosophy, the issue of problematic translations from the Dutch original is immediately there again. The 1948 French edition calls the chapter 'L'ontologie des Bantous' and the 1959 version follows with 'Bantu Ontology'. In the Dutch original, it says however 'De wezensopvatting der Bantu', which could be best translated as: 'The Understanding by the Bantu of Essential Being'. 'Wezen' can be translated into English as 'essence' or 'being', but as it is clear from the text, Tempels doesn't use it to indicate the basic characteristic of a group of things, like in 'the essence of the human being is his reason' or 'the essence of stone is its solidity'. Rather he uses it to indicate the unique and most inner aspect of every individual being, its deepest ontological aspect; thus, a translation with 'essential being' captures this meaning better. What characterises Bantu metaphysical understanding now, according to Tempels, is that this deepest ontological aspect is not considered to be static, as is common in Western metaphysics (being qua being), but rather dynamic. Essential being is that energy that flows from the divine into all that is, and that gives every being its individual character while simultaneously making it inherently dependent on others.

The above paragraph gives my best explanation of what Tempels found to have blown his mind when he started to understand the foundations of Bantu life and thought. He captured it by opposing it (too strongly, it may be argued) to Western ontological understanding of essential being. The one is dynamic, the other static. Bantu ontology has 'levenskracht' or 'vital force' as its core indicator, whereas Western ontology uses 'substance' or 'being as such', in itself or 'in se', as Tempels has it (Tempels 1945, 30). The Western concept is non-relational, disconnected; the African is connected and inherently relational. Tempels' contemporaries had a hard time understanding what he meant, and why they should go along with it, to the frustration of the author himself, as we find in his letters to Possoz, where he vents his frustration and combines this with attempts to clarify things further.[18]

The most encompassing clarifications are to be found in a letter written in June of 1946 that is entirely dedicated to answering misunderstandings, even by Possoz himself, of the formula that Tempels came up with as 'wezen = kracht', traditionally translated as 'being = force'. Tempels starts by saying that Possoz misunderstands Bantu ontology when he claims that Tempels takes the concept of essential being away from the Bantu and only leaves them the concept of force. Thus, he would leave them a dynamism which is not truly ontological, that is, of course, measured against the standard Aristotelian-Thomistic ontology. To make himself better understood, Tempels quotes from a sympathetic writing of a Monsignor Cleire, who translates Bantu ontology in purely Aristotelian terminology:

> But if I [Cleire] understand you [Tempels] well, you mean by 'being' nothing but 'essence'- *id quodest*, and by energy nothing but nature, i.e. *principium agendi*, what is defined by Aristotle as: *principium et causa motus et quietis in quo est primo et per se et non secundomaccidens*.[19]

Cleire comes up with the word 'energy' rather than 'force' to better explain what is meant by 'kracht' in Dutch and relates it to the 'principle of agency' of Aristotle, which is the principle of movement and rest that is necessary and not accidental to being.[20] Tempels adds:

> For the negro it is not about accidental force, it is about the essence of the being itself and as such, and the essence or the nature of the being as such is determined by us as that what is, by the Bantu as force that is, or as existing force.[21]

His point is that the catholic thinkers of his day were so convinced of Thomistic-Aristotelian ontology, that they could not accept any alternative understanding of the deep structure of reality. So here we find this (for a catholic theologian and philosopher) strange and interesting step Tempels makes – to go from a monistic to a pluralistic understanding of ontology:

> Everything is captured in us, or at least many Western scholastic philosophers, claiming: the essence is the static element that is the foundation of the accidental aspects among which is force, whereas the negroes claim: well now, that static element of yours we take to be dynamic, we think that it is in its nature to be force [...].[22]

This dynamic element, the principle of agency, is intrinsic to every being, whether alive, as in organic or animal life, or not, as in non-organic being. It takes different forms, however, depending on the type of being, and finally also differs per individual being. Tempels uses this ontology to understand all phenomena of African life which he has come to know – and this is made clear only fully in the empirical part of Bantu Philosophy, where we find, as it were, the full-color print of the ontological theory of vital energy. Here a full understanding is made possible of how this energy works, what it is doing, so to speak. Or: what is the 'agency' in this 'principle'.

The principle of agency of this vital energy resembles ebb and flow (my words, A.R.), as every single thing can increase or diminish in vital energy. Since everything is connected, we have to understand that no increase nor diminution happens without the influence of some other thing's energy. As we can increase someone's potential to flourish by encouraging them, we can hinder the same potential by putting them down and making them insecure. Vital energy works, further, in hierarchical ways: a being that is stronger in its essential being has more power to influence one that is weaker. A parent normally has more power over a child than vice versa, and a human over a non-human animal. An animal overpowers organic matter, and a divine being does the same to humans. And in all cases, in more or less direct ways, this energy flows from God. From the empirical part, where Tempels also explains the clan and family structure of African societies, we can deduce that the relation between God and all other beings is somewhat like the intimate link between a parent and a child, a bond that can never be completely severed. This understanding then differs greatly from the European understanding of this relation as one between a powerful and distinct Super Being creating other distinct beings that stand on their own (as substances).

The many very concrete examples of practical life, of legal institutions, of emotions and interactions that Tempels gives in his book make clear that the 'dynamism' derided by his scholastically trained contemporaries, can only be understood as an intimate and deep relatedness of all there is, as in one big and super-diverse family. A family consisting of lizards and bugs, but also of mountains and winds, of majestic elephants and of scorching sunlight. Seen from this ontological standpoint, it makes no sense for the human being to disconnect itself from the rest of nature, to try to rule or transform nature without taking into account that we are part of it, since we are just a knot in a web. To say it in Tempels' own words:

> Nothing moves in this universe of energies without moving or potentially moving the entire whole. This world of energies is like a spider's web, of which one can move no single thread without making the entire web quiver with it.
>
> (Tempels 1945, 32)

This quote captures well the core of why Bantu Philosophy is a meaningful text to reflect on what African Philosophy can bring to Deep Ecology. The deep and intimate relatedness of everything provides an ontological foundation to ecological life and thus goes deeper than any moral calls to an ecologically responsible life for humans as in the activist approach of deep ecology that founds itself on the intrinsic value of more-than-human life.

The Dynamic Web We Live in

In this final section of my chapter, I will go into some problems with Tempels' characterisation of the workings of the web of nature and how they can be resolved to make his ontology better fit an indigenous ontological understanding. To do so, we will focus on the hierarchical characterisation that Bantu Philosophy gives of its dynamic and relational ontology, a characterisation that will be critically discussed with reference to a recent article on Bantu language and the understanding of African ontology (Monga-Kasimba 2022). I will conclude that a more appropriate interpretation of this core element of hierarchy and the dynamic relational ontology can help us provide a fruitful African founding for Deep Ecology as ecological philosophy or ecosophy.

Tempels uses the idea of hierarchy to explain that, according to the Bantu, reality is not a 'chaotic whole of forces that wriggle through each other and bump into each other' (Tempels 1945, 36), but an orderly whole that they understand according to their metaphysical principles. Tempels understands these principles to include a fixed hierarchy. At the end of his ontological chapter, he summarises the workings of forces according to three fixed metaphysical laws (which he proposed as a hypothesis):

1. The (living or deceased) human being can empower or diminish another human being in their essential being.
2. Human vital energy can directly influence lower (animal, vegetal, or mineral) energies in their essential being.
3. A being with reason (spirit, deceased, or living human being) can influence another being with reason indirectly by exerting vital influence on a lower energy (animal, vegetal, or material) and letting that energy work upon the other being with reason.

The third principle attracts our attention, as it is introduced by Tempels to solve a problem with his principle of a hierarchical flow of vital energy through the world we live in. He mentions that some authors claim that the Bantu call some lower energies (rocks, trees, plants) '*bwanga*', meaning exerting influence upon anything or anyone that comes near them. This, if true, would disturb the hypothesis of Bantu Philosophy that there is a unidirectional hierarchy in the world flowing from God towards the ancestors, the spirits, the human beings, the animals, etc. Tempels solves this problem by saying that his conversations with Bantu people do not confirm this aberration, as the Bantu will always in the end admit that such a powerful tree or rock will be under the influence or guardianship of a (higher) spirit. Even though Tempels can thus defend the consistency of the hierarchy of being,[23] we should question if this part of his ontological hypothesis best captures the foundational structure of indigenous thought, as in his 'case study' of Bantu philosophy. The idea of hierarchy does not even harmonise with the spider's web metaphor, where *any* force can and will influence any other force.

As central and possibly problematic as the idea of hierarchy in Bantu ontology is, there has not been much academic discussion of it. Mostly, it is taken as just part and parcel of Tempels' theory and not as part of a philosophical hypothesis that can be improved upon and changed. In a recent article, however, written by Congolese scholar Patrick Monga-Kasimba, we do find a critique of the idea of hierarchy. (Monga-Kasimba 2022). The author, knowledgeable in several Bantu languages, proposes how a deeper discussion of the meaning of Kiluba concepts should further refine the systematisation of Bantu philosophical thought. He investigates how the Luba understood the hierarchy of being and shows thereby that it was not just a European idea projected by Tempels on their understanding of the world. Quoting a Professor Mbuya Mukombo, the grandson of Maman Lenge,[24] we read how Tempels travelled to learn about the different Luba peoples, such as the Baluba of Katanga and the Luba Shankadi:

> It was in his quest that Tempels encountered Maman Lenge. He received many insights about the world of the Baluba of Katanga, their relations to things, to human beings and to God. After completing gathering insights on his pilgrimage that led him right up to lake Tanganyika, Tempels published his work turned famous: Bantu Philosophy [...].
>
> (Monga-Kasimba 2022, 210)

According to Mbuya, there are six hierarchical levels of beings: (1) God , (2) divine spirits or divinities, (3) ancestor-spirits, (4) the recently deceased, (5) the living, (6) and finally all the other beings: animals, plants, and minerals (6). Monga-Kasimba says this confirms Tempels' description in Bantu Philosophy of a hierarchy of beings, but he denies the vertical directness of this hierarchy, as explained below.

More problematic aspects of the hierarchy of beings become clear when Monga-Kasimba turns to Afro-centric research (especially the work of Joseph Mabika Nkata) that claims the connection of creation and God/creator to be very different in African and in European traditions of thought. Whereas in the European world, the universe is created by a God who remains external to it, in the African world, the universe is a whole that includes God, all beings, and all non-being, together in their connectedness.

> It is in this sense that the energies can influence each other by virtue of their internal nature, according to the Baluba Shankadi as well as all negro-African peoples. [...] The universe, for the negro African, is a whole, a TM. The '*tem*' is a relational, dynamic reality.
>
> (Monga-Kasimba, 214)

This understanding of the universe as a relational whole aligns with Tempels' description of a holistic understanding of all beings as connected through the vital energy that flows from the creative source that is called God.

Monga-Kasimba then returns to the Luba understanding of reality and argues how exerting a vital influence in Bantu ontology should be understood as mutual, circular, direct, and indirect rather than mono-directional from a higher to a lower being. God is not only influencing all (lower) beings that depend on the vital energy flowing from him, but can also *be* influenced from below – by the pleading of intermediate spirits sent by humans. Similarly (lower) beings such as animals or trees can exert vital influence on humans, who are considered higher in the vertical articulation of the hierarchy of vital force that Tempels seems to have copied from a traditional Aristotelian model. This leads Monga-Kasimba to argue:

> The Tempelsian hierarchy of Being should therefore be reexamined in light of the negro-African mental framework. According to the Aristotelian philosophy of the participation of beings in Being, the influence between beings is vertical, from higher to lower. The negro-African practice however doesn't work this way.
>
> (219)

To support his claim to re-articulate the workings of vital energy in an African relational ontology, Monga-Kasimba gives the example of a ruler who uses the skin of a powerful wild animal to add to his vital energy, which means that the animal is not considered to have a fixed lower place than the human concerning their vital energy. If the animal could not be seen to be higher with respect to hunting power, for instance, it would not make sense for the ruler to wear its skin. Tempels relates similar cases, but interprets them in a complicated manner to maintain his vertical hierarchical understanding of the workings of vital energy. In his interpretation of such cases as depending on a higher spirit that works through the animal skin, Tempels, according to Monga-Kasimba, has fallen victim to the Aristotelian-Thomistic scheme of a fixed hierarchy. One could add that Tempels, when discussing a similar case, fails to uphold his hierarchical model, as he claims that in this case, the Bantu do not uphold their own foundational philosophy well. (Cf. Tempels 1945, 45–47)

The critical claim of the core of Bantu ontology, as described by Tempels, is not rejected by Monga-Kasimba. However, he claims that certain aspects of it, especially the idea of a fixed hierarchy of being, may not be well-founded in actual Bantu thinking and thus distort the core of Bantu philosophy.[25] The conclusion, then, must be that the 'web of nature' is better understood as a circular whole (like a web) in which influences, direct and indirect, are mutual and flow from higher (meaning closer to the divine source) to lower, as well as vice versa. So a certain type of hierarchy in the relational web of vital energies could be maintained, but its Aristotelian, static, and monodirectional articulation should be rejected. Instead, the relational ecological whole of which the human being is a part, as articulated in the Bantu and African wisdom tradition, may be more multi-directional and dynamic in its workings than Tempels realised when he wrote his book.

Our final question is now what contribution this repaired Bantu ontology can bring to the present-day real-world problems that deep ecology has attempted to address? Bantu ontology, understood as multidirectional and more-than-human in its dynamic of energies, brings God as well as the human being down from their supposed positions higher up in the moral and ecological order. The conclusion from such an ontology should be that God will not save us, but we can also not save ourselves. According to an indigenous understanding of reality, things are more complex. A philosophy that captures an indigenous approach to nature well cannot avoid bringing us closer to a 'pagan' or 'shamanic' thinking and practice.[26] Such an approach entails that the human position is simultaneously humble *and* ambitious. Dependent on so many forces larger than us, respectful of their intrinsic value, but also still desiring to change the world to our needs, we are torn between the promise of technology on the one hand, and humility and '*gelassenheit*' on the other hand. There just simply is no technical solution to our environmental problems. Reality, as understood in an indigenous ontology, tells us we can maybe try to find a technical solution, but we should never forget what the others (spirits, animals, plants, divinities) are doing for themselves and maybe for us. Humbly asking for the collaboration of the more-than-human energies in maintaining and repairing our shared habitat, as in shamanic practices, might be understandable and sensible from the point of view of a relational holistic ontology, but could hardly be understood if we only acknowledge the intrinsic value of more-than-human beings and stress human moral responsibility for them and the ecology. From such a holistic relational perspective, one could say: we are in a world larger than we can capture, and should transform our practice accordingly to a proper ecological being in relation.

Notes

1 The ontological status of value is central to the work of the German phenomenologist Max Scheler (1874–1928), as argued in the dissertation of James Hackett (Hackett 2013). Only recently we have seen an increased interest in Scheler's work, as most of it was, for a long time, not translated and overshadowed by the fame of thinkers in the Husserlian school of phenomenology.

2 Taylor has described his position in slightly different ways in his subsequent works. For the idea of moral ontology, his *Sources of the Self* is the major source (Taylor 1989).

3 Cf. Mudimbe 1988.

4 In the DRC, French is the official language, and there are four national languages: Kikongo, Lingala, Swahili, and Tshiluba. The many other, smaller languages are not included in educational programmes, etc. Professor Nkulu-N'Sengha kindly pointed out to me that while Kiluba is not recognised as a national language, the main written and printed texts in the language are religious (the Bible and related matter).

5 Among the translations, which exist in German, Italian, Spanish, Portuguese, French, and English, I will focus only on the French (1948) and mostly on the English (1959) translations, as these are the ones that have formed the basis for most scholarly study and reactions to Bantu Philosophy.

6 In his book *The Hermeneutics of African Philosophy*, Serequeberhan, however, misquotes Bantu Philosophy to support this, cutting out a part of a sentence where Tempels critically describes the attitude of colonials (in the King version, this quote is on pp 171–172) as a view Tempels himself would have held.

7 And thus no longer the private possession of the Belgian king Leopold II, which was called Congo Free State and is now known for the most brutal pillaging for rubber, making the king excessively wealthy. Adam Hochschild described in his 1998 book how the Congo Free State came about and was finally given up to the Belgian government after increasing criticism of the cruelties perpetrated there on behalf of the king. The placement under the Belgian government may have led to the abolishment of the most horrific methods to enforce slave labour but didn't lead to the dismantling of a racist settler colonial system of occupation and export of valuable resources to build wealth in Europe.

8 See the biography by A.J. Smet: https://www.aequatoria.be/tempels/bio.htm.

9 There is very little literature on Possoz in English. He published his works in French, and they have been mostly ignored later on by scholars in the historiography of colonial studies in law. A brief introduction to his work and relationship with Tempels in English can be found in Clement 2020.

10 The first chapter was published as a separate essay in *Equatoria* in 1944 (Vol VII), chapters 2 through 6 were published in *Band* (February, March, July, October, and November) in 1945, numbered as 1 through 5, and the final chapter 7 as chapter 6 (conclusion) in

Band in January 1946. I thank Professor Michael Meeuwis for providing me with copies of these publications.

11 This was first noticed by Henk Haenen in his dissertation, cf. Haenen 2006. Willem Storm noted the same about the English translation in 1993.

12 The archive in Leuven holds the printed 1945 version with what seem to be Tempels' corrections.

13 Convinced of his own good intentions, of his intellectual integrity, and increasingly involved in what would become the main project of his life: the transformation of Catholic Church theology and practice on the ground of the encounter between Europeans and Africans, Tempels often felt misunderstood, and his text not treated as it should have been. This led him to add an eighth chapter even later, to further explain the intentions of the book. This chapter was never contained in any of the standing published versions of the book, but is included in the internet-published French critical edition of Smet. See: https://www.aequatoria.be/tempels/FTCriticalEditon.htm.

14 All translations of texts quoted, that were published in Dutch or French, are by the author of this chapter.

15 In fact, where I translate with 'indigenous' the original has 'primitieven', so Tempels uses the term current in his time but denies its normally denigrating meaning in the same sentence. That is why I chose the more neutral 'indigenous'. For a substantial discussion on the use of these two terms in the frame of the Study of Religions, see James L. Cox, 2007.

16 Letter from 8/3/1947, KADOC BE/942855/815 – 14.

17 Letter from 31/3/1947, which in this case is even very Flemish, mentioning the famous Belgian version of French fries: 'Wie die terminologieopvat in westerschfilosofischen zin, die zit noodzakelijk in de patatten […]'. KADOC BE/942855/815 – 14.

18 It was not Tempels' style to argue things out in academic journals, which would have solved a lot of misunderstandings of his work. He rather sent all his thoughts to Possoz, in Dutch, with rare exceptions, such as in the studies done by A.J. Smet, who was a missionary and professor of African Philosophy in Congo (now the DRC, previously named Zaïre); these works have remained there, waiting. Smet's work is mostly available (in French) on the website dedicated to Tempels' work: https://www.aequatoria.be/tempels/index.htm.

19 Letter from 23/6/1946, KADOC BE/942855/1856.

20 I think Cleire, cited below, uses the more apt translation of 'kracht', which is also more unequivocal in English and French, namely, 'energy' instead of 'force'. Thus we could translate Tempels' formula as 'essential being = energy'. Or to make it more snappy: 'essence = energy'.

21 Idem note 12.

22 Ibidem note 12.

23 This originally Neo-Platonic idea briefly regained popularity through the work of the philosopher and historian Arthur Oncken Lovejoy, *The Great Chain of Being*, that was published in 1936. It is not exceptional at all that a catholic thinker who aims to move beyond scholastic thought, like Tempels, would feel the appeal of its main idea.

24 Apparently, Maman Lenge was a recognised wise woman and could be called an indigenous philosopher.

25 Monga-Kasimba instead follows Mabika's adoption of certain concepts of A.N. Whitehead next to ancient Egyptian ones to explain the Bantu universe more accurately.

26 'Pagan' has commonly been used by Christians to indicate ideas and practices that are outside the realm of the governance of the one God they recognise as the all-creator and judge of all that humans do. It is a philosophically problematic concept, for how can there be something outside the realm of the one God? For my definition of the concepts 'shamanic' and 'shamanistic', see Roothaan 2019, xvi, footnote 2.

References

Note: all archival sources were accessed at KADOC in Leuven, Belgium, and folder references are given in the footnotes.

Britannica, The Editors of Encyclopaedia. "Great Chain of Being". (1936)*Encyclopedia Britannica*, 10 December 2021, https://www.britannica.com/topic/Great-Chain-of-Being. Accessed 19 July 2024.

Bujo, Bénézet (2016 [2000]). *Foundations of an African Ethic: Beyond the Claims of Western Morality*. The Crossroad Publishing Company, New York.

Clement, Piet (2020). "Tempels Revisited. The Conversion of a Missionary in the Belgian Congo, 1930s–1960s". In Vincent Viaene, Bram Cleys, and Jan de Mayer (eds.), *Religion, Colonization and Decolonization in Congo, 1885–1960*. Leuven University Press, Leuven, 241–160.

Cox, James L. (2007). *From Primitive to Indigenous: The Academic Study of Indigenous Religions*. Ashgate, Aldershot and Burlington.

Diagne, Souleymane Bachir (2013). "Préface". In *Placide Tempels, La Philosophie Bantoue*. Paris: Présence Africaine , I–IV.

Drengson, A., Devall, B., & Schroll, M. A. (2011). The deep ecology movement: Origins, development, and future prospects (toward a transpersonal ecosophy). *International Journal of Transpersonal Studies*, 30(1–2), 101–117. http://dx.doi.org/10.24972/ijts.2011.30.1-2.101

Eze, Michael Onyebuchi (2017). "Humanitatis-Eco (Eco-Humanism). An African Environmental Theory". In A. Afolayan and T. Falola (eds.), *The Palgrave Handbook of African Philosophy*. New York: Palgrave Macmillan. 621–632

Hackett, James Edward (2013). "Scheler's Phenomenological Ontology of Value: Implications and Reflections for Ethical Theory." Dissertations. Paper 679. http://opensiuc.lib.siu.edu/dissertations?utm_source=opensiuc.lib.siu.edu%2Fdissertations%2F679&utm_medium=PDF&utm_campaign=PDFCoverPages

Haenen, Henk (2006). *Afrikaans Denken: Ontmoeting, Dialoog en frictie. Een filosofisch onderzoek*. Buijten & Schipperheijn, Amsterdam.

Hochschild Adam (1998). *King Leopold's Ghost: A Story of Greed, Terror, and Heroism in Colonial Africa*. Houghton Mifflin Company, Boston and New York.

Hountondji, Paulin J. (1996 [1976]). *African Philosophy: Myth & Reality*. Indiana University Press, Bloomington and Indianapolis.

Ibanga, Diana-Abasi (2023). "Ndu-Mmili-Ndu-Azu ('Live-and-Let-Live'): Ekwealo's Version of Environmental Ethics." *Environmental Ethics*, 45(2), 147–173.

Katcheckpele, Léonard Amossou (2019). *D'Osiris au Bantou theologian. Théologie africaine et vie en abondance, undegénéalogiehistorique et critique*. L'Harmattan, Paris.

Matolino, Bernard (2022). "Placide Tempels's Legacy in the Bantustanisation of Philosophy in Africa." In Frans Dokman and Evaristi Magoti Cornelli (eds.), *Beyond* Bantu Philosophy*: Contextualizing Placide Tempels's Initiative in African Thought*. Routledge, London and New York, 65–77.

Monga-Kasimba, Patrick (2022). "La hiérarchie tempelsienne de l'être à la lumière de la culture Luba Shakandi." In Charles Kantenga Kasongo, Delphin KibambaMbayo, and Olivier NkuluKabamba (eds.), *La Philosophie bantoue du Père Placide Tempels. Relecture diversiforme présentée dans le cadre du centenaire du diocèse de Kamina (1922–2022)*. Paris: L'Harmattan, 193–222.

Mosima, Pius Maija (2016). *Philosophic Sagacity and Intercultural Philosophy: Beyond Henry Odera Oruka*. African Studies Centre, Leiden.

Mosima, Pius Maija and Nelson Shang (2023). "We Live in Paradise. Beautiful Nature in African Tradition." In Bolaji Bateye, Mahnoud Masaeli, Louise Müller, and Angela Roothaan (eds.), *Beauty in African Thought: Critical Perspectives on the Western Idea of Development*. Lexington Books, Lanham, Boulder, New York and London, 33–50

Mudimbe, V.Y. (1988). *The Invention of Africa. Gnosis, Philosophy, and the Order of Knowledge*. Indiana University Press, Bloomington and Indianapolis; James Currey, London.

Mungwini, Pascah(2022). "Indigenous (African) Philosophy and the Fecundity of Expression." In Frans Dokman and EvaristiMagotiCornelli (eds.), *Beyond* Bantu Philosophy*: Contextualizing PlacideTempels's Initiative in African Thought*. Routledge, London and New York, 51–64.

Naess, Arne. (1973). "The shallow and the deep, long-range ecology movement: A summary". *Inquiry: An Interdisciplinary Journal of Philosophy*, 16(1–4). 95-100.

Roothaan, Angela (2019). *Indigenous, Modern and Postcolonial Relations to Nature. Negotiating the Environment*. Routledge, London and New York.

Roothaan, Angela (2022). "Terug naar Bantoe Filosofie. Tekst en betekenis van het werk van Placide Tempels". *Filosofie & Praktijk*, 43, 77–91.

Serequeberhan, Tsenay (2012 [1994]). *The Hermeneutics of African Philosophy: Horizon and Discourse*. Routledge, New York ānd London.

Smet, A. J. (2001). La Philosophie Bantu de PlacideTempels;Augmentée du huitiemechapitreinédit; Avant-propos et Edition critique de la Traduction française par le Prof. Emer. A.J. SMET C.P.: https://www.aequatoria.be/tempels/FTCriticalEditon.htm

Storm, Willem (1993). "Bantoe-Filosofie vs. Bantu Philosophy". *Quest* VII(2), 66–75.

Taylor, Charles (1989). *Sources of the Self. The Making of the Modern Identity*. Cambridge University Press, Cambridge, NewYork and Melbourne.

Tempels, Bantu Philosophy. Translated by Colin King. France: Présence Africaine 1959 [1945].

2 The Environment in Yorùbá Collaborative Ontology

Abidemi Israel Ogunyomi

Introduction

The environment has been a major subject that has attracted the attention of scholars in recent times. Several scholarly articles and books have been written, from philosophical perspectives, on human– environment relation and the need to conserve the environment in the contemporary world. This is in view of the anthropocentric orientation of human beings towards the environment and the negative effects such orientation has for other organisms (non-human animals) in the environment. Human beings have, for long, been (mis)led to think that other entities (living or nonliving) in nature are for their comfort, consumption, and use; hence, they should be engaged and directed to their service. Accordingly, human beings have overexploited their environments and have, by so doing, directly exposed other living beings to the danger of extinction and non-living beings to degradation through their endless constructive, scientific, and technological activities.

How to save the environment from the many terrible effects which the activities of human beings have left on it is, therefore, a major problem for philosophers of the environment and other environmentally inclined scholars. The realisation of the fact that the environment is not only for human beings but also for all living organisms calls for a giant caution in the activities of human beings in the environment. This is because, whether directly or indirectly, the activities of human beings affect not only other organisms but also human beings themselves in the environment. This is evident in the upsurge of environmental crises like erosion, flood, sun-scorch, famine, global warming, and other environmentally related crises that persistently threaten human existence. The implication of this is that a review of human attitude becomes imperative in order to ensure general environmental balance. The predominantly parasitic and anthropocentric orientation of human beings needs to be substituted for eco-friendlier, bio-friendlier, and complementary orientations which will help in the conservation and preservation of other life forms within the environment. This has given rise to the proliferation of environmental theories such as biocentrism, ecocentrism, deep ecology, animal right theories, eco-bio-communitarianism, and other environmental theories of relation. However, while other environmental ethical theories have been explored from the African perspective, little has been done in exploring deep ecology from an African (Yorùbá) perspective. This chapter aims to fill this gap by discussing

DOI: 10.4324/9781003480914-3

the notion of collaborative ontology in Yorùbá Eco-philosophical thought and underscore how it is imperative for deep ecology. This chapter is divided into three major sections, aside from the introduction and the conclusion. In the first section, I undertake an explanation of the Yorùbá notion of the environment, their creationistic and teleological worldview, and how it gives credence to their belief in the interconnectedness of reality. I also discuss the notion of collaborative ontology in Yorùbá thought, where non-human animals are actively featured in the process of creation. This, as I shall explain, gives credence to the view that all things emanated from the same source to serve specific purposes and that, by implication, they are offspring of the same father. Something related to this idea has been explained as a theory of sociation in sociology and anthropology by Akinsola Akiwowo and other scholars. Akiwowo used the notion of "*Asuwada*," meaning "beings that exist as a group," as a theory of sociation to explain the reason human beings must continue to live in a society and cooperate with one another towards achieving a common goal.

He developed and popularised this theory of sociation to justify why human beings must continue to exist in a society and in groups. On the basis of this *Asuwada*, he further expanded on other theories of sociation such as *Alajobi* (those who are tied up in the same web of birth) and *Alajogbe* (co-tenants) to emphasise the necessity of sustaining human society. However, in this section, I build upon Akiwowo's theory of sociation and consider its applicability to human–environment relations through a conceptual expansion of the notion of "collaborative ontology,"[1] which features non-human entities in the making of reality and the myth about the creation of the universe in the metaphysics of the Yorùbá as an indigenous theory that is imperative for deep ecology. This will be instrumental to the explanation of the relationship between human beings, non-human animals, and non-living things, with the intention to present a sample of the African perspective on deep ecology. This collaborative ontology conveys the impression that nothing in the world ought to be mistreated or exploited as mere instruments for the realisation of the interests of other beings because, in the final analysis, they originated from the same root and they participated jointly in the process of being.

In the second section, I undertake a brief discussion of the major principles of deep ecology as an alternative scholarly perspective that emphasises the inherent value of all life forms and advocates the moderation of human activities in order to protect other beings in the environment. In the third section, I explore the indigenous worldview of the Yorùbá through the analysis of some verses of Ifá on what should be the disposition of human beings toward the environment. I underline how this indigenous explanation is based on the idea of collaborative ontology in Yorùbá thought and how it is imperative for the practice of deep ecology.

The Environment and the Idea of Collaborative Ontology in Yorùbá Thought

The environment, in Yorùbá language, is called *àyíká*. The word *àyíká* also means "things that surround us." Accordingly, it is a holistic term that refers to physical entities (human beings, non-human animals and non-living things in the universe) and non-physical entities (spirits, apparitions, deities, divinities, and forces). The Yorùbá notion of the environment can best be understood in relation to their worldview on the nature of reality, which compels me in this session to briefly explain the metaphysics of the environment in Yorùbá thought, since metaphysics, as a cardinal branch of philosophy, takes the fundamental structure of reality as a whole and the nature of existence, essence, and being to be its preoccupation (Lowe 2002, 2–3; Teffo and Roux 2003, 3; Wilfred 2017, 19–20).

Generally, African metaphysics is holistic and deeply teleological in nature (Teffo and Roux 2003, 196–197). This is particularly the case in Yorùbá worldview: their ontology and cosmology rest on creationist and teleological explanations where every being, whatever its nature, is believed to have been created in order to serve some purpose. This explains, in the first instance, the reason such notions as destination, determination, and predestination are prominent when it comes to the explanation of the purpose of human existence in the Yorùbá worldview (See Gbadegesin 2003, 220–228; Makinde 2010, 120–137.) It is not, however, restricted only to human beings. Other beings in the environment, whether living or non-living, are believed to have been created to serve some specific purposes.

Òrìsàńlá (also called Ọ̀btálá), the creativity deity, is believed to be directly responsible for the task of creation in Yorùbá thought, on the instruction of Olódùmarè, the apex being in the Yorùbá pantheon of gods (Idowu 1962, 71–75; Awolalu 1979, 21–22.) However, other deities such as Ògún, Àjàlá, and Olódùmarè also featured in some aspects (Ogunyomi 2024, 91.) This being the case, everything created by Òrìsàńlá is believed to be divine in itself. In Yorùbá myth of creation, he was said to have been sent by Olódùmarè to establish the earth. He descended with a parcel of soil and some non-human animals, specifically a hen and a chameleon, which also featured prominently in the task of creation (Abimbola 2006, 120.) However, he was unable to carry the task out because he got drunk at Òkè-Àrà (mountain of wonders) where all the deities descended prior to the making of the earth. Nevertheless, his younger brother, Odùduwà, carried out the task. After the application of the soil on the pre-existing vast body of water, it was the hen that spread it on the surface of the water until it began to coagulate and extend. This later reformed into the tangible space referred to as the land. After the hen had already carried out the task of extending the soil, the chameleon, being a fragile and light being, tested it for the first time to check whether or not it was safe to be trekked upon (Abimbola 2006, 120, 124.)

It was after all these activities and the attestation of the chameleon to the solidity and firmness of the newly created land that all other deities began to descend on the earth, and trees, plants, forests, and other things began to spring from the newly created earth. This ontology that features deities, animals, and non-living things is not only cooperative but also collaborative in nature. This informed the notion of "collaborative ontology" in this paper.

Beside the above explanation, Akinsola Akiwowo, while presenting a theory of sociation, cited the Ifá verse of *Ìròsù-Wòrì,* which is significant for the understanding of the relationship between human beings and the environment and also imperative for the practice of deep ecology. The verse goes thus:

> *Nígbàtí awọn ọmọOdùduwà péjọ, Àwọn tí Odùduwàyàn láti kó ire náàwá òde ayé li à uń pè li Ènìyàn. Àwọn ọmọ Odùduwàni:igi, ọ̀pẹ̀, àtiẹranko àti ekòló, ènìyàn, àti àwọn ohun míràn bẹ̀ẹ̀ bẹ̀ẹ̀.*
>
> When the children of Oduduwa assembled before Him, those whom He chose to convey goodness to earth are those we refer to as Ènìyàn (human beings) The children of Odùduwà include: the trees, the palm trees, beasts (animals), earthworms, human beings and such like things.
>
> (Akiwowo 1983, 11)

From the above, Akiwowo further postulated a socio-anthropological theory of *Alájọgbé* (co-tenants/cohabitants) and *Alájọbí* (those who are tied up in the same web of birth). Nevertheless, these theories, when critically considered, are not only significant for the explanation and understanding of what should be an ideal relationship among human beings in society but also significant for the explanation and understanding of what should be an ideal relationship between human and non-human beings in the environment.

The notion of collaborative or cooperative ontology further strengthens Akiwowo's theory of sociation and makes it applicable to the understanding of human–environment relation. While Akiwowo's theory of sociation, when extended to the environment, conveys the impression that human beings and non-human beings are all co-tenants and co-occupants of the environment, collaborative ontology goes further to contend that both human and non-human beings are responsible for bringing the environment about, due to the reason that they all participated jointly in the task of creation. In other words, the making of physical reality, part of which the environment is, is a combined effort of human and non-human beings in Yorùbá ontology.

The above gives credence to the belief in the interconnectedness of reality where everything there is, is considered to be related and connected to other things. Although, strictly, there is a dualistic explanation of reality along a physical and non-physical divide. However, the non-physical is believed to inhere or inhabit in the physical, thereby conveying the impression that they are inseparably interconnected, interpenetrated, and interdependent. For instance, the spirits, which are supposed to be non-physical beings, are believed to inhabit and reside in physical things like the forests, mountains, rivers, trees, rocks, and groves.

Against the above background, Mbiti (1969, p. 76) maintains that the divinities are associated with natural phenomena and objects, as well as with human activities and experiences in Yorùbá thought – Ògún is associated with iron, war, metallurgy, hunting, and smithy; Èsù is associated with stones; Sango is associated with lightning and thunder; Òrìsàńlá is associated with snails; Ọya is associated with rivers, and so on. In addition, Bolaji Idowu contends that human beings are co-tenants with other beings in the environment (Idowu, 1978, p. 206). Also, Omosade Awolalu maintains that the forest, the hills, trees, rocks, rivers, and mountains are considered to be spiritual, divine, and sacred because they are believed to serve as abodes for other categories of beings in the environment (Awolalu, 1979, pp. 48–49.)

Moreover, there is a blurry relation among the categories of being referred to above and human beings in Yorùbá thought. John Mbiti maintains that while animals and plants constitute food for human beings, they also have religious associations with human beings within the African context, which may, sometimes, account for how both human beings and animals originated at the same time from the same place (Mbiti, 1969, p. 50.) This is especially the case in Yorùbá worldview. While the Yorùbá kill some animals for food, some animals, such as the leopard (Ẹkùn), the giant rat (Òkété), and the parrot (Odídẹrẹ́/Aiyékòótọ́), are believed to share relational bonds with some clans, villages, and towns such as the Ògbùró, Ìkòyí, and Ìwó respectively. Accordingly, those animals are considered to be close associates, and in some cases, matriarchs and patriarchs of such lineages (see Babalola, 1966, p. 23; Ajuwon, 1980, p. 67; Yemitan, 1988, pp. 34–35, pp. 37–38.) Adebayo Babalola presents a compelling instance of this in his documentation of a chant of salutation to Ògúnwálé, an individual elderly hunter from the Yorùbá lineage of Ẹlẹ́rìn (Ẹ̀rìn- Ilé) in Kwara State, Nigeria. The chant goes thus:

Ọ̀rọ̀ ḿbẹ lónìí àjọsọ.
Ng ó re'lé Ògúnwálé, gbogbo jànmọ́ọ̀ wa.
Igún orí Àkó ni nwọ́n,
Àkàlàmàgbò ọrùn Ọsè.
Tẹ̀ǹtẹ̀rẹ́, ẹiyẹ́ orí Ọ̀pẹ yẹ̀kú;
Akúwáàgbawà.
Akúfẹ́iyẹrápó;
Ọmọ Olóriogun.

Today, there is much for us to talk about.
I will chant a salute to *Ògúnwálé,* all you people gathered here.
He is associated with the Vulture on *Àkó* tree top,
And with the Hornbill on the Baobab tree.
As well as the Egyptian plover,
A bird whose haunt is the branches of a young palm-tree.
He who at death received a shield.
He whose death gives birds a chance to see the quiver;
Warriors of the Generalissimo (Babalola, 1966, pp. 198–199.)

Apart from indicating that *Ògúnwálé* shares a relational tie with some birds such as the Vulture, Hornbill, and Egyptian plover, the chant above also underlines the relationship of those birds (and by implication, *Ògúnwálé*) to some trees such as the *Àkó,* Baobab, and Palm trees. Besides, praise poetries and panegyrics are composed to celebrate non-human animals, plants, and non-living entities (see Babalola, 1966, pp. 87–116; Oni, 2003, pp. 94–96; Ajibade, 2005, pp. 161–166.)

Again, some powerful figures are believed to have transformed into animals and non-living entities such as rivers, stones, and trees in the past. Because of this, the descendants of these powerful people worship and offer sacrifices to some of these non-living entities.[2]

In addition, it is believed that the spirits of human beings can inhabit some animals like cats, insects like cockroaches, and birds (Yemitan, 1988, pp. 12–14, pp. 48–49; Ajibade, 2005, p. 159). This, however, raises the metaphysical problems of identity, disembodiment, and transmigration of the soul, which deserve a special focus in another outlet in order to be thoroughly considered. Some animals and birds, in addition to having negative causal implications for whoever kills them, are natural environmental sanitisers and also instrumental in the making of rituals and sacrifices in Yorùbá religion; hence, they are forbidden to be killed. Some of these birds include the vulture (*Igún*), the hornbill (*Àkàlàmàgbò*), and the parrot (*Odídẹrẹ́/Aiyékòótọ́*) (Ajibade, 2005, p. 157; Abimbola, 2015, pp. 65–67).

Other birds that are considered to be naturally sacred in Yorùbá thought are *Àgbìgbò, Wọ́sọ̀-Wọ́sọ̀, Tẹ̀tẹ̀rẹ̀, and Kowéè*. The significance of *Kowéè* among the Yorùbá deserves special attention. It is a mysterious bird that breaks the news of death to the relatives of the deceased irrespective of whether they are in the town where the deceased lived or not. Many of these birds and animals also feature prominently in several chapters and verses of Ifa. In addition to drawing normative or didactic inferences from the behaviour of these animals, in most cases, the relationship between those non-human animals and human beings is usually strongly underlined in Ifa verses (See Abimbola, 2015, pp. 65–67, pp. 111–130, pp. 171–192). For example, in some verses of Ifa, twins are said to share the same ontology with a species of monkey called *Ẹdun* and *Àíyá*. Accordingly, they are forbidden to kill or eat monkeys. There are other verses underlining the relationship between human beings and non-human beings in the Ifa corpus (See Abimbola, 1976, pp. 195–231).

The above is predicated on the conviction that all beings, whatever their nature (corporeal or incorporeal), are interconnected. The environment, on this basis, is viewed in a holistic sense in such a way that if anything negatively affects a certain part of it, it automatically affects the remaining part as well. Thus, human beings are not seen as the most important beings in the universe; other entities, physical and non-physical, living and nonliving, are as important as human beings. While this holistic conception of reality takes cognisance of the different categories of being, it disallows a total separation among them. This is observable in the extension of intrinsic values to some non-human animals and non-living entities, including the land.

The land itself, on which all things in the environment dwell and make their abode, is considered to be a force. This is evident in the fact that it is offered proprietary sacrifice in Yorùbá land (see Adeoye, 1985, pp. 356–360) and it is used in making oath or covenant by the Yorùbá with a strong belief that it possesses the natural power to punish defaulters, traitors, betrayers, or oath breakers (*Ọ̀dàlẹ̀*) and vindicate the victims of oath breaking. That is why the following sayings: *Ọ̀dàlẹ̀bá'lẹkú, ẹni tóbá dalẹ̀, ábá'lẹ̀ lọ* (traitors die with the land, whoever betrays goes with the land);

bí ènìyàn bá ún yọ́'lẹ̀ ẹ̀ dà, ounburúkú á máa yọ̀rúu wọn se (whoever betrays in secret suffers from terrible things secretly) in Yorùbá, bear heavy causal implications.

The land is believed to mete out retributions or punishment to those who break the principles of secret agreements or contravene the tenets of an oath or covenant initially sworn to concerning certain things in the past. Now, it does not matter whether the defaulter intentionally or unintentionally breaks the rule; anyone who breaks the rule is called a traitor *Ọ̀dàlẹ̀*). Such a person automatically invites the appropriate sanctions or punishment(s) that are attached to it from the land on whose surface the oath or covenant was taken – and which, again, *witnessed* that very act of betrayal.

The Idea of Deep Ecology

Deep ecology can be described as a biocentric environmental movement concerned centrally with the preservation of all life forms within the environment. This movement attributes intrinsic value to all life forms and advocates a rejection of any attempt to valorise the supremacy of human beings over other entities in the environment. Deep ecology, in particular, discourages anthropocentrism, which, in the first instance, is the basis for the trivialization of the worth of other life-forms in the environment.

While discussing the conceptual difference between deep ecology and other environmental philosophies, Merley Jacob avers that a distinction is often made in environmentalism between mainstream (shallow) and deep ecology. She, however, considers the distinction to be biased and inadequate on the grounds that it sweepingly generalises a uniform content for all mainstream perspectives and exaggerates the necessity of curtailing the human-centered temperament of the mainstream perspectives for the achievement of the kind of changes required for the control of the environmental crisis, which is the position of deep ecology (Jacob, 1994, p. 447).

However, Jacob's position can either be considered plausible or otherwise only when one understands the subject matter and the core claims of deep ecology. According to George Bradford, deep ecology as an environmental theory is traceable to Arne Naess, who described it in the 1970s as an "attempt to ask deeper questions" about the prevalent orientation of human beings towards the environment and to originate a new philosophy of nature that rejects the anthropocentric attitude of human beings and recommends an eco-friendly attitude that recognises the intrinsic value and independent worth of other life forms in nature as a substitute (Bradford, 1989, pp. 8–11). The following are often listed as the basic tenets or core principles of deep ecology:

1. All life (human and non-human) on Earth has intrinsic value.
2. The richness and diversity of life forms contribute to the realisation of these values and are also values in themselves.
3. Human beings are not privileged in any way and should not reduce this richness and diversity except to satisfy vital needs.

4. The flourishing of human life and cultures, as well as non-human life, requires a substantial decrease in human population.
5. The present human interference with the non-human world is excessive, and the situation is rapidly deteriorating.
6. The present policies affecting basic economic, technological, and ideological structures must be changed; the resulting state of affairs will be deeply different from the present.
7. The ideological change is mainly that of appreciating life quality (dwelling in situations of inherent value) rather than adhering to an increasingly higher standard of living; there will be a profound awareness of the difference between big and great (Jacob, 1994, p. 480; Ibanga, 2017, pp. 102–110.)

From the basic principles in deep ecology identified above, the following could be considered the central concerns of deep ecology: a strong defence of the welfare of human beings in addition to the welfare of non-human animals and non-living entities in the environment, a celebration and protection of different life-forms as having inherent values, a rejection of anthropocentrism that places only human beings at the centre of the universe and reduces other life-forms to the position of tools or instruments to be deployed in the service of human beings, control of human population, interests, desires, and goals to ensure the thriving and survival of other life-forms, and the making and enforcement of relevant and necessary eco-friendly laws and regulations to keep human actions in check.

Collaborative Ontology: An Imperative of Indigenous Yorùbá Eco-Philosophy for Deep Ecology

Having briefly articulated the worldview of the Yorùbá about the environment, which hangs on the notion of collaborative ontology, the meaning of collaborative ontology itself, and the idea of deep ecology, I undertake in this section the task of indicating how the notion of collaborative ontology in Yorùbá eco-philosophy is imperative for the practice of deep ecology as one of the persistently blooming aspects of environmentalism or eco-philosophy in scholarship. In doing this, I shall also investigate the indigenous thought of Yorùbá people concerning what is called "deep ecology" as conceived in Western environmentalism.

The creationist and the teleological worldview of the Yorùbá, as explained in one of the previous sections, and their belief in the interconnectedness and interpenetration of reality can be identified as the first basis for deep ecology. The reason for this is that, apart from accommodating the basic principles of deep ecology outlined above, they also embody them both in thought and in practice. The reservation of intrinsic value, albeit relational in their context, for non-human animals and non-living things based on the view that they are naturally sacred in Yorùbá thought, conforms to the basic principles of deep ecology.

In the previous section, I explained that both living and non-living things, such as animals, birds, mountains, groves, rivers, forests, trees, and land, are worshipped and propitiated, in some cases, by the Yorùbá people. This explains the level of sacredness and reverence they reserve for these beings. Besides, there are taboos guiding against the desecration of these entities, and there are punishments for the contravention of those taboos, respectively.

The oral literature of the Yorùbá people concerning their indigenous epistemology on human–environment relations also exemplifies the reverence they have for other life forms and the pity they express when these forms run into extinction or are summarily violated. For instance, Yorùbá hunters, in their *Ìrèmọ̀jé* (funeral dirges), embark on strong eco-lamentation that decries, with an evocative effect, the degree of degradation and mismanagement of the environment by human beings. This oral eco-lamentation advocates deep ecology. To further showcase the Yorùbá indigenous perspective on deep ecology, the consideration of some documented oral materials by Ifayemi Elebuibon about Ọbàtálá, the creativity deity, is imperative.

Ifayemi Elebuibon presents a profound elucidation of the panegyrics of Ọbàtálá, where the ancient people of Ife were said to have appealed to him through sacrifice to him when he visited all life-forms in the land with death by withholding rain due to the reason that the people made him angry. He was propitiated to preserve all life-forms in the environment and prevent them from going into extinction. It goes thus:

Àgbẹ̀dẹníífọ'rin jí;
Ifánioníkòló ótún kòló sín.
Ijọ́ọbatiún jẹ,
A òrírúọbabáyìirí,
Àfi Ọbàtálá, Ọbàtárìsà,
Ọba pátápátátííbá wọn gbélé Ìrañjé.

Òun lókóokùn ejiso,
Ó dáọsẹ̀rẹ̀ sẹ Òòrùn sílẹ̀,
Oníkírúeku ó máará,
Kírúẹja ó máará,
Kírúẹyẹ ó máará,
Kírúẹran ó máa rá,
Kírú èèyàn ó máa rá,
Èji Ẹlẹ́mẹ̀rẹ̀gbàwá o,

Kírúeku kó máràá,
Kírúẹja kó máràá,
Kírúẹran kó máràá,
Kírú èèyàn kó màràá,
Èji Ẹlẹ́mẹ̀rẹ̀gbàwáo.(Elebuibon, 1998, pp. 100–101.)

Blacksmiths are the purifiers of metals;
Ifá advises those with unkept beards to adjust them.
Ever since kings have been installed,
We have never seen this type of king before,

Except for *Ọbàtálá, Ọbàtárìsà,*
The supreme king that resides in the city of *Ìrañjé.*

He withheld the rope of rain
And ignited the ray of the Sun
He cursed the race of rats to vanish;
He cursed the race of fishes to vanish;
He cursed the race of animals to vanish;
He cursed the race of birds to vanish;
He cursed the race of human beings to vanish.
Èji Ẹlẹ́mẹ̀rẹ̀ (Ọbàtálá,) please, come to our aid.

Do not let the race of rats vanish;
Do not let the race of fishes vanish;
Do not let the race of animals vanish;
Do not let the race of birds vanish;
Do not let the race of human beings vanish;
Èji Ẹlẹ́mẹ̀rẹ̀ (*Ọbàtálá*), please, come to our aid. (my translation)

The significance of the above verse for deep ecology cannot be overemphasised. Besides underlining the importance of restoring and maintaining environmental order, it underscores the huge weight of value the Yorùbá attribute to the preservation of all life-forms in the environment. This is reflective in the pragmatic effort of offering propitiations to the deity of creativity in order to prevent all life-forms from going into extinction. In another panegyric of *Ọbàtálá*, deity of creativity, he was reported to have been angry with tree fellers, herb fetchers, deforesters, and hunters of *Àpà* and *Ìrañjé* for violating his instruction not to intrude upon a particular forest that he conserved by himself. It goes thus:

...Ó (Ọbàtálá) kìlọ̀ọ̀rọ̀kan,
Wọn ògbàl'Ápà.
Ká sọ̀rọ̀fún ni ká mágbà,
A máa dùn mítítí.

Óní sẹ́wé sẹ́wé Ìjàó,
Ẹ mád'Ápà sẹ́wé mọ́n.
Bẹ́ ò fẹ́ni o, Àlápàní ósinmi ókó ewésísẹ́ mọ́n;

Sẹ́gi sẹ́gi Ìrànjé,
Óníẹ mád'Ápà sẹ́gi mọ́n;
B'ó fẹ́, ó kọ̀, Àlápàní ósinmiigisísẹ́ mọ́n.

Àwọn Ọdẹ Àpà, nìkánlókọ'tíọ̀gbọnin;
Àwọn Ọdẹ Àpà, wọ́n ya'láìgbọrọ̀
ỌrẹỌdẹ, ÌmùnlẹỌdẹ,

Wọn kìí b'Ọ̀dẹ̀ẹ́jàl'Ápà;
Arọdẹwọ̀gún, aréníibáỌdẹ̀ẹ́ se.

Ìgbàtí óún báỌdẹjàl'Ápà,
Asọinání baba e bora.
Ógbọ́'rọ̀kan, ó jó mùkúmùkẹ.
Òkùsùtíibá Ọdẹ̀ẹ́ j'ágẹ̀rẹ̀ wọjà.
Ikú tíí b'ọdẹ ẹ́ pínwóọgbọọgba;
Ikútíí s'ọdẹ l'ófòọfà
Ó b'Ọ́dẹ mẹ́fàl'Ápà, ó pẹ̀ta nínú wọn,
Ó dé'jìní 'gbèkùn l'ókè Àpà;
Óníọ̀kanyókù órelé lọ̀ọ́wí...[3] (Elebuibon, 1998, pp. 144–146.)

....He (*Ọbàtálá*) gave a warning,
But they did not adhere to it at *Àpà*.
When one refuses to adhere to instructions,
It pains me so deeply.

He warned the herb fetchers of *Ìjàó*
Never again to intrude the forest of *Àpà* to fetch herbs.
Whether they liked it or not,
They had to stop fetching herbs at the forest of *Àpà*.

He warned the wood fetcher of *Ìrànjé*
Never again to fetch wood at *Àpà*.
Whether or not he liked it,
He had to stop fetching wood at the forest of*Àpà*.

Only the hunters of *Àpà* refused his warning
For they were very disobedient.
Friends and associates of hunters
Do not fight with hunters at Àpà
Arọdẹwọ̀gún (*Ọbàtálá*) also played with the hunters.

But when he eventually fought with the hunters at *Àpà*,
He wrapped himself with a garment of fire.
He heard an information to which he danced:[4]
For he is *Òkùsú*[5] that dances to *Àgẹ̀rẹ̀* drum with the hunters to the market.

He is death that shares money equally with hunters.
He is death that wastes hunters' weapons.
He met six hunters in the forest of *Àpà* and killed three;
He tied two hunters down at the mountain of *Àpà*
And sent one to inform others at home. (my translation)

The above verse, taken from the panegyrics of *Ọbàtálá*, is a strong indication of how dissatisfied he could be with the mismanagement of the environment, being a revered deity of creativity. It is also a pointer to the fact that the Yorùbá are not oblivious to the practice of environmental conservation and deep ecology, having been fully aware of the displeasure of the deity of creativity in the misuse of the environment. The killing of the disobedient hunters of *Àpà* by *Ọbàtálá,* due to the reason that they refused to adhere to his instruction concerning the conservation of *Àpà* forest, serves as an indigenous and didactic deterrent for the abuse of the environment.

This is imperative for the practice of deep ecology because it advocates some of the basic tenets and principles of deep ecology, such as environmental justice, rejection of anthropocentrism, regulation of human activities in order to ensure the survival of other life-forms in the environment, and so on. The view of the Yorùbá on the creation of the universe itself, which gives credence to participatory, cooperative, and collaborative ontology, in which other life-forms are actively featured, is the basis upon which their oral and indigenous knowledge system concerning what ought to be the ideal relationship between human beings and the environment, as discussed above, is based. Yorùbá collaborative ontology is, therefore, imperative for the practice of deep ecology. The only exception to the aforementioned discussion is when the vital needs of human beings are concerned. However, this may be a general problem for all cultural civilisations. The reason for this is that the exact scope of what constitutes (or ought to constitute) the "vital needs" of human beings has yet to be clearly determined even in the basic principles of deep ecology. This constitutes a little difficulty for all environmental theorists, which future research may hope to address.

Conclusion

In this chapter, I discussed the worldview of the Yorùbá on the environment. I explained how their understanding of the environment rests on a teleological and creationist basis and the holistic nature of this understanding of the environment. I explained how the creationist account of the Yorùbá about the origin of the world features cooperation and active participation of deities and non-human beings. This was described as collaborative ontology. I presented a brief account of the basic claims of deep ecology as an environmental movement. I explained the indigenous view of the Yorùbá about what should be the ideal relationship between human beings and the environment, and I underlined how this explanation rests on the notion of collaborative ontology earlier mentioned and how it is imperative for deep ecology as conceived in Western environmentalism.

Notes

1 I am grateful to Dr Elvis Imafidon of the School of Oriental and African Studies, University of London, for suggesting this phrase. It was apt!

2 Many powerful people in Yorùbá land are reported to have turned to some non-living entities in the past. Examples are Ọya, wife of Sàngó, and Onídèyí, a prin-

cess of Olúkọsin, who both were reported to have turned into rivers. Kúrunmí, the ÀàrẹỌnàKankanfò (Generalissimo) of Yorùbá land, was reported to have turned into a palm tree when he was about to be captured by the forces of Ibadan under the command of Bashọrun Ògúnmọlá, after the murder of his seven sons during the Ijaye war, ordered by Aláàfin Adelu. This was said to have infuriated Ògúnmọlá, who wished he had captured him alive. For this reason, he was said to have ordered his soldiers to fell all the palm trees in that particular spot. Others, such as Oníkòyí and Ológbùró, were reported to have turned into a giant rat and a leopard, respectively. Their descendants, by implication, attribute intrinsic value to these animals and forbid them to be killed or eaten in their lineages.

3 Baba Oluwole Fawole, an Ifa priest of Òkètasẹ̀ in Ile-Ife, was greatly helpful in providing further useful information on certain unclear aspects of this line in Ifayemi Elebuibon's work.

4 Of course, the information referred to here was that the hunters of *Àpà* were hunting in the forest against the warning of the deity of divinity.

5 Name of a river animal that hunters must not kill. It is believed that any hunter that kills this animal will also die within a couple of days.

Bibliography

Abimbola, K. (2006) Yorùbá *Culture: A Philosophical Account.* Birmingham: Iroko Academic Publishers.

Abimbola, W. (1976) *An Exposition of Ifa Literary Corpus*. Ibadan: Oxford University Press.

Abimbola, W. (2015) *Sixteen Great Poems of Ifá*. Ibadan: University Press Plc.

Adeoye, C. L. (1985) *Ìgbàgbọ́ àti Ẹ̀sìn Yorùbá*. Ìbàdàn: Evans Brothers Nigeria Publishers Ltd.

Ajibade, G. O. (2005) 'Animals in the Traditional Worldview of the Yorùbá', *Folklore: Electronic Journal of Folklore*, 30, pp. 155–172. www.folklore.ee/folklore.

Ajuwon, B. (1980) 'The Preservation of Yorùbá Tradition through Hunters' Funeral Dirges', *Africa: Journal of the International African Institute,* 50(1), pp. 66–72.

Akiwowo, A. (1983) *Ajobi and Ajogbe: Variations on the Theme of Sociation*, Inaugural Lecture Series 46. Ile Ife, Nigeria: University of Ife Press.

Awolalu, J. O. (1979) *Yorùbá Belief and Sacrificial Rites*. London: Longman Group Limited.

Babalola, S. A. (1966) *The Content and Form of* Yorùbá *Ìjálá*. London: Oxford University Press.

Bradford, G. (1989) *How Deep Is Deep Ecology?* New York: Times Change Press.lib .anarhija.net.

Elebuibon, I. (1998) *The Adventure of Obatala Part 2.* Osogbo: Ara Ifa Publishing.

Gbadegesin, S. (2003) 'Ènìyàn: The Yorùbá Concept of a Person', in L. Teffo and A. Roux (eds.) *African Philosophy Reader.* London: Routledge, 208–228.

Ibanga, D (2017) 'Is Deep Ecology Inapplicable in African Context: A Conversation with Fainos Mangena', *Filosofia Theoretica: Journal of African Philosophy, Culture and Religions,* 6(2), pp. 101–119.

Idowu, E. B. (1962) *Olódùmarè: God in Yorùbá Belief.* London: Longman Group Limited.

Idowu, E. B. (1978) *African Traditional Religion.* Oyo: PP Publications.

Jacob, M. (1994) 'Sustainable Development and Deep Ecology: An Analysis of Competing Traditions', *Environmental Management,* 18, pp. 477–488.

Lowe, E. J. (2002) *A Survey of Metaphysics*. Oxford: Oxford University Press.

Makinde, M. A. (2010) *African Philosophy: Demise of a Controversy*. Ile Ife, Nigeria: Obafemi Awolowo University Press.

Mbiti, J. S. (1969) *African Religions and Philosophy*. London: Heinemann.

Ogunyomi, A. I. (2024) 'An Existential Interpretation of Evil: A Critique of Ebun Oduwole and Kazeem Fayemi on the Philosophical Problem of Evil in Yorùbá Thought', *Journal of Comparative Literature and Aesthetics*, 47(1), pp. 87–101.

Oni, K. (2003) *Kike Oloburo: Akojopo Ewi Igbalode*. Osogbo: Sumob Publishers. ISBN: 978-2992-49-6.

Teffo, L. and Roux, A. (2003) 'Metaphysical Thinking in Africa', īn L. Teffo and A. Roux (eds.) *The African Philosophy Reader*. New York: Routledge, pp. 192–208.

Wilfred, L. (2017) 'African Metaphysics: Traditional and Modern Discussions', in I.E. Ukpokolo (ed.) *Themes Issues and Problems in African Philosophy*. Switzerland: Palgrave Macmillan. pp. 19–59.

Yemitan, O. (1988) *Ìjálá Aré Ode. Ibadan*: University Press Limited; New York: Random House.

3 Radicalizing Ubu-ntu

Some Critical Thoughts on Mogobe Ramose's Philosophy of Ubu-ntu and a Proposal for Its Desuperiorisation

Björn Freter

Introduction

Mogobe Ramose has developed an incredibly sophisticated philosophy of ubu-ntu. Ramose's influence is already significant, but the true depth of his philosophy has not yet been adequately appreciated. His ideas are so rich, and the philosophical community has much to glean from them.

In our contribution, our first task must be to understand his philosophy of ubu-ntu. During this work, we will eventually come across the problem that ubu-ntu philosophy normatively overemphasises umu-ntu, i.e., the *homo loquens et sapiens (the talking and rational human being)*, the human be-ing, by assuming that it is this human be-ing (and only this human be-ing) who brings the epistemology and ethics of ubu-ntu into reality. We will analyse the tension that arises from this imbalance, from this potential superiorism in ubu-ntu philosophy. We will respond to this tension by proposing a desuperiorisation of umu-ntu and the ubu-ntu maxime *umuntu ngumuntu ngabantu (a human being is a human being through human beings)*.

The Fragmentation of Being

Ramose unfolds important parts of his ubu-ntu philosophy with the help of grammatical observations and etymologies, primarily the etymologies of ubu-ntu and umu-ntu[1]. These words, ubu-ntu and umu-ntu, however, are not mere linguistic entities, arbitrarily connected to their extra-linguistic denotations. The words ubuntu and umuntu are non-accidental aspects of the ontology of ubuntu and umuntu. The grammatical-etymological investigation is an ontological investigation. Ubu-ntu and umu-ntu *are* ubu-ntu and umu-ntu and the *words* ubu-ntu and umu-ntu are parts of their respective beings. An attempt to understand Ramose must always keep this in mind. Ramose's "body of work is grounded on an ontology" (Hallen, 2019, p. 14) and this ontology is "foundational" (Hallen, 2019, p. 14; see Kimmerle, 2006, p. 190). This becomes obvious when Ramose introduces his idea of the rheomode language.

This rheomode language opposes "the subject-verb-object understanding of the structure of language" (Ramose, 2005, p. 37) which is based "upon fragmentative thinking" (Ramose, 2005, p. 37). The subject-verb-object understanding of

DOI: 10.4324/9781003480914-4

language "posits a fundamental irreconcilable opposition in be-ing becoming. On the basis of this imputed opposition be-ing becoming is fragmented into pieces of reality with an independent existence of th[ei]r own" (Ramose, 2005, p. 37). This fragmentation is to be undone with the rheomode language. We have to take "the verb as its point of departure" (Ramose, 2005, p. 41) and no longer the noun. This allows us to preserve "the incessant flow of motion as be-ing" (Ramose, 2005, p. 41). However, since "there is always the doer in the do-ing", the verb in the rheomode language has to be understood as a "verbal noun, that is to say, [a] gerund" (Ramose, 2005, p. 41).

The rheomode language:

> is a critique of a thought and language structure which assumes and imposes a strict divide and a necessary sequence in terms of subject-verb-object. It is an appeal for the understanding of entities as the dimensions, forms and modes of the incessant flow of simultaneously multi-directional motion.
>
> (Ramose, 2005, p. 40)

This rheomode language is "against the fragmentation of be-ing through language" (Ramose, 2005, p. 41), it aims to preserve "be-ing as a whole-ness" (Ramose, 2005, p. 41).

When we philosophise about ub-untu – in Ramose's understanding – we have to do so in a rheomode language, as the "logic of *ubu-ntu* is distinctly rheomodic in character" (Ramose, 2005, p. 41). Let us now try to understand ubu-ntu in Ramose's sense

Ubu-ntu

Ubu- evokes be-ing, it brings be-ing to our consciousness, to our attention. Ubu- is enfolded be-ing, it is un-manifested or un-concrete be-ing, it is *wholeness*. When the enfolded ubu- unfolds, it comes to our reality in a concrete form of ex-istence, i.e., as a particular thing. That which exists is that which ex-ists, i.e., that which ex-ists is that which emerges – *sister* (causative of *stare* meaning "to stare") out of – *ex* – not-being, nothing.

Ubu-, Ramose continues to explain, "is always oriented towards -ntu" (Ramose, 2005, p. 271). Ubu- and -ntu are both aspects of the be-ing. While ubu- evokes the enfolded be-ing as be-ing and as nothing else, it "may be said to be distinctly ontological" (Ramose, 2005, p. 36). -ntu, on the other hand, evokes the unfolding of be-ing as "being […] a specific entity subsisting in a particular mode in space-time" (Ramose 2007, p. 124), as "an apparent (that is, observable) concrete form or mode of be-ing" (Dladla, 2019, p. 157), and this "may be said to be distinctly epistemological" (Ramose, 2005, p. 36). Ubu-ntu thus constitutes "the fundamental ontological and epistemological category in the African thought of the Bantu-speaking people. It is the indivisible one-ness and whole-ness of ontology and epistemology" (Ramose, 2005, p. 36). That which is, in its most general (or abstract, see Oosthuysen, 2016, pp. 41–43) sense – this is the ontological aspect – unfolds into a (particular) entity for us to understand, to know, to recognize (see Ramose, 2005, p. 37) – this is the epistemological aspect.

It is important to note that ubu-ntu is "always a -ness and not an -ism" (Ramose, 2005, p. 36). Even though the suffixes -ness and -ism both indicate nouns, their meanings are quite different, and, as we need to remember, this is not an arbitrary linguistic observation; this is of immediate ontological relevance. The suffix -ism indicates that its base word is, for instance, a movement or a belief. The suffix -ness indicates a state, a condition, or a quality. Ramose does not want ubuntu to be understood as an -ism, i.e., he does not want it to be understood as an instance, movement, or belief, but as a state or a quality. This would be an etymo-ontological misunderstanding. Ubu-ntu is not a dogmatic and thus static belief or a school of thought; it evokes one-ness and whole-ness and their endless interactions. In fact, "[o]ur fixation to -ism is ethically untenable since it ultimately leads to unjustified violence in defence of absolutism and dogmatism" (Ramose, 2017, p. 39).

It is precisely the onto-epistemological coincidence of whole-ness and one-ness that is brought to our attention by the hyphenation of ubu-ntu on the one hand and by its characterisation as a verbal (or gerundial) noun on the other, i.e., a word in verb form functioning as a noun (as, for instance, in "the thinking"). Ubu-ntu is a philosophy of the *ultimate dynamic unity of all that is*. The hyphen always reminds us that ubu-ntu (and all other hyphenated terms) "have to be understood as processes or in a dynamic sense" (Kimmerle, 2006, p. 82). Ramose's ubu-ntu philosophy is a philosophy of *the one-ness and* whole-ness; it is "fundamentally holonistic" (Ramose, 2005, p. 42), it is, in one word, *Onto-Holoness* (but not Onto-Holon*ism*).

Ramose elucidates further that at the "ontological level" – that is, the level of the enfolded be-ing – "there is no strict and literal separation and division between ubu- and -ntu" (Ramose, 2005, p. 36). However, ubu- must continuously unfold itself with time into -ntu and become a particular one-ness. In more emphatic words, ubu- *desires* its unfolding to ex-ist as a particular entity, and this unfolding never stops. Ramose writes: "On this reasoning, *ubu-* may be regarded as be-ing becoming and this evidently implies the idea of motion" (Ramose, 2005, p. 36). In fact, "motion is the principle of be-ing for *ubuntu*" (Ramose, 2005, p. 36).

As the reader will have noticed, Ramose hyphenates not only ubu-ntu but be-ing and ex-istence as well. He seems to develop similar etymo-ontological ideas for be-ing as he has done for ubu-ntu. We can explain it like this: be-ing consists of the root be- and the suffix -ing. Be- is a root with extensive meaning. Most importantly, be- can be used to say that an entity (derived from the Latin *ens*, the present participle of *esse*) *is*, *exists*, or *lives*. In be- we find, like in ubu-, a most general reference to being, that is to say: be- evokes being *as* being.

The meaning of the suffix -ing depends on whether a particular word is an adjective, a verb, or a noun. The meaning of the verbal -ing is certainly the most important here. The verbal use always has to do with something in action, in movement. "Philosophizing" refers to an ongoing action, "gathering" refers to an instance of an action, "writing" to the result of an action, "sawing" to an action connected with a specific thing. While be- refers to being *as* being (similar to ubu-), -ing refers to a being becoming or a being having become (similar to -ntu). The prefix be- and suffix -ing brought together as one word thus evoke be-ing in its most general sense (be-ing as be-ing) and being in its dynamic particular sense (be-ing as becoming).

Nonequilibrium

Interestingly, Ramose points out that due to its generality, ubu- "is marked by uncertainty" (Ramose, 2005, p. 45) This is of foundational, existential importance, as it immediately determines our *conditio humana*: "The experience of nonequilibrium is [...] the basic problem of human existence" (Ramose, 2005, p. 39). And thus the central question of our lives emerges: "how does one respond to the fundamental instability of be-ing?" (Ramose, 2005, p. 46) Ramose provides the answer right away: "In *ubuntu* philosophy a human being in the world of the living must be *umuntu* in order to give a response to the challenge of the fundamental instability of be-ing" (Ramose, 2005, p. 46). In fact, "be-ing human is not enough. One is enjoined, yes, commanded as it were, to actually become a human being" (Ramose, 2005, p. 37). Let us now find out what umu-ntu means.

Umu-ntu, Human Being, and Human-ness

Ramose goes on to investigate one of the particularizations of ubu-, i.e., umuntu. -ntu is now interacting with umu-. However, Ramose writes, "[w]hereas the range of *ubu-* is the widest generality, *umu-* tends towards the more specific" (Ramose, 2005, p. 36; see Oosthuysen, 2016, pp. 25–26). Compared to ubu-, umu- is more specific; however, both still denote:

> the highest [to be understood in an elative sense] level of generality in the sense that they [scil. ubu- and umu-] specify nothing in particular. They are [...] dangling and will remain in that position until they are grounded by a particular suffix.
>
> (Dladla, 2019, p. 157)

Umu- is now to be grounded by -ntu, "[j]oined together with *-ntu* then *umu-* becomes *umuntu*" (Ramose, 2005, p. 36). This is a very important step in Ramose's argument. He goes on to explain that,

> *[u]muntu* means the emergence of homo loquens [i.e. the speaking human being] who is simultaneously a *homo sapiens* [i.e. the thinking or rational human being]. In common parlance, this means the human be-ing: the maker of politics, religion, and law. *Umuntu* then is the specific concrete manifestation of *umu-:* it is a movement away from the generalized to the concrete specific.
>
> (Ramose, 2005, p. 36)

Umu-ntu is thus "the embodiment of the ontology and epistemology of *ubu-*" (Ramose, 2005, p. 45).

All this allows for a translation of the two vital terms: umu-ntu and ubu-ntu. Umu-ntu can be translated as *human being*; it is the particular entity making laws or conducting philosophy. Ubu-ntu can be translated as *human-ness*; it has to be "thought of as describing the more general and abstract human-ness or be-ing human" (Dladla, 2019, p. 157).

Umu-ntu, like ubu-ntu, is always in motion, continuing to "conduct an inquiry into being, experience, knowledge, and truth" (Ramose, 2005, p. 36) which "is an ongoing process impossible to stop unless motion itself is stopped" (Ramose, 2005, p. 36).

We can further refine the understanding of umu-ntu. It refers to the *living* human being. The living human being is to be distinguished from those "[human] beings departed from the world of the living through death", i.e., the "living dead (*abaphansi*)" (Ramose, 2005, p. 45) and the "yet-to-be-born" (Ramose, 2005, p. 45). These are the "three levels of human ex-istence, [...] the onto-triadic structure of be-ing" (Ramose, 2005, p. 45).

The Particular Importance of Umu-ntu

Umu-ntu, the living human be-ing, is obviously of particular importance. Ramose remarks that "[w]ithout the speech of *umuntu* [i.e. the speech of the *homo loquens et sapiens*]" (Ramose, 2005, p. 37). ubu- has to remain silent. Ramose expresses this in surprisingly harsh words. He says that "*Ubu-* is condemned to unbroken silence" (Ramose, 2005, p. 37). Finding out that ubu- is *condemned* to silence is, after all we have learned about ubu-, rather surprising. Why is ubu- condemned? Who condemned it? And for whom is ubu- silent? For umu-ntu, for any particularisation of -ubu? Ramose does not seem to answer these questions. He does, however, instruct us what to do about the damnation to silence. This silence must be, again a tonally rather surprising word, *broken*.

We can understand the intensity of language when we realise that the fundamental existential problem of the human being is addressed here. If we are not umu-ntu, i.e., that entity that makes ubu- speak, we will never be able to find an answer to the question of how to "respond to the fundamental instability of be-ing" (Ramose, 2005, p. 46). Therefore, ubu- must not be silent, and thus umuntu must speak. But how should we understand this speech of umu-ntu?

Let us remember: Ubu-, human-ness, is always oriented towards -ntu, its concretization. Now we learn that the speech of umuntu is breaking the silence of ubu-, and that it is thus "anchored in, revolves around, and [...] ineluctably oriented towards *ubu-*" (Ramose, 2005, p. 37). The speech of umu-ntu "directs and focuses the entire epistemological domain towards the ontology of *ubu-*" (Ramose, 2005, p. 37). In fact, the "language of *umuntu* 'relevates' [derived from Latin *relevare*, to highlight, to lighten], i.e., it directs and focuses the entire epistemological domain towards the ontology of *ubu-*" (Ramose, 2005, p. 37).

This revelation through the language of umuntu is done "by the contemporaneous and indissoluble coupling of *ubu-* and *umuntu* through the maxim *umuntu ngumuntu ngabantu* [in isiXhosa] (*mothokemotho ka batho* [in Setswana])" (Ramose, 2005, p. 37). This connection of ubu- and the maxim is by no means self-evident – as the reader will have noted, the maxim does not contain the term ubu-ntu, but only the word umu-ntu – twice in its singular form and once in its plural form (*bantu*).

We would like to note that connecting this maxim and ubu-ntu was introduced relatively late in ubu-ntu philosophy in general. It can be found for the first time in the foreword to the American edition of Augustine Shutte's book *Philosophy in Africa* published in 1995 (the South African edition was published in 1993 and does not contain the remark) (see Shutte, 1995, p. vi; Gade, 2011, pp. 313–315).

But we will continue to focus only on Ramose: Ubu- is, however, present in Ramose's particular understanding of umuntu. Ramose thus connects the maxim and ubu-ntu intentionally and with the obvious assumption that these two express the same onto-epistemological principle. Ramose translates the maxim like this: "a human be-ing is to affirm one's humanity by recognising the humanity of others and, on that basis, establish humane relations with them" (Ramose, 2005, p. 37).

Let us take a brief look in some more detail at the language of the maxim. We can translate it rather literally, in the following way:

umuntu	human being
ng-	is
umuntu	human being
nga-	through
bantu	human beings

Or, to make it more fluent in English:

umuntu	[A/One] human be-ing
ngu-	is
muntu	[a/one] human be-ing
nga-	[nga- understood here as a prefix of bantu] through
bantu	[other] human beings

And, to express the maxim again in Ramose's dynamic Onto-Holoness of ubu-ntu, we can say:

umuntu	A [particular] human be-ing
ng-	is [affirming]
umuntu	[the humanity of themselves]
nga-	through
bantu	[recognising the humanity of all other] human beings [and by relating to them]

We learn: it is umu-ntu affirming umuntu by means of (*nga*) bantu. It is umu-ntu relevating and (potentially) practising ubu-ntu. It is umu-ntu breaking the silence of ubu-ntu. But does this not introduce a strange imbalance?

Umuntu-Superiorism?

Dladla comments very accurately, although he does not seem confused by the imbalance, that in "philosophical terms, *umuntu* precedes Ubuntu in practice and, by virtue of such precedence *umuntu* is the progenitor of the epistemology of Ubuntu" (Dladla, 2019, p. 159; see also Bewaji, 2003, p. 392). In a later paper, Ramose even writes: "The epistemology of *ubu-ntu* presupposes the ex-istence of *umu-ntu*, the human being. Without *umu-ntu* neither the epistemology nor the ethics of *ubu-ntu* can come into being" (Ramose, 2017, p. 40).

Umu-ntu may not bring ubu-ntu into reality, but it does so for its epistemology. Umu-ntu is not the creator of the uncreated ubu-ntu (its ontology), but umu-ntu is the particularised entity that makes the *episteme* (gr. ἐπιστήμη meaning "knowledge") of ubu-ntu accessible for other human be-ings (*bantu*). Umu-ntu is the entity that makes it possible for other human be-ings to recognize, to know, to learn about ubu-ntu. This expresses, even if it may not be intended, a strange hierarchy, perhaps even an *umu-ntu-superiorism*. Superiorism can be understood as an incapability, an inability and/or unwillingness to accept otherness *sui generis*, and to take the other (being) as the other (being) seriously, in our case: the other living be-ings. All non-umu-ntu be-ings seem marginalised. The human be-ing, the *homo loquens et sapiens*, brings ubu-ntu epistemology and ethics into reality – it is *not* any other species, not any other living be-ing. This is a strange – and, unless something has escaped our attention, unexplained – fixation on the human being, a strange absence of all the other be-ings (see Horsthemke, 2015, p. 82, and a critique of Horsthemke in Etieyibo, 2017). Ramose remarks that "[h]umanness regards being, or the universe as a complex wholeness involving the multi-layered and incessant interaction of all entities. [...] Herein lies the ecosophical dimension of the indigenous African concept of *Ubuntu*" (Ramose, 2009, p. 309, see also Ramose, 2009, . p. 312). But just three sentences later, he notes that "human beings are indeed an intrinsic part of physical nature although possibly a privileged part" (Ramose, 2009, p. 309). Something is wrong here. This does not fit together.

Let us remember the maxim *umuntu ngumuntu ngabantu*. We have learned that ubu-ntu is contained in the maxim, as it is umu-ntu who will bring ubu- epistemologically into existence. And we have learned that a human be-ing is a human being by affirming the humanity of others. Again, we have to wonder, where are all the other be-ings? Whatever they are, they are not in a position to lighten ubu-ntu; this happens only through the *lingua humana*. It seems that a *lingua elephanti*, a *lingua panthera,* or a *lingua hippopotami* are not to be considered; these *linguae* seem to have no relevance whatsoever for bringing the epistemology of ubu-ntu into reality. But why not? Does the trumpeting of an elephant, the clicking of a dolphin, the grunting of a hippopotamus, the rattling of a termite, the rustling of a tree, the vibration of a mountain, or the sound of the ocean's waves not contribute to the revelation of aspects of ubu-?

Ramose writes that African philosophy has a "rheomodic or oceanic character" (Ramose, 2017, p. 37), that the "African philosophic conception of the universe is not only pantareic [derived from Greek πάντα ῥεῖ, meaning 'everything flows'] but it is musical as well" (Ramose, 2005, p. 42). Music is the "dance of be-ing" (Ramose, 2005, p. 42), and "the invitation of the dance of be-ing is undeclinable since it is understood as an ontological and epistemological imperative" (Ramose, 2005, p. 42), to "dance along with be-ing is to be attuned to be-ing" (Ramose, 2005, p. 42). But can all the be-ings outside of the "three levels of human ex-istence", outside of "the onto-triadic structure of be-ing" (Ramose, 2005, p. 45) not dance along, not attune themselves to the dance of be-ing? Is their part in the dance of be-ing not required? Do all the non-umu-ntu species, or even their (often forgotten) individuals (see Baratay, 2022), have nothing to shed light on? Is umu-ntu revolving around itself? Why, of all the living be-ings, is the human be-ing in this privileged position? Is it the human being's ability to reason that means that it alone can fight against the fragmentation of be-ing? If it is umu-ntu who relevates be-ing *inmodo humano*, why should *indlovu*, the elephant, not do the same *inmodo elephanto*? Could there not be something in the ontology of ubu-ntu that um-untu cannot relevate, something that um-ntu cannot know that can be known *per naturam elephanti*?

We learned the meaning of *umuntu ngumuntu ngabantu*. Is there also a meaning, an importance, a relevance to a sentence like, for instance, *indlovu yindlovungezindlovu* (an elephant is an elephant through elephants) or perhaps even *isilwaneyisilwanengezilwane* (an animal is an animal through animals)? Is the human being not an animal as well? Is this not fragmentation of be-ing (see Mutwa, 1997, p. 13; see Horsthemke, 2018)? The ubu-ntu interpretation of *umuntu ngumuntu ngabantu* does allow the affirmation of the Other, provided that this Other is another human be-ing.

Should the human be-ing really be able to affirm itself and its human other without any reference to a non-*homo loquens et sapiens*? Do we not need the non-human to understand us and to affirm us and the non-human other as well? Is this not fragmenting be-ing? Does ubu-ntu rest on an umu-ntu-superiorism? Does ubu-ntu rest on an umu-ntu-superiorist fallacy? Is it really the case that this liberating, caring, affirming, unifying philosophy of ubu-ntu is like this? Have we missed something? If so, we simply cannot find it.

Speciesism?

Unfortunately – for now – we thus have to agree with Clive Tendai Zimunya and Joyline Gwara, "that in the *ubuntu* discourse there is no reference to non-human creatures. In all its major formulations, *ubuntu* essentially relates to human beings. [It is] essentially and exclusively about humans' relationship to other humans" (Zimunya and Gwara, 2018, p. 27). It seems as if there is a certain contempt for non-human be-ings in this discourse, if only insofar as the non-umu-ntu be-ings have little, perhaps no relevance for the unfolding of ubu-. There seems to be the deepest divide, in Ramose's philosophy, between the human be-ing and all other be-ings. Perhaps one must indeed speak of speciesism or even, as Steven Best put it (even though he was not referring to ubu-ntu specifically), of *species apartheid* (Best, 2017, p. 48)? Peter Singer – following Richard Ryder (see Ryder, 1989) – has described speciesism as "a prejudice or attitude of bias in favor of the interests of members of one's own species and against those of members of other species" (Singer, 2009, p. 6; see Freter, 2019). This is somewhat related to the problem we came across in ubu-ntu philosophy. However, the term cannot encompass umu-ntu in Ramose's sense, since umu-ntu refers not only to the (sub)species *homosapiens* but to the *living homosapiens* at one of the "three levels of human ex-istence" (Ramose, 2005, p. 45), i.e., the yet-to-be-born human be-ings, the living human be-ings, and the living dead. The highly controversial (see Wilkins, 2009) biological concept of species is certainly not broad enough for the onto-triadic structure of be-ing, and thus the term speciesism is not applicable here either. However, it seems that the problem of the overemphasis, the apparent superiority of the onto-triadic structure over other existences remains.

Superiorist Fallacy?

Um-ntu-superiorism suffers from the same problem as all superiorisms, i.e., the *superiorist fallacy*. The superiorist fallacy occurs precisely at the moment when a philosophy introduces a *normative difference between beings*, precisely at the moment when one being becomes the (sole) normative point of reference for the other being. The superiorist fallacy happens when a being, in our case, the human being, *arbitrarily superiorises itself and arbitrarily inferiorizes all those who are not (enough) like itself*. The superiorist fallacy is a fallacy because it *arbitrarily connects the phenomenon of beings being different to these beings being of different value*.

The superiorist fallacy cannot be defended – even though it still can, unfortunately, become a practical reality. The only requirement for the factual reality of the pseudo-conclusions of the superiorist fallacy is power, and power does not have to commit to truth.

The problem that comes to light in superiorism is that of misunderstanding difference. This misunderstanding is twofold:

1. Difference is not Normative Difference

   It is no challenge to find that two things are different from each other. In fact, everything is different from everything else – perhaps we can even say that with the one exception of a particular thing being identical to itself, nothing is ever not different from everything else. There is no reason to assume that these differences are of any normative relevance. Of course, it is possible that, for instance, one human being might prefer one thing to another. However, preference does not question the value of one thing over the other *per se*, but only for this one human being. It is a strange, apparently hard-to-resist philosophical temptation to examine all that exists over and over again to find out what is *per se* better and what is *per se* worse. And it seems an equally hard-to-resist philosophical temptation to ascribe dominion to what has been found to be better over what has been found to be worse – which, in turn, will deepen the posited power differential and produce further differences that will be misunderstood as further proof of an alleged normative difference. We like to call this the cycle of superiorism.

To reject normative difference by no means implies a nihilistic denial of values. Living beings, for instance, are different but not of different values. Norwegian Deep Ecologist Arne Næss pointed out that "all living creatures have their own intrinsic value, that is to say, a value irrespective of the use they might have for [hu]mankind [at alle levendevesenerharegenverdi, det vilsiverdiuavhengig av den nytteverdisom de måtte ha for menneskene]" (Næss, 2000, p. 15, 2002, p. 6). This does not deny difference of beings, but it does deny difference of their respective values.

2. Arbitrary Attribution of Relevance to Differences

Let us think back to speciesism. We learned with Singer that speciesism is a bias "in favor of the interests of members of one's own species and against those of members of other species" (Singer, 2009, p. 6). To be able to be prejudiced against members of another species, one is required to understand the difference between the superior – human – species and all the other species that might exist. There is no denying that the human being is different from all other beings. But why would that which constitutes the difference between human beings and all other beings be the one relevant factor? A ring-tailed lemur is also different from all other living beings; it is even considered to be the only species in the genus Lemuridae. There is no other living being that has stink-fights like ring-tailed lemurs. Or think of a pronghorn: there is no other living being that grows branching horns. Why are these not the all-decisive normative differences? Anthropocentric positions or positions of human exceptionalism will immediately point out that only the human being is rational, autonomous, self-conscious, moral, and so on. These traits, anthropocentrists claim, are what make human beings unique. Even if we ignore the fact that this can be doubted (see Chapman and Huffman, 2018) and accept for the sake of argument that these traits are unique to human beings, we still run into a foundational problem. Uniqueness is not unique; "uniqueness is expressed in each species" (Wallner, 2019, p. 2). If uniqueness were a criterion for superiority, we would have to conclude that, since every living being and every species is unique, every living being and every species is superior – which means that none is.

Why would human beings' uniqueness be so uniquely unique that it surpasses all other uniqueness? There is and there can be no explanation as to why traits that might be unique to human beings would have to be considered more unique or even superior. Are termites deficient because they lack

> humanness? Are humans deficient because they lack termiteness, because we do not communicate through vibrations and pheromones? Why does the difference between a beluga whale and a Kermode bear not imply superiority, but the difference between a human being and a beluga whale or a Kermode bear does? The answer is simple: it does not. We posited our superiority, we have enforced it violently (see Joy, 2010; Best, 2017, and the excellent, but unfortunately largely overlooked study by Karr, 2015), and we worked for centuries on making ourselves believe that we have done so for good reasons. However, we did not do it for good reasons. Speciesism is "as such [...] a selfish emotional argument rather than a reasoned one" (Ryder, 2010, p. 2) The most important *argument against* human exceptionalism, speciesism, or umu-ntu-centrism remains: there is *no argument for* human exceptionalism, or speciesism, or umu-ntu-centrism.

We are in an unpleasant situation. Should we really assume that Ramose missed all this? We cannot find substantial textual counter-evidence, but we remain somewhat uncertain. We would like to conclude by making a productive suggestion on how to counter anthropocentrism: Umu-ntu-superiorism needs to, and can easily, be desuperiorised.

Desuperiorisation and Radical Ubu-ntu

We have proposed understanding superiorism as an incapability, an inability and/or unwillingness to accept otherness *sui generis.* Desuperiorisation is accordingly to be understood as a philosophical method that allows us to develop the capability and willingness to accept otherness sui generis, to become capable, able, and willing to take the other as the other seriously (see Freter, 2020, 2022). Desuperiorisation means to consciously, intentionally, and practically unwant alleged superiority, whether it a superiority that we claim for ourselves or that someone else for themselves.

To desuperiorise a philosophical thought means practically:

1. to investigate whether a philosophy commits the superiorist fallacy;
2. to investigate whether this philosophy necessarily depends on the superiorist fallacy;
3. to investigate whether the philosophy in question can still be consistently thought when normative equality replaces the initially assumed normative difference.

We have already shared our thoughts on point 1. of the above list. We came, albeit with some reservations, to the conclusion that we can identify a superiorist fallacy.

The next step can be taken with some ease, since we can refer to some of the considerations above. Ubu-ntu philosophy does not necessarily depend on an assumption of the superiority of the human being. Ramose made the point that

"*Ubu-* is condemned to unbroken silence" (Ramose, 2005, p. 37) and that only *homo loquens et sapiens* could bring ubu-ntu's epistemology and ethics into reality. Desuperiorisation does not require giving up this thought *per se*. What is required, however, is to add a qualifier to this statement: only *homo loquens et sapiens* and its onto-triad can bring a particular aspect of ubu-ntu epistemology and ethics into reality: Ubu-ntus' umu-ntu ontology can only become (epistemological) reality if umu-ntu does what is necessary, i.e., adopt the rheomode of ubu-ntu, "dance along with be-ing" (Ramose, 2005, p. 42), ex-ist in Onto-Holoness. The same applies *mutatis mutandis* for anything else, the ocean, the baobab, the puff adder, or the bluespotted stingray. This has another consequence that leads us to propose extending the ubu-ntu maxim *umuntu ngumuntu ngabantu*. We propose (in isiXhosa):

> *umuntu ngumuntu ngabantukwaye abo bangengobantu (umzekelo: imvelo, umhlaba, nakonke okunempilo.*
> *A human being is a human being through human beings and through those who are not human beings(for instance, nature, the world, all that has life).*

Or perhaps:

> *umuntu ngumuntu ngabantu, nakonke okuphilayo.*
> *A human being is a human being through human beings and through all living beings.*

Would it not be possible to claim that a human being is a human being through human beings *and* through the sunbird, Table Mountain, or the Hawaan Forest? Would this not be much more plausible than the isolated circular determination of the human being through the human being? No doubt, it is of the highest importance to highlight the interrelationality and equality of *all* human beings. But it also seems important to highlight that those who are umu-ntu cannot understand themselves circularly, in relation only to themselves. One of the suggested addenda would allow us to bring the relationality of ubu-ntu to full bloom. The uniqueness of umu-ntu would *not* be denied, while at the same time *not* denying the uniqueness of all that which is not umu-ntu but is in relation with umu-ntu.

This brings us to the last step. It seems possible to think of Ramose's ubu-ntu philosophy consistently including the proposed addendum without any major problems. We would consider the plausibility of ubu-ntu as indeed to be significantly increased by the appreciation of all beings. Perhaps this would allow for the unfolding of the full radicality that is contained in the idea of ubu-ntu; perhaps this would allow a progression towards equilibrium and unfragmented be-ing.

Conclusion

Philosophy owes a lot to Ramose's tireless work. His philosophy is of a poetic density, a depth of thought, and a unifying power that is rarely found. It is to be hoped that the reader was able to notice the deep appreciation for Ramose's philosophy – and that the critical uncertainty that we detected and the desuperiorisation that abounded in the addendum of the um-ntu maxim were always borne by the need to truly understand Ramose, to take the thoughts of this great philosopher radically seriously.

Note

1 I would like to express my deep gratitude to Mogobe Ramose for his support and all that I have learned from him. I also would like to thank the editors Kenneth Uyi Abudu, Kevin Behrens, and Elvis Imafidon for their support and patience. My thanks also go to Makhozasana Lunga, Mthunzi Thwala, and Thando Thwala for the help with some isiXhosa problems. Finally, I would like to thank Yvette Freter for her valuable comments on the draft versions. This paper is dedicated to her.

References

Baratay, É. (2022). *Animal Biographies: Toward a History of Individuals*. Athens: The University of Georgia Press.

Best, S. (2017). 'Species apartheid and the killing fields of Africa', in: R. Ebert and A. Roba (eds.) *African and Her Animals: Philosophical and Practical Perspectives*. Pretoria: UNISA Press, pp. 46–66.

Bewaji, J.A.I. (2003). 'Beyond ethno-philosophical myopia. Critical comments on Mgobe B. Ramose's *African Philosophy Through Ubuntu*', *South African Journal of Philosophy*, 22(4), pp. 388–401.

Chapman, C. A. and Huffman, M. A. (2018). 'Why do we want to think humans are different?', *Animal Sentience* 23(1), pp. 1–8.

Dladla, N. (2019). 'Ramose's conception of ubuntu as a philo-praxis of liberation', in: H. Lauer and H. Yitah (eds.) *The Tenacity of Truthfulness: Philosophical Essays in the Honour of Mogobe Bernard Ramose. Ugumuwadhanayaukweli: insha za kifalsafakumuenzi Mogobe Bernard Ramose*. Dar-es-Salaam: Mkukina Nyota, pp. 145–169.

Etieyibo, E. (2017). 'Anthropocentrism, African metaphysical worldview, and animal practices: a reply to Kai Horsthemke', *Journal of Animal Ethics* 7(2), pp. 145–162.

Freter, B. (2019). 'Veganismus als Anti-Nihilismus', *Zeitschrift für Kritische Tierstudien* 2, pp. 93–99.

Freter, B. (2020). 'Decolonization of the west, desuperiorisation of thought, and elative ethics', in Elvis Imafidon (ed.) *Handbook of African Philosophy of Difference*. Cham: Springer, pp. 105–127.

Freter, B. (2022). 'Decolonial philosophical praxis exemplified through superiorist and adseredative understandings of development', in Jonathan Okeke Chimakonam, Edwin Etieyibo, and Ike Dimegwu (eds.) *Essays on Contemporary Issues in African Philosophy*. Cham: Springer, pp. 209–226.

Gade, C.B.N. (2011). 'The historical development of the written discourses on ubuntu', *South African Journal of Philosophy*, 30(3), pp. 303–329.

Hallen, B. (2019). 'Be-ing and being ramose', in: H. Lauer and H. Yitah (eds.) *The Tenacity of Truthfulness: Philosophical Essays in the Honour of Mogobe Bernard Ramose. Ugumuwadhanayaukweli: insha za kifalsafakumuenzi Mogobe Bernard Ramose*. Dar-es-Salaam: Mkukina Nyota, pp. 13–20.

Horsthemke, K. (2015). *Animals and African Ethics*. New York: Palgrave Macmillan.

Horsthemke, K. (2018). '*Isilwane: The Animal – Ubuntu, ukama* and environmental justice', in: R. Ebert and A. Roba (eds.) *African and her Animals. Philosophical and Practical Perspectives*. Pretoria: UNISA Press, pp. 3–21.

Joy, M. (2010). *Why We Love Dogs, Eat Pigs, and Wear Cows: An Introduction to Carnism*. San Francisco: Conari.
Karr, S.M. (2015). *Verbundenheit. Zum wechselseitigen Bezogensein von Menschen und Tieren*. Berlin: Neofelis.
Kimmerle, H. (2000). 'African philosophy through ubuntu, by Mogobe B. Ramose [Review]', *African Philosophy* 13(2), pp. 189–197.
Kimmerle, H. (2006). 'Ubuntu and communalism in African philosophy and art', in: H. van den Heuvel, M. Mangaliso, and L. van de Bunt (eds.), *Prophecies and Protests: Ubuntu in Glocal Management*. Amsterdam: Rozenberg, pp. 79–91.
Mutwa, C. (1997). *Isilwane: The Animal. Second Impression*. Cape Town: Struik Publishers.
Næss, A. (2000) [1998]. *Livsfilosofi: Et personligbidrag om følelserogfornuft*. Second edition. De norskeBokklubbene.
Næss, A. (2002) [1998]. *Life's Philosophy: Reason and Feeling in a Deeper World*. Translated by Roland Huntford. : USA: The University of Georgia Press.
Oosthuysen, J.C. (2016). *The Grammar of isiXhosa*. Stellenbosch: Sun Press.
Ramose, M.B. (2005) [1999]. *African Philosophy Through Ubuntu*. Revised edition. Harare: Mond Book Publishers.
Ramose, M.B. (2007). 'African philosophy as the bridge to intercultural philosophy', *International Journal of the Humanities*, 4(9), pp. 123–131.
Ramose, M.B. (2009). 'Ecology through Ubuntu', in: M. F. Murove (ed.) *African Ethics: An Anthology of Comparative and Applied Ethics*. Scottsville: University of KwaZulu-Natal Press, pp. 308–314.
Ramose, M.B. (2017). 'Wiping away the tears of the ocean: Ukusulaizinyembezizolwandle', *Theoria: A Journal of Social and Political Theory*, 64(153), pp. 22–57.
Ryder, R.D. (1989). *Animal Revolution: Changing Attitudes Towards Speciesism*. Cambridge: Cambridge University Press.
Ryder, R.D. (2010). 'Speciesism again: The original leaflet', *Critical Society* 2, pp. 1–2.
Shutte, A. (1995). *Philosophy for Africa*. American edition. Milwaukee: Marquette University Press.
Singer, P. (2009) [1975]. *Animal Liberation*. New York: Ecco.
Wallner, B. (2019). 'Of course, humans are not unique!', *Animal Sentience* 23(25), pp. 1–2.
Wilkins, J.S. (2009). *Species: A History of the Idea*. Berkeley, Los Angeles and London: University of California Press.
Zimunya, C.T. and Gwara, J. (2018). 'The ethics of *ubuntu/hunhu* in the animal rights discourse in Africa', in: R. Ebert and A. Roba (eds.) *African and Her Animals: Philosophical and Practical Perspectives*. Pretoria: UNISA Press, pp. 22–32.

4 Interrogating Deep Ecology within the Frame of Ubuntu Philosophy

Dennis Masaka

Introduction

In efforts to proffer an emblematic idea of how African peoples, especially through their philosophy of ubuntu (and its variants), relate to the environment, there have been indications of it as something which is overly different from how peoples from other traditions conceive such relations. The scenario created is that of complete difference where African ecology is presented as deeply relational (Terreblanche, 2018, p. 181), and this reflects African ecological thinking (even when one can still hold that such ascription of the same moral status has overly human interests at stake). This is a belief that the environment has essentially the same status as the one extended to humans, while ecologies of other traditions, such as those of the Euro-North American world, are posited as starkly anthropocentric. This is implied in Mathew Bukhi Mabele, Judith E. Krauss, and Wilhelm Kiwango's claim that '*Ubuntu* disengages from western ways of knowing about human–environment interactions, as it is predicated on promoting the many links between humans and nonhumans' (Mabele, Krauss, and Kiwango, 2022, p. 92).There is no explicit articulation of what these links would look like in order to justify the claim of difference. However, as I will argue, this does not take away the strong appeal of an ubuntu-inspired ecology in giving intrinsic worth to the environment and what exists in it, just as the one extended to humans. That said, this 'narrative of opposites' points to the existence of different strands of deep ecology. While there are different strands of deep ecology such as radical ecology, there has not been a clear and deliberate attempt to delineate and define the strand and form of deep ecology that emerges from Ubuntu philosophy. This chapter delineates such a strand as qualified or moderate deep ecology. This is an under-appreciated yet, to me, a closer estimation of a typical African ecology, one that is at the intersection of radical deep ecology and an anthropocentric position.

I see ubuntu-inspired ecology as promising a more compelling ecology than its contender positions that lean more towards an anthropocentric stance. With this view in mind, I seek, in this chapter, to challenge the claim that ubuntu is characteristically supportive of deep ecology, as some of its defenders might want to claim (Le Grange, 2015, p. 301). I argue that while ubuntu might express the need to preserve the environment for present and future generations, it does so essentially

DOI: 10.4324/9781003480914-5

through qualified intrinsic lenses for it leverages human survival on respect for the environment as worthy of inherent value.

To say that the environment has stark intrinsic worth could be something that can rarely be spoken of in relation to humans because humans tend to think of the environment partly in terms of the benefits that it might extend to them. Added to that, whatever commitments humans have for a fairer and just ecology are expressly human imaginations of how the environment can best be used or treated. It is not at all something on which the environment has input, assuming that it has the wherewithal to partake in such deliberative processes leading to an agreed position on sound ecology. By so doing, I seek to show why it is less compelling to take ubuntu as a plausible base for grounding radical deep ecological thinking, but rather as a normative position that, on balance, guarantees a human-mediated intrinsic ecology.

In this chapter, the first aim is to muse over what deep ecology (Singer, 1994, p. 280) means in extant literature, and how a typical African community would view its relations to the environment. The intention is to prepare the ground for the second aim of the chapter, whose task is to demonstrate the central claims of ubuntu and what they might mean concerning how humans extend the moral community to the environment. Thereafter, I show that ubuntu can best be viewed as embodying a qualified or moderate notion of deep ecology. This is in opposition to a strand of thinking which might want to perceive it as overly committed to deep ecological thinking.

Deep Ecology and the African Space

Deep ecology has often been presented as a game changer in a world where activities and interactions with the environment are patently human-centered. This means that they show a lack of concern for the environment as an entity that is important in itself and never to be seen in instrumental terms. That deep ecology seeks to reverse the injustice of seeing the environment as embodying instrumental value has made it an intervention that is attractive to reconfigure ecology in ways that bring a different and fresh sense of relations that could thwart the environmental disasters of our times. In this section, I aim first to show how extant literature has understood deep ecology (Singer, 1994, p. 280) especially in the light of present-day ecological or environmental catastrophes, with Africa being a point of focus. In the second part, I argue that the enduring injustice of colonial extractivism (*see* Moti, 2019, p. 483) has generally meant that expressions of interest to rethink the relations of humans and the environment are still far from yielding desired results. With this in mind, I show, in the ensuing section, that a notion of deep ecology that does not engage the problematic of the colony, which is at the centre of some of the ecological disasters in Africa and other parts of the world, might not contribute much to the efforts to attain the idealistic deep ecology.

The thinking that seeks to change the way humans relate to the environment from one that sees it in an instrumental view to one that takes it as embodying inherent or intrinsic status has been called deep ecology (Singer, 1994, p. 280;

Naess, 2005, pp. 4–5). This obviously is a promise of difference in terms of how humans relate to the environment. The 'deep' which is appended to or prefixed to 'ecology' signals this commitment to change not only in terms of a move towards an equality of status of humans and the environment but also in terms of giving premium to the preservation of the environment for the benefit of present and future generations. The change sought is a reaction to the challenges of our times whereby the lust to fulfil human needs has led to the unsustainable exploitation of the environment and its defilation in ways that have posed a threat to the very existence of humans (Bond, 2006, p. 5). In a way, one might see the reaction of humans to ecological catastrophes that are consequent upon their careless use of the environment as primarily motivated by a threat to their own very existence. It is not because it has just dawned on them that, after all, the environment has inherent value which entitles humans not to treat it as merely expendable or that the future generations are owed a share of it. The deep ecology movement has been widely discussed in the Euro-North American world in response to the ecological disasters that have been a consequence of their accelerated exploitation and defilation of the environment in their insatiable quest for wealth generation, industrialisation, nuclear power, and the arms race more generally.

The pursuit of wealth through industrialisation has seen, especially, some members of the Asian and Euro-North American world turning to other parts of the world in search of new sources of natural resources and farmlands (*see* Ikuenobe, 2014, p. 1). The manner of exploitation of the environment that fulfils the profit motive and power consolidation motive will, in the end, contribute to the unsustainable use of the environment. In so doing, the ecological disasters escalated through Asian and Euro-North American-driven industrialism have been 'imported' to other parts of the world such as Africa. Through the unjust wars of conquest and imperial scramble for cheap natural resources, the ecological disasters driven by human pursuit for profit and power have become a worldwide phenomenon (see Ramose, 2009, p. 313). Africa has borne the bane of ecological disasters, which are mainly caused by fortune seekers coming from other regions, who have mainly seen the environment in instrumental lenses. Such an attitude to the environment is in opposition to the ecocentric stance for which some typical African normative positions, such as ubuntu, lean towards (*see* Terreblanche 2018, 181).

Talk of a more just and considerate relations between humans and the environment needs to be understood in the context of realisation of the unmitigated use of the environment through plunder and its defilement if humans were to continue on this trajectory. However, for Africa, pleadings from other spaces for a more just ecology seem to mean something other than the intended message. The reason is that such expressions of desire for a just ecology have not spared Africa from the problem of imperial plunder of natural resources that have been its albatross since the inception of unjust wars of conquest. If anything, there has been more sustained imperial plunder of African resources by power holders of this world for refinement and processing in their metropoles at a time when the call for a just ecology has gained currency across the world (Hountondji, 1990, p. 8). Indeed, this has often been done with the blessings of some corrupt African political elites who, by association, are complicit in the continued defilation of the environment and the unsustainable depletion of Africa's resources even when there have been murmurs of countries in Africa having seemingly transited the colony.

With that in mind, it would seem to make more sense to ponder what could be a plausible way to transition the 'old order' of overly instrumental use of the environment and enter an era of a just ecology after addressing the problem of the lingering colony. For one, talk of a decolonial or just ecology goes against the agenda of the colony. This agenda is principally motivated by the imperial plunder of the environment's resources, without due regard not only for the nominal benefits that accrue to host countries but also for the ecological effects such activities would have on present and future generations. As Patrick Bond aptly points out concerning the plunder of Africa's natural resources,

> the looting of Africa dates back many centuries, to the point at which value transfers began via appropriation of slave labour, antiquities, precious metals and raw materials. Unfair terms of trade were soon amplified by colonial and neocolonial relations. These processes often amounted to a kind of 'primitive accumulation', by which capital of Northern countries grew by virtue of looting Africa.
>
> (Bond 2006, 135–136)

This shows that the colony in Africa is still very much in force in its neocolonial mode. In this regard, the primitive extroverted extraction of Africa's natural resources by foreign organisations continues to prevail over concern for the welfare of present and future generations (*see* Hountondji, 1990, p. 8). This coloniality of the environment has created obstacles to efforts to attain the somewhat elusive ecological justice, which seriously considers future generations into account. The dominant presence of organisations and peoples from other regions in the extraverted exploitation of the environment in Africa cannot be thwarted by simply invoking the deep ecology narrative.

The point to be raised is that the decolonial ecology cannot be decolonial in the sense we typically understand it if it is not underwritten by a serious effort to transition from the colony. Instead, there must be bold attempts to overturn the colony in which the ecological disasters that countries in Africa are currently experiencing have their historical sources. Such a move is necessary because any efforts to bring into play an ecology that estimate to the idealistic deep ecology are sure to be pitted against the interests of dominant forces of capital. It then becomes a contest between the economic interests of forces of capital and the efforts to set in place a fairer and just ecology. It would appear that in light of the uneven balance of power between the forces of capital and countries in Africa, for which the former is favoured, efforts to consummate a fairer and just ecology can be hindered. This serves to show the destabilising effects of the power of the colony in which much of the countries in Africa are still very much immersed (Ikuenobe, 2014, p. 13). Be that as it may, the colony and its unsustainable hunger for African resources must be overcome if a decolonial ecology that estimate to the idealistic deep ecology is to successfully take hold.

Ubuntu and Deep Ecology

In the light of the ensuing ecological crisis gripping the African continent and the world at large, there have been various efforts to overcome it. This, it is thought, can be done by drawing from what would count as typical normative directions that would transform human–environment relations in a fairer and just way; or something that inclines more comprehensive ecological relations. In search, then, is such normative grounding that would typically help in consummating relations that align with more radical ecological thinking, or deep ecology, as some thinkers call it. It is to this that ubuntu has been variously spoken of as an ideal normative position to reorient in a decolonial fashion relations of humans and the environment so that the intrinsic value of the latter is accepted by the former. With this in mind, I show in this section what ubuntu is, and how it has been thought to be a plausible intervention to thwart the current ecological crisis facing Africa and the world at large. I then show how it can make a significant foothold in environmental debates if it were to be complemented with the decolonisation of the environment in Africa from colonial tentacles. Based on how ubuntu has been invoked in environmental debates mainly in Africa, I then attempt to show why it can be said to measure up to the demands of deep ecological thinking.

Ubuntu is a normative dispensation that takes as its key underlying principles the intrinsic value that each individual has as a constitutive ontological property and the importance of interpersonal acceptance of the fact that such value can be reinforced and reinvigorated by acquiescing to its unquestionable presence in all humans, or simply the communing scenario (Tutu, 1999, p. 29). These communing scenarios help to build and cherish the importance of interpersonal or inter-entity relationships and that interdependency is a reality of existence (Crippen, 2021, p. 240). Individual humans have what it takes to be creatures of dignity and to be owed any such rights and privileges that go with being human. However, such rights and privileges can be more assured in a scenario where every other human being sees the importance of living in reciprocal and respect-inspired relations with others. The social contract-like community in the mind of Ubuntu philosophy requires mutual acceptance of the importance of upholding the intrinsic status of all members of the moral community. Seeing a community constructed around ubuntu thinking as one that treasures relational and interdependent existence (*see* Behrens, 2010, p. 470) over and above the intrinsic worth of an individual human being is meant to ring-fence the entitlements of individuals and, more importantly, showcase the high value of cooperative existence.

But more importantly, this seemingly human-only moral community is not all there is if one were to consider how ubuntu is broadly conceived in the African space. It also concerns itself with how humans relate to the environment and what there is in it. As Polycarp A. Ikuenobe correctly notes, 'Africans consider nature as an end in itself which should be respected for its own sake: it is sacred and divine, it is in harmony with humans, and it has an inherent moral worth which derives from its role in cosmic and ontological order and its moral and religious primacy' (Ikuenobe, 2014, p. 10). When ubuntu is expanded to embrace the environment and its constitutive segments, ubuntu by the same measure extends the embrace of moral community and the values according to which members of the expanded moral community are to be viewed.

This expansive idea of moral community imbedded in the philosophy of ubuntu accounts for the inclusion of the environment as an entity that shares relationships and moral status with humans. Taken in this sense, the ubuntu community would then theoretically and practically show no bias in terms of how members of this expanded moral community are to be treated and to treat each other, for each member or segment ought to be seen as important in itself. Cooperative, interdependent, or relational existence helps to fortify the intrinsic worth of members of the community made of humans and the environment. Nevertheless, such an ubuntu-grounded community has often come under scrutiny from scholars, with Kevin Behrens (2010, 476) being one such scholar who raises what I would consider to be a merited doubt about when non-animate reality might equally fall under moral consideration as is extended to non-human animate reality. A detailed engagement with the problematic of an expanded moral community will be the subject of the last section of this work.

Because of its somewhat organic idea of an expanded moral community, ubuntu has been variously touted as good enough to fulfil the expectations of deep ecological thinking. Recall that deep ecology takes interest in an expanded moral community where inherent value is extended to its members and segments. According to many of the ubuntu defenders, there possibly could be no other persuasive normative position than ubuntu that can best embrace human, animate, and inanimate reality of the environment into a coherent inherent value-laden community (Murove, 2004, p. 196). Some of the claims made about ubuntu to show its qualities that are thought to measure up to the equality of status between humans and the environment are: 'ubuntu is regarded as a deeply bio- or ecocentric ethics' (Terreblanche 2018, 181); 'there is no division between human society and other living things. Human existence is only meaningful when seen as a continuum with all else that exists' (Murove 2009, 324); and '*ubuntu/botho* is an ecosophy that is aligned with the principles of the deep ecology movement' (Le Grange 2015, 301). Seen in the context of alternative theoretical grounds for an attractive human–environment relationship, these expressions are meant to showcase the merits of ubuntu in offering a moral community that thwarts historical asymmetries in terms of how the environment weighs against humans. This serves to show that for some thinkers (*see* Le Grange 2015, 301) that ubuntu can well embrace deep ecological thinking.

Let's recall that in the context of countries in Africa, the ecological crisis being experienced is taking place within the context of neo-colonialism or coloniality, that is, a scenario where the colony has outlived claims of independence. This then means that the ecological crisis in Africa has colonial connections. This explains why efforts to search for a normative position to ground a fairer and just ecology in Africa make reference to the failings of imposed approaches of human–environment relations. With this in mind, the choice of ubuntu as one such ideal normative position stems from its qualities to thwart the ecological crisis being experienced in Africa and the world more broadly. Its resources, such as relationality, interdependence on grounds of sameness of inherent value, solidarity, and fellowship between humans and the environment, have been its trump card. Ubuntu sets a decolonial path meant to rescue the environment from human conquest through the instrumental weight extended to it. The restoration of the intrinsic value of the environment from its conquest is the liberatory path that ubuntu is thought to engender, thus leaning towards what other traditions refer to as 'deep ecology'.

Ubuntu philosophy scholars have made explicit this decolonial turn that is meant to secure the environment from further plunder and the threat of extinction on account of its assumed instrumental value. Mogobe Bernard Ramose (2009, 312), while noting that perhaps humans are the 'privileged part' of the expanded moral community, argues that the African concept of harmony, which is a key pillar of ubuntu, requires that a fair balance be struck among humans themselves as well as between humans as a genus and the environment. To add, 'the quest for harmony is thus the striving to maintain a comprehensive but specific relational condition among organisms and entities. It is the constant striving to strike, and then maintain, a balance between human beings and physical nature' (Ramose 2009, 309). Harmony, to me, goes hand in hand with certain entitlements to those things brought into communing, such as respect for their intrinsic and non-instrumental status. This would entail that in living in harmony, humans and the environment are engaged in a relation built on respect and equal status of existence for members of this expanded community.

Perhaps, as part of a strong commitment to ecological communing drawn from a reading of ubuntu philosophy, some African communities have in place ingenious strategies to promote fairer and just relations between humans and the environment. What comes to mind are taboos that come in the form of stern threats of misfortune, illness, or even death that would visit anyone who engages in actions considered injurious to the environment or some of the things that belong to it (Masaka and Chemhuru, 2011, p. 142–144).These prohibited actions include cruelty to non-human animals, defilation, and unmitigated use of the environment or some of its segments or members for human self-centered goals. A complementary strategy to engender respect-filled relations with the environment is humans' adoption of totems. This enjoins humans to extend intrinsic importance and respect to members or parts of the environment to which they are in totemic relations. Defilation, harm, or use of totem animals or objects for human ends is, by extension, an invitation for trouble to those in totemic relations, which no sane human will be prepared to endure, so goes the thinking (Chemhuru and Masaka, 2010, p. 123). Put together, these strategies show how an ubuntu-driven community would go in creating an expanded moral community and grounding it on mutual relations.

Besides that, ubuntu is thought of as a philosophy that engenders a strong sense of 'mutual interdependence' between humans and the environment (Murove 2009, 325).

Invoking African relational thinking, best represented by the concept of ukama which also manifests in ubuntu thinking, Munyaradzi Felix Murove takes a well-lived life to be one which, besides living in harmony with the environment, also shows that 'concern for the past is inseparable from concern for the future' (Murove 2009, 330). I would also add that the concern of the present is inseparable from the well-being and welfare of the future. This is what measures to ecological communing, whereby there is equal dispersion of inherent value to both humans and the environment that together forms an expanded present and future moral community. Because of this tradition of relationality borne out of the equality status of members of the expanded moral community, concern for future generations becomes very much the responsibility of present generations. Understood thus, African relational thinking very well takes as its utmost concern interconnectedness, interdependence, and inherent value of members of the expanded moral community, which also include generations to come.

In the light of what ubuntu defenders have put forth as its merits in transforming relations between humans and the environment, it is important to show how it might be conceived to be in agreement with what has been elsewhere named 'deep ecology'. Recall that deep ecology wants humans and the environment to establish relations that underscore that both have inherent value, and that future generations' interests must be upheld by present generations. Ubuntu makes similar, if not more persuasive, claims to an egalitarian community where humans and the environment are thought to be one grand community, both having inherent value. This obviously would constitute a decolonial venture or departure from seeing the environment in instrumental terms.

For its commitment to a radical change that brings humans and the environment into germane and inherent value-filled communion, an ubuntu-driven moral community has been seen as a typical intervention that would thwart the ensuing environmental crisis facing Africa and the world at large and save it from further damage. To some, the goal of deep ecology has long been part of ubuntu philosophy, from time immemorial (*see* Le Grange 2012, 333–334). With this in mind, ubuntu would take the mantle of a decolonial initiative that also takes on board the central claim of deep ecology. In essence, the expanded moral community which ubuntu constructs is non-anthropocentric, for it takes both humans and the environment as having inherent value. It is its radical non-anthropocentric stance and ecocentric orientation that make it lean towards the central claim of what some philosophic traditions take as deep ecology.

Ubuntu and Qualified Defence of Deep Ecology

Having shown how ubuntu leans towards an inherently value-inspired expanded moral community, or what has elsewhere been called 'deep ecology', it is important to adjudicate the merits of this claim. This is necessary because for ubuntu to claim to have decolonial credentials meant to overcome the coloniality of the environment, it must show an explicit departure from an overly instrumental view of the environment for which the colony in Africa is infamous.

From the views of prominent writers on ubuntu and the environment, what a typical ubuntu community does is to take the environment to be of equal status to human counterparts (*see* Terreblanche, 2018, p. 181). Thus conceived, an ubuntu-inspired moral community would be different from other contending human–environment relations, such as one that has been associated with the colonial period in Africa. However, it is important to establish whether ubuntu can really be said to embrace the environment as an inherently value-filled member of the moral community, as ubuntu defenders allude (*see* Le Grange, 2015, p. 301). To my estimation, an ubuntu-backed moral community can ideally be conceived as something akin to a qualified defence of deep ecology. With this in mind, I demonstrate why it is less compelling to take ubuntu as a plausible normative position for grounding deep ecological thinking,particularly when considered in its radical posture – when it is viewed as an inherent value-giving and assuring social contract between between humans and the environment. Relatedly, I probe what a 'relationship' entails conceptually when it is used in connection with the harmony of humans and the environment (*see* Chigangaidze, 2023, p. 1849). Such an inquiry is important in building a case that ubuntu can best be understood as offering a qualified defence for deep ecology.

Whenever we talk about how humans might best relate to the environment, it is clear that the whole discourse is a human creation. Following from this, whatever value and rights are extended to the environment can be described as humans' impressions on how best they weigh it 'for its own sake' and the importance of safeguarding it for present and future generations. As I have argued elsewhere, any ascription of inherent value to the environment is simply a human estimation of what would constitute a fairer and just relationship between humans and the environment (Masaka, 2019, pp. 227–228). To read claims of equality of moral status otherwise would be to miss the point that such ascriptions are wholly human inventions on what humans take to be the weight of status that can be extended to the environment.

While it is clear that ascription of moral status to the environment is a human 'benevolent act', it is often couched in the language that gives an impression of a social contract between humans and the environment. Alternatively, it would appear that the idea of a moral community that ubuntu has in mind is a product of a 'conversation' of some kind between humans and the environment, culminating in some form of social contract. It is this social contract-like scenario that would give effect to an ubuntu-inspired decolonial moral community. Connected to this is the talk of 'human *relations with*[1] the natural environment' (*see* Murove 2009, 316), that is, humans' relationship with the environment. Such talk implies agency of those things or entities that are said to be in a relation. Expectedly, such agency in the construction of a social contract cannot be spoken of the environment and its segments. It is something that humans might construct in order to overcome the instrumental use of the environment by using language in such a way that it would appear as if there are, as it were, conversations between them and the environment whose outcome is a social contract-like moral community.

An objector might rightly retort that the source of equality-of-moral-status drive might not be an important issue here. What is at issue is simply the normative position that has the wherewithal to break the cycle of plunder of the environment in Africa and the world at large, and install within humans a sense of responsibility in their dealings with the environment, for it is a member of the moral community with inherent worth just like any other member. Such an objector's point is merited because a fairer and just ecology is a concern emanating from some humans' admission that their relationship 'with' the environment has been an exploitative one to the latter. For that reason, any normative position that realistically takes the environment into the moral community as a member with equal moral status with humans, to me, is something that will make a great difference in the fight against the environmental crisis of our times (*see* Terblanché-Greeff, 2019, p. 106). With this in mind, ubuntu's idea of a moral community made of humans and the environment existing in a relational web underpinned by equal ascription of inherent value is something worthy of commendation even though it is a wholly human construct.

Looking at some accounts of an ubuntu-inspired moral community, it seems clear that an ecology it embodies leans more towards considerations of human interests (Murove 2004; Ramose 2009). It does so by ascribing inherent value to the environment as a way of proscribing actions that pose an existential threat to it, which may, in turn, pose an existential threat to present and future human generations. This is possibly out of the realisation that human interests can best be served if the environment is accorded non-instrumental status. Take, for example, Ramose's claim that:

> through autopoietic activity the human being is constantly interacting with its environment, with nature in pursuit of 'self-preservation'. However, the pursuit of self-preservation is simultaneously a direct or indirect, immediate or mediated, propinquitous or more remote preservation of the universe as wholeness.
>
> (2009, 310)

Similarly, Murove expresses something I think is closer to the idea of a qualified deep ecology that I have in mind when he states that Africa has a normative position best suited to overcome the ecological crisis unfolding in the present world more broadly. To this, Murove has the following to say, 'Africa [yet] possesses in its own traditional culture the roots for an ethical paradigm to solve the current environmental crisis. This is an ethic of the interdependence of individuals within the larger society to which they belong and to the environment on which *they all depend*"[2] (Murove 2009, 315).

To me, the views expressed by Ramose (2009) and Murove (2009) demonstrate that in the pursuit of human preservation, humans feel the impulse to take the environment as part of the moral community that enjoys similar entitlements as those extended to humans. This is an ubuntu strategy that seems to work well in mitigating the unsustainable use of the environment by conceiving it as a worthy member of the moral community. While, in these views, there is a clear reference to the interdependence of humans and the environment, one can also notice the urge that human interests seem to have over propping the idea of an environment with intrinsic value. For this reason, the deep ecology idea that can be drawn from an ubuntu-inspired moral community might best be described as a qualified one.

I take qualified deep-ecology to mean extension of inherent value to the environment not solely out of the realisation of its importance in itself but also motivated by what unmitigated use of it might mean to the welfare and well-being of present and future generations. My take is that deep ecology in its radical form does not seem to be attainable. The reason is that even in those normative positions such as ubuntu that present a persuasive case for the respect of the inherent value of the environment, there is always an explicit pronouncement of what failure to do so would mean to the interests of humans. Claims of *interdependence* of humans and the environment that constitute an ubuntu moral community point to an undeniable reality: they are bound in a relational grid (Nana, 2016, p. 165), where the welfare and well-being of one depends on equal respect for the welfare and well-being of the other.

But, then, being a 'consensus position' drawn by one part of the moral community when ideally it should be an outcome of contributions from both parts is most likely to tilt the scale in favour of the drafters of the 'consensus' position. Nevertheless, the relationship that is an outcome of an ubuntu moral community is very much a departure from one where the environment is seen overly in instrumental terms. Following from this, such a relationship that the ubuntu moral community instils is not anthropocentric, nor can it neatly fit the bill of deep ecology. It leans more towards a qualified sense of deep ecology because humans extend inherent value to the environment, noting that seeing it in an instrumental fashion poses an existential threat to the survival of present and future human generations.

Concluding Remarks

What I would say is that even in its qualified sense, the deep ecology that the ubuntu moral community embodies makes something akin to a decolonial move regarding the relations between humans and the environment. By moving away from an instrumental view of the environment that has been implicated in the environmental crisis that is presently unfolding across the world and proposing a normative position that substantively changes the outlook, the ubuntu moral community wants to put a stop to the unmitigated and unsustainable use of the environment. This means that an ubuntu moral community helps to thwart the asymmetrical relations between humans and the environment, which some normative positions seemingly abet. Saving the environment from further damage ought to be genuinely a global concern. Mere theoretical pronouncements on the need to avert it will not be helpful.

A normative position in the mould of ubuntu would help to reconfigure human–environment relations in ways that would save the environment were it to influence human–environment relations not only in Africa but throughout the world. An ubuntu moral community might be interpreted as leaning towards what we have called 'qualified deep ecology' and thus on some points short of the idealistic optimism of deep ecology. However, it still has what it takes to offer a persuasive intervention to debates on the environmental crisis based on its notion of a moral community of inherently valuable entities. By leveraging human survival on the protection of the environment from plunder and unmitigated use (Ikuenobe, 2014, p. 10), an ubuntu moral community helps the environment from real threats of extinction.

Notes

1 My emphasis.
2 My emphasis.

References

Behrens, K. (2010) 'Exploring African Holism with Respect to the Environment', *Environmental Values*, 19, pp. 465–484.

Bond, P. (2006) *Looting Africa: The Economics of Exploitation.* Pietermaritzburg: University of KwaZulu-Natal Press.

Chemhuru, M. and Masaka, D. (2010) 'Taboos as Sources of Shona People's Environmental Ethics', *Journal of Sustainable Development in Africa*, 12(7), pp. 121–133.

Chigangaidze, R. K. (2023) 'Environmental Social Work Through the African Philosophy of Ubuntu: A Conceptual Analysis', *International Social Work*, 66(6), pp. 1845–1856.

Crippen, M. (2021) 'Africapitalism, Ubuntu, and Sustainability', *Environmental Ethics,* 43(3), pp. 235–259.

Hountondji, P. (1990) 'Scientific Dependence in Africa Today', *Research in African Literatures*, 21(3), pp. 5–15.

Ikuenobe, P. A. (2014) 'Traditional African Environmental Ethics and Colonial Legacy', *International Journal of Philosophy and Theology*, 2(4), pp. 1–21.

Le Grange, L. (2012) '*Ubuntu, ukama*, Environment and Moral Education', *Journal of Moral Education*, 41(3), pp. 329–340.

Le Grange, L. (2015) 'Ubuntu/Botho as Ecophilosophy and Ecosophy', *Journal of Human Ecology*, 49(3), pp. 301–308.

Mabele, M. B., Krauss, J. E. and Kiwango, W. (2022) 'Going Back to the Roots: *Ubuntu* and Just Conservation in Southern Africa', *Conservation and Society*, 20(2), pp. 92–102.

Masaka, D. (2019) 'Moral Status of Non-human Animals from an African Perspective', in M. Chemhuru (ed.) *African Environmental Ethics: A Critical Reader*. Cham: Springer Nature Defense of Moderate Anthropocentric Thinking, pp. 223–237.

Masaka, D. and Chemhuru, M. (2011) 'Moral Dimensions of Some Shona Taboos (*Zviera*)', *Journal of Sustainable Development in Africa*, 13(3), pp. 132–148.

Moti, U. G. (2019) 'Africa's Natural Resource Wealth: A Paradox of Plenty and Poverty', *Advances in Social Sciences Research Journal*, 6(7), pp. 483–503.

Murove, M. F. (2004) 'An African Commitment to Ecological Conservation: The Shona Concepts of *Uhama* and *Ubuntu*', *The Mankind Quarterly,* XLV(2), 195–215.

Murove, M. F. (2009) 'An African Environmental Ethic Based on the Concepts of Ukama and Ubuntu', in M. F. Murove (ed.) *African Ethics: An Anthology of Comparative and Applied Ethics*. Scottsville: University of KwaZulu-Natal Press, pp. 315–331.

Nana, A. E. (2016) 'The Impact of Ecology in African Traditional Religion', *International Journal of Theology and Reformed Tradition*, 8, pp. 164–174.

Naess, A. (2005) 'The Deep Ecology Movement: Some Philosophical Aspects', *OpenAirPhilosophy.org*, 20, pp. 1–16.

Ramose, M. B. (2009) 'Ecology *through* Ubuntu', in M. F. Murove (ed.) *African Ethics: An Anthology of Comparative and Applied Ethics*. Scottsville: University of KwaZulu-Natal Press, pp. 308–314.

Singer, P. (1994) *Practical Ethics*. Cambridge: Cambridge University Press.

Terblanché-Greeff, A. C. (2019) 'Ubuntu and Environmental Ethics: The West Can Learn from Africa When Faced with Climate Change', in M. Chemhuru (ed.) *African Environmental Ethics: A Critical Reader*. Cham: Springer Nature, pp. 93–109.

Terreblanche, C. (2018) 'Ubuntu and the Struggle for an African Eco-socialist Alternative', in V. Satgar (ed.) *The Climate Crisis: South African and Global Democratic Eco-Socialist Alternatives*. Johannesburg: Wits University Press, pp. 168–189.

Tutu, D. M. (1999) *No Future Without Forgiveness*. New York: Doubleday.

5 An African Theory of Moral Status

A Relational Alternative to Individualism and Holism[1]

Thaddeus Metz

Introducing African Relationality

English-speaking philosophical accounts of moral status are usually either holist or individualist. For example, claiming that a being warrants moral consideration in light of its capacity for reason or pleasure implies that moral status is a function of something internal to an individual. In contrast, views that ascribe moral status to ecosystems and to species are holist accounts, deeming moral status to inhere in groups of certain kinds. Much of the contemporary debate about moral status concerns whether it is adequately captured by either individualism or holism alone, and which specific versions of these general perspectives are most attractive.

There is, however, a third perspective on moral status available, one that grounds moral status on *relational properties*. My aims in this chapter are to develop a new relational theory of moral status and to bring out its often appealing implications in a wide variety of contexts. While I do not have the space to argue that it is more justified than all the views that currently dominate English-speaking discussion, I do mean to demonstrate that it should be taken just as seriously. Of particular interest is that the theory promises to solve some long-standing conundrums that continue to plague Anglo-American discussion of moral status, e.g., why animals and humans might both have moral status that is of the same kind but different in degree, why even a severely mentally incapacitated human being might have a greater moral status than an animal with identical internal abilities, and why a newborn infant might have a greater moral status than a mid-to-late stage foetus.

I call the theory I develop 'African' since it is grounded on a central strand of sub-Saharan ethical thought, which places relationality at the core of morality. One will find individualism and holism held by African moral philosophers, but the relational theory I articulate arguably fits with much more of the field *qua* communal. One will also find some relational ideas in Western philosophy, particularly in the ethic of care, but it is far from a prominent contender, probably in part because it and similar views have been overly parochial. My goal is to spell out a relational account of moral status that is informed by sub-Saharan ethical reflection, avoids major problems facing the relational accounts that one currently finds in the literature, and, above all, plausibly accounts for many intuitions about moral status.

DOI: 10.4324/9781003480914-6

The Afro-relational account may be viewed as one attractive interpretation of the key deep ecological point that aspects of the nonhuman world morally matter for their own sake, ultimately entailing that killing animals for sport, harming them for cosmetic testing, and eating them when unnecessary for health are wrong (Metz 2024). That view is in contrast to the 'shallow' environmentalism according to which all of nature exists merely to be used for human purposes, albeit in a long-term or 'sustainable' manner.

However, there are other views sometimes associated with the tradition of deep ecology that my favoured account of moral status does not support, including the claim that literally all living beings have an equal non-instrumental value. I believe this much stronger position is in serious tension with the claim that meeting the vital needs of human beings can justify actions such as killing carrots to feed a person or saving the life of a person instead of that of a cat (supposing one cannot save both and must choose). In fact, much of what motivates the Afro-relational view is that, even if all living beings have some non-instrumental value, there are intuitively gradations of moral status, with some beings mattering more than others from a moral point of view.

I begin by briefly defining what I mean by 'moral status' and differentiating between individualist, holist, and relational theories of it (The Concept of Moral Status). In the next section, I articulate the view that, roughly, the more a thing is capable of being part of a certain communal relationship, the greater its moral status (section 3). Then, I apply the modal-relational theory to many cases, noting respects in which it easily entails and reasonably explains a variety of firm judgements about moral status, including several that have evaded theoretical capture up to now by the most prominent, viz., Kantian and utilitarian traditions (section 4). I conclude that, in light of the African theory's ability to account for many widely shared intuitions, it warrants no less attention than individualist and holist accounts as a promising form of monism about why beings matter morally (section 5).

The Concept of Moral Status

The concept of moral status is the idea of something being the object of a 'direct' duty, i.e., owed a duty in its own right, or the idea of something that can be wronged. Contrast this idea with that of an object of an 'indirect' duty, where an agent has a moral reason to treat it a certain way, but not ultimately because of facts about it. For a standard example, I could have a duty not to destroy a football, where this duty would be indirect, something owed not to the ball, which cannot be wronged, but to its owner.

The way I understand moral status, it in principle admits of degrees. Some philosophers do not understand the concept this way (Gruen, 2003; cf. DeGrazia, 2008), but there are strong reasons for siding with those who think otherwise (such as Miller, 1988; Van De Veer, 1995; Warren, 2003).[2] For one, the existence of degrees of moral status best explains many intuitions about forced trade-offs amongst the urgent interests of different beings. Suppose, for example, that you are driving a bus and have no alternative but to run over either a mouse or a normal, adult human being. The right thing to do would be to run over the mouse, which seems best explained by the idea that its moral status is not as great. For another, differential moral status also accounts best for uncontroversial judgements about how to treat beings that have already been killed or otherwise died. If an animal is dead for whatever reason, many find it permissible not to let its body go to waste, but such a practice applied to humans would be horrific. In any event, I shall work with a sense of 'moral status' that is not an all-or-nothing matter; I suppose that something can have either full moral status, which normal, adult human beings are typically thought to have, or partial moral status.

An individualist account of moral status is the view that properties intrinsic to an entity ground the capacity to be wronged or to be the object of a direct duty. An intrinsic property, as understood here, is a property that is internal to an individual and that includes no essential connection to any other being. Influential forms of individualism include the views that moral status is solely a function of: being a living organism (biocentrism), exhibiting the capacity for autonomy or rationality (Kantianism), being able to recollect the past and satisfy desires for future states of affairs (subject of a life theory), and having the capacity for preference realization/frustration or for pleasure/pain (welfarism). All these properties can be purely internal to an individual.

A holist account of moral status, as I construe it here, is the view that the bearers of moral status are groups, where a group is a discrete collection of entities that are near, similar to, or interdependent with one another. Aldo Leopold's (1968) land ethic is a clear version of holism, as are accounts that accord moral status to species and not merely to their individual members.

A relational account of moral status is the view that it is constituted by some kind of interactive property between one entity and another. It therefore stands 'in between' individualism and holism. Similar to individualism, a relational account implies that moral status can inhere in things as they exist apart from their membership in groups. A relational theory implies that something can warrant moral consideration even if it is not a member of a group, or, more carefully, for a reason other than the fact that it is a member. Similar to holism, though, a relational account accords no moral status to organisms on the basis of their intrinsic properties. A relational theory implies that a being warrants moral consideration only if, and because, it exhibits some kind of intentional (attitudinal) or causal (behavioural) property with regard to another being.

One might reasonably wonder what the difference is between holism and relationalism, for, after all, is not a group just a kind of relationship? As I understand 'group', something could constitute a group without it being a 'relationship' in the relevant sense. For example, the species of elephants is a group, but there is no *interaction* between elephants in Kenya and those in India, and hence no relationship. Furthermore, there could be a 'relation' between organisms that do not constitute a group, say, a person whom tourists pay to ride on an elephant. In short, relationships require interaction of some kind between organisms, while groups do not, and groups require some kind of discreteness, while relationships do not. To be sure, there are *some* groups that are constituted by relationships, e.g., ecosystems and biotic communities within them, but the properties of being a group and of being a relationship are nonetheless conceptually and metaphysically distinct and ground differing accounts of moral status.

A Modal-Relational Account of Moral Status

A natural way to appeal to relationality to ground moral status would be to maintain that being in a relationship with a certain agent is necessary and sufficient to confer moral status on it. However, such an approach would suffer from the grave problem of failing to entail that certain beings that intuitively have moral status in fact do. If something had moral status only if it were in actual relation with an agent, then anything that is not would not count at all from the agent's moral point of view. Too many beings that uncontroversially have moral status would be excluded. To avoid such problems of arbitrariness and parochialism, my approach is to keep the characteristically African idea that morality is a function of community in some way but to focus on a novel relational property. My proposal is that a being has moral status roughly insofar as it is *capable* of being part of a communal relationship or of entering into community with us. In the rest of this section, I spell out the nature of communality as I construe it, ground a conception of moral status on the capacity for it, and briefly bring out its sub-Saharan pedigree. I apply the theory only in the following section.

What is a morally attractive sort of communal relationship? Elsewhere, in the context of formulating a principle of right action grounded on African mores, I have articulated and defended the following analysis of communality: a relationship in which people identify with each other and exhibit solidarity with one another (Metz, 2007). To identify with each other is largely for people to think of themselves as members of the same group, i.e., to conceive of themselves as a 'we', as well as for them to engage in joint projects, coordinating their behaviour to realise shared ends. For people to fail to identify with each other could involve outright division between them, i.e., people not only thinking of themselves as an 'I' in opposition to a 'you' or a 'they', but also aiming to undermine one another's ends. To exhibit solidarity with one another is mainly for people to engage in mutual aid, to act for the sake of one another. To act for the sake of another is, in the first instance, to seek to improve some other organism's objective quality of life for non-instrumental motives. Solidarity is also a matter of people's attitudes such as affections and emotions being invested in others, e.g., by feeling good consequent to when their lives flourish and bad when they flounder. For people to fail to exhibit solidarity would be for them to be either indifferent to one another's welfarist (and perhaps additionally perfectionist) good or downright cruel.

An equivalent way of phrasing my preferred analysis of communality is in terms of relationships in which people share a way of life and care for one another's quality of life. Note that the combination of sharing a way of life and caring for others' quality of life is basically a relationship that English-speakers call 'friendship' or a broad sense of 'love'. Hence, one way to put my favoured conception of moral status is this: the more a being is capable of being part of a friendly or loving relationship with normal humans, the greater its moral status.

By the present view, a being that is capable of being both the *subject and object* of such a relationship has *full* moral status, whereas a being that is capable of being merely the *object* of such behaviour has *partial* moral status. Being a subject involves identifying with others and exhibiting solidarity with them oneself. A being can be a subject of the relevant communal relationship insofar as it can think of himself as a 'we', seek out shared ends, sympathise with others, and act for their sake. In contrast, a being can be the object of a friendly relationship insofar as characteristic human beings could think of it as part of a 'we', share its goals, sympathise with it, and benefit it. Note that having the capacity to be an object of such a relationship does not imply that a being would or even could *respond* to any friendly engagement by another (*contra* Noddings 1984).

This modal view is neither that a communal relationship itself has moral status, nor that only those who are in such a relationship have it, but rather that those who could be part of such a relationship have it.[3] To be 'capable' of being part of a communal relationship means being able *in principle,* i.e., without changes to a thing's nature. Contingent obstacles to being a subject or object of a communal relationship are not constitutive of a thing's 'capacity' to be a part of it in the relevant sense.

Examples of contingent inabilities to be the object of a communal relationship, irrelevant to a being's moral status, include the facts that the being is: unknown, the object of false beliefs, and the object of fear or disgust on the part of an individual. While these might hinder a being's actually becoming the object of identity and solidarity with one of us, they are not relevant to determining its moral status, for it nonetheless could become such an object in light of its nature and ours. A being that 'cannot' be the object of a communal relationship in the relevant sense would, for instance, be one that utterly lacks the ability to have a better or worse life, say, a rock.

Examples of contingent inabilities to be the subject of a communal relationship include being asleep, having drunk too much alcohol, electing not to sympathize, being too sick to help others, and being ignorant of what would benefit others. While these might hinder a being's actual ability to become the subject of identity and solidarity, they are not relevant to determining its moral status, for it nonetheless could become such a subject in light of its nature. A being that 'cannot' be the subject of a communal relationship in the relevant sense would be one that is, say, genetically unable to act for the sake of others.

I mean two things when I say that the *greater* the capacity for communal relationship (of identity and solidarity), the *greater* a being's moral status. In the first place, I am indicating, as I have said, that being capable of both subject and object of a communal relationship constitutes a higher status than merely being the object. In the second place, though, I mean that large, and not merely incremental, differences in degrees of ability to be either a subject or an object of a communal relationship constitute differences in moral status. So, for example, if, by virtue of the nature of human beings, dogs, and mice, humans were *much* more able to identify with and exhibit solidarity with dogs than with mice (upon full empirical information about both), then dogs would have a greater moral status than mice. The qualification about 'much' is important. Of those who believe in degrees of moral status, most implicitly think of it in a way that can be captured well, I submit, with an orbit metaphor. Supposing that dogs were only marginally more capable than, say, cats of entering into the relevant communal relationship with us, cats would intuitively get 'pulled into' the same moral status orbit that dogs have.

Thinking of moral status in terms of the capacity for relationships of identity and solidarity coheres well with a principle of right action prescribing respect for such relationships, which *prima facie* captures several salient (not essential) facets of behaviour and thought south of the Sahara desert.[4] For example, sub-Saharans often think that society should be akin to family; they tend to believe in the moral importance of greetings, even to strangers; they typically refer to people outside the nuclear family with titles such as 'sister' and 'mama'; they frequently believe that ritual and tradition have a certain degree of moral significance; they tend to think that there is some obligation to wed and procreate; they usually do not believe that retribution is a proper aim of criminal justice, inclining toward reconciliation; they commonly think that there is a strong duty for the rich to aid the poor; and they often value consensus in decision-making, seeking unanimous agreement and not resting content with majority rule. I have the space merely to suggest that these recurrent (not invariant) practices are plausibly entailed and well explained by the prescription to respect people insofar as they can both share a way of life and care for one another's quality of life. I am not suggesting that this principle has been believed by even a majority of Africans; my point is rather that it promises to capture in a *normative theory* several salient aspects of a communal way of life that have been widespread in particularly the large sub-Saharan region, and hence that it and its companion account of moral status qualify as an 'African' philosophical ethic.[5]

Although I believe it apt to call the theory of moral status I have constructed 'African' since it is grounded on features of folkways and worldviews that are salient amongst peoples indigenous to the continent, it ultimately matters little what it is called. What matters most is how well it accounts for comparatively firm judgements about what has moral status and to what degree.

Applying and Defending the Modal-Relational Account

Here I consider what the account of moral status from the previous section entails in a variety of contexts, bringing out its many strengths and putative weaknesses in relation to dominant Western approaches. I first apply the theory to human beings, then to non-human animals, and finally to other things such as plants and ecosystems.

When it comes to normal, adult human beings, exemplary instances of beings with full moral status, it follows from the modal-relational theory that they have it. This theory implies that full moral status inheres in beings that can be both subjects and objects of communal relationships, and typical humans not only are capable of being identified with and being cared for, but also can identify with others and care for them. So far, the theory does as well as Kantianism and utilitarianism at *entailing* the intuitive moral status of human beings.

Furthermore, there are reasons to think that the African account of human moral status is *explanatorily* more attractive than at least some of the Western individualisms. Utilitarians deny that dignity is a genuine moral category; while they readily think in terms of a being warranting moral consideration, they do not believe that any being by its nature has a superlative non-instrumental value respect for which grounds basic human rights. However, many of those who believe that human beings have a moral status also think that they have dignity, and modal-relationalism can account for that intuition. It is naturally extended to say that humans have dignity if they have the full ability to participate in communal relationships. If one is driving a bus and must run over either a mouse or a human being, the adherent to the present view can plausibly contend that the reason to run over the mouse is that it lacks a dignity that the human being has when it has the capacity to love and be loved by us.

Friends of Kant's and Kantian views, of course, are also friends of dignity. Although Kant himself maintained that the capacity for moral reason is definitive of dignity and hence of moral status (1797, Ak. pp. 423, 434–435), contemporary Kantians usually ground it in a being's more general capacity for reasoned deliberation or self-governance. The African theory is an attractive alternative to Kant's view and is competitive with regard to recent forms of Kantianism. Kant himself cashed out the ability to act morally in terms of a capacity to act on the basis of representations that are beyond space and time in a noumenal realm, which involves a less than compelling metaphysics that the African theory does not invoke. And unlike contemporary Kantianism, the modal-relational theory entails that a rational being that can, say, neither think of himself as a 'we' nor act for the sake of others lacks a full moral status, which many (though, I accept, not all) will find intuitive. The suggestion is not that individuals who have acted wrongly have a lower moral status, but rather that individuals who are utterly incapable of any other-regard do. In contrast, the Kantian (though probably not Kant) is committed to the view that a rational being capable only of self-regard has a dignity equal to one capable also of other-regard.

Let us consider the case of the purely self-regarding rational agent in more detail. Although the African theory probably cannot claim an advantage relative to the Kantian with respect to this sort of case, I submit that the former is not much, if any, worse off than the latter. First off, note that the modal-relational theory need not entail that autistic humans, or at least 'higher' functioning ones, such as the well-known Temple Grandin, lack a dignity or full moral status. While these individuals appear to be comparatively incapable of empathetic awareness and sympathetic emotion, the evidence indicates that they are capable of additional forms of other-regarding behaviour (Kennett, 2002). They are able to coordinate their actions with others, to do what is likely to make others better off, to act for the sake of others, and so on. Given the orbital account of degrees of moral status I suggested above, the African theory entails that many autistic individuals have a dignity or a moral status equivalent to what we have.

However, the African theory does appear to entail that *severely* mentally incapacitated human beings and *extreme* psychopaths lack dignity, for they are incapable of being subjects of a communal relationship. The Kantian, of course, is also committed to the view that severely mentally incapacitated human beings lack dignity and indeed a moral status altogether. So, the nub of controversy between the Kantian view and the African one is the extreme psychopath. Now, even though the extreme psychopath, we imagine, has the capacity only for being an object of a communal relationship, that capacity is *much greater* than that had by other beings such as mice and dogs, and hence the modal-relational view entails that, even though he lacks dignity, his moral status is greater than that of animals. Compared with animals, normal human beings are more able to include 'deformed' humans such as psychopaths, as well as the mentally incapacitated, in a 'we', cooperate with them, act in ways likely to improve their quality of life, exhibit sympathetic emotions with them, and act for their sake. We do much more for the psychopathic and the mentally incapacitated than we do animals, which is evidence of a greater ability to make them an object of a friendly relationship. And even with respect to psychopaths, we readily feel for them and are disinclined to punish them (while remaining firm about their confinement, to protect others) when we learn that many of them suffered horrific abuse as children.

The modal-relational theory is interesting and worthy of consideration for its ability to ascribe a 'middle' moral status, i.e., neither highest nor lowest, to both the extreme psychopathic and the severely mentally incapacitated. Neither Kant nor the Kantian would grant any moral status to the severely mentally incapacitated; the Kantian would ascribe full moral status – indeed, a dignity--to the psychopathic; and the utilitarian would naturally grant full moral status to both, as they are capable of pleasure or preference satisfaction.[6] The African theory presents a plausible alternative of granting to both more moral status than animals but less than 'normal' humans.

Furthermore, the modal-relational account is to be credited with providing a reasonable explanation of why members of two different species that have the same internal properties, viz., animals and severely mentally incapacitated humans, could nonetheless have different moral statuses. It provides a genuine reason to doubt the influential 'argument from marginal cases' (e.g., Singer, 1993; Dombrowski, 1997), the rationale that at least some animals are entitled to as much moral status as humans, since the two are biologically similar. The African theory entails that even if there is no *intrinsic* difference between two beings, there could be a *(modal-) relational* difference between them, *qua* capacity to have a life that is shared with, and cared for by, normal human beings, that grounds differential degrees of moral status. The idea that humans have a greater moral status than animals is a persistent intuition, and invoking the property of degree of capacity for communal relationship is a more attractive way to account for it than is the speciesist one of the bare fact of human life.

The last category of human beings that I address is the very young. According to the modal-relational theory, gametes, zygotes, blastocysts, and embryos lack moral status, and, if they would warrant protection, it would be only on indirect grounds.

They cannot be subjects of a friendly relationship, for they lack the abilities for conceptualization, intentional action, emotion, and the like that are constitutive of it. Furthermore, they cannot even be objects of such a relationship, since they are not yet organisms capable of engaging in goal-directed activity and of being better or worse off.

Of course, the embryo is alive, and so there may be a sense in which it would be 'better' if it were to stay alive, but not in the relevant sense of being 'better off', i.e., having an improved quality of life. And while it is true that some people, particularly pregnant women, might be inclined to think of themselves as a 'we' with extraordinarily young human beings, the tendency for us to identify with them increases substantially as they develop. Typical humans much more readily think of themselves as a 'we' with mid-to-late stage foetuses and newborns than they do with a mere clump of rudimentary cells, at least upon a clear awareness of their nature. Furthermore, foetuses, at a certain point, are organisms with the capacity to feel pain and hence to be an object of solidarity. Although they still cannot be subjects of a relationship of identity and solidarity, they can be objects of them to some meaningful degree. Similar remarks go for newborn infants. So, the African theory appears to entail that human beings younger than two months old – when the very large majority of abortions take place – lack a moral status, but that they acquire a moral status upon becoming a mid-to-late stage foetus.

Comparing mid-to-late stage foetuses and newborns now, the former would have less moral status than the latter if normal humans could more easily make the latter the objects of a friendly relationship, upon information about both kinds of beings.[7] Newborns and foetuses, while sometimes identical in biological terms, are not identical in respect of essential functions, and partly for this reason are not identical in respect of capacity to be the object of identity and solidarity with characteristic human beings. Babies cry, look people in the eyes, ingest food through their mouths, are out in the world, have desires, and engage in goal-directed behaviour, which enables them to be more readily conceived as a 'we', sympathised with, and helped by normal humans. The point is not that newborns are more similar to us than are foetuses, but rather that the latter are, *by their nature, cloistered* in such a way as to make communal relationship with them much more difficult. If that metaphysical claim were indeed true, then it would follow from the African theory that newborns have a greater moral status than foetuses.

Again, the modal-relational theory is to be credited for promising to account for a long-standing and widely shared intuition that has not been firmly captured up to now, namely, that moral status increases along with the development of a being from embryo to foetus to newborn. Utilitarianism cannot entail or explain that intuition, for once a foetus is capable of pain or preferences, there is no difference in moral status between it and a newborn, such that, notoriously, if abortion is routinely permissible, infanticide often is too. Kantians also have difficulty accommodating the differential moral status of foetuses and infants, as either they both lack moral status for being non-rational, or they both have moral status for exhibiting the potential for rationality. Only a *relational* theory of the present sort has any real chance of differentiating between these beings. And *nothing else could*, given their intrinsic similarity.

I now turn from focusing on human beings to saying more about animals. In my view, a glaring weakness amongst dominant theories of moral status is that they cannot entail all the following: (1) both humans and animals have moral status; (2) humans have a greater moral status than animals; (3) both humans and animals have moral status for a unitary reason. On (1), I find it incontrovertible that at least some animals have a moral status, and even those who question the existence of animal *rights* and defend practices such as eating animals and experimenting on them invariably admit at some point that they do (e.g., Frey, 1983). The best explanation of why it is wrong to put a live cat in a microwave merely for the thrill ultimately has to do with facts about the cat, not solely the agent or other persons; Kant's ethic and contemporary Kantianism are simply mistaken.

Although utilitarianism accounts for (1), it fails with respect to (2). It entails that many animals have the same moral status as human beings, for both kinds are capable of pleasure and pain or of preference satisfaction and frustration. And Tom Regan's (2004) subject of a life criterion, according to which a being has moral status roughly insofar as it has propositional attitudes towards the past and the future, as well as Paul Taylor's (1986) biocentric view, which prescribes respect for all living organisms, are also well-known for entailing the equal moral status of humans and many animals, something that I, and I presume most readers, find counterintuitive (The Concept of Moral Status).).

Finally, there are accounts of moral status that entail (1) and (2) but that are pluralistic and hence fail (3). For instance, Christopher Stone (1985) and Mary Anne Warren (2003) are known for thinking that several features ground moral status of differing degrees, ranging from rationality to sentience to actual relationships to landscapes to species. However, all things being equal, a view is more desirable, the greater its unity and simplicity, meaning that a monistic theory is preferable.

Modal-relationalism promises to account well for (1), (2), and (3); it appeals to a singular property to ground moral status that entails differing degrees of it between humans and animals and between various classes of animals as well. As I have explained, since normal humans are capable of being both subjects and objects of harmonious relationships, they have full moral status. There is evidence that some animals, such as chimpanzees and gorillas, are capable of being subjects to an extent, at least insofar as they can coordinate their behaviour, exhibit sympathy, and act consequent to it (e.g., de Waal, 1996), and so they would have the highest moral status in the animal kingdom. Below them would be animals that are utterly incapable of being subjects of a friendly relationship, but would have a substantial ability to be objects of one, namely, most other warm-blooded creatures. Further down would be animals that are capable of having a better or worse life and so could be objects of a friendly relationship, but would be less able to be objects than other animals due to either their comparative lack of sentience or the dispositions of human nature, say, molluscs.

I now address beings other than humans and animals, and, not coincidentally, respond to two major objections that are likely to be made to this African-based theory of moral status. To some readers, the glaring weaknesses of this view are that it is, first, anthropocentric and, second, unable to account properly for deep ecological concerns. I take up both objections in turn before concluding.

The modal-relational theory might appear to be anthropocentric in that it cashes out moral status in terms of certain human capacities. To be able to be an object of a communal relationship, on this view, is analysed in terms of a capacity to relate to normal human beings in a certain way. So, there is an irreducible appeal to humanity in its conception of moral status.

However, the modal-relational theory is not anthropocentric in the most clearly questionable forms. First, it does not entail that natural objects have merely an instrumental value for human beings, for it instead entails that many animals have moral status, i.e., they are the proper objects of ethical concern in their own right. Second, the view is not speciesist, for it does not imply that human life *qua* human has a basic moral value. The theory is capacity-based, making it an empirical matter to determine not only which beings have the capacity to identify with and exhibit solidarity towards others (which chimpanzees and gorillas may well have), but also which beings can be the objects of such a relationship with us.

Even so, some friends of deep ecology will find the present theory unsatisfactory. For one, it entails that plants – even Redwoods – lack moral status, so that any duties with regard to them will be indirect. Although there is a sense in which a tree can 'flourish' (as Stone 1972 helped make clear), that is a matter of being able to be a good instance of its kind, and not of being able to live a good life, which is necessary to be an object of solidarity. For another, groups such as species and ecosystems obviously are incapable of being subjects of friendship and also cannot be objects of it, as they are not *themselves* beings with goals and lives that can be better or worse off.

In reply, here is some reason to think that such ecological concerns might be plausibly captured without ascribing moral status to plants and groups. To put it tersely, not all final value is moral value. As Robert Nozick points out (1989, pp. 163–164, 170–171), scientific theories and the game of chess are organic unities and, for that reason, plausibly have a value that is not merely instrumental, but it is implausible to think that they are ever the direct object of moral obligations. Even if the modal-relational theory of moral status denies it to plants and groups, there could be other respects in which these things are finally (not merely instrumentally) valuable and warrant respect and protection, even at some cost to beings capable of communal relationship (supposing, as is plausible, that morality is not invariably overriding).

Consider how an agent (or at least one not yet in the grip of a theory) would likely feel if she destroyed a Redwood, landscape, or an ecosystem. I submit that, insofar as she would feel guilt, she would most likely feel guilt with respect to the animals or humans, including future generations, affected by this behaviour, and she would probably not feel *guilt* with respect to the tree, landscape, or the ecosystem *as such*. She would be more likely to feel *shame* with respect to the way her behaviour has affected those things, but guilt is a much more reliable marker of moral status than is shame. People often feel shame about their unattractive appearance, or their impoverished condition, or their having been the victim of wrongdoing, conditions unrelated to their having violated moral norms. There is, therefore, *prima facie* reason to exclude plants and groups from the domain of moral consideration in their own right, and to continue to try to account for any non-instrumental value they have in other terms.

Conclusion

Those deeply committed to holism and individualism, or even a combination of them, may well not be convinced by this discussion. Diehard holists will reject the idea that anything other than a group can ground moral status, while pure individualists will reject the recurrent suggestion that two beings that are internally identical (foetus versus neonate, severely mentally incapacitated human versus animal) could differ in their moral status. However, my aim has not been to convince anyone to change her mind or even to provide a complete justification for doing so. My goals have instead been the more limited ones of articulating a new, modal-relational account of moral status grounded on ideas salient in sub-Saharan moral philosophy, demonstrating that it avoids the severe parochialism facing actual-relational accounts, and showing that it accounts better than standard Western theories for a variety of widely shared intuitions about what has moral status and to what degree. Many of these intuitions are captured by neither holism nor individualism and have lacked a firm philosophical foundation up to now. Of importance here is the African theory's promise to underwrite the ideas that humans and animals have a moral status because of the same property that differs in degree, that severely mentally incapacitated humans have a greater moral status than animals with the same internal properties, and that a human's moral status increases as it develops from the embryonic to foetal to neo-natal stages. It is too early to conclude that one has most reason to believe the modal-relational theory of moral status, particularly in light of criticisms that I have not yet had the space to address,[8] but I submit that the field would be misguided to ignore this unique and promising view in future debate.

Notes

1 Acknowledgements: This chapter first appeared as an article in Ethical Theory and Moral Practice, 15, (2012): 387–402, and is reprinted in an abridged version, with modifications, and with the permission of Springer.
2 The rest of this paragraph borrows from Metz (2010: 308–309), which includes some evidence beyond what is mentioned here.
3 There are a very small handful of other modal views in the literature, restricting moral status either to the capacity to be caring (Jaworska, 2007) or to the capacity to be an object of sympathy (Mercer, 1972, pp. 129–133).
4 As I have argued in Metz (2007).
5 For a discussion of geographical labels and why calling something 'African' does not imply either that it is utterly unique to or exhaustive of that locale, see Metz (2015).
6 Some utilitarians argue that people's 'global' desires, with regard to their lives as a whole, are more important than 'local' ones, giving them priority over the severely men-

tally incapacitated in cases of conflict. However, what if a normal person, such as Galen Strawson, lacked such desires?

7 This line of reasoning is inspired by some remarks from Slote (2007, pp. 17–19).

8 Does modal-relationalism entail that cuter animals have a greater moral status than ugly ones, that a late-term human foetus has a lower moral status than a chimpanzee, that a stereotypical Mother Teresa has a greater moral status than us, or that someone actually part of a communal relationship with us has a greater moral status than someone who merely could? If so, are these implications counterintuitive?

References

de Waal, F. (1996) *Good natured*. Cambridge, MA: Harvard University Press.

DeGrazia, D. (2008) 'Moral status as a matter of degree?', *Southern Journal of Philosophy*, 46, pp. 181–198.

Dombrowski, D. (1997) *Babies and beasts*. Chicago: University of Illinois Press.

Frey, R.G. (1983) *Rights, killing and suffering*. Oxford: Basil Blackwell.

Gruen, L. (2003) 'The moral status of animals', in E. Zalta (ed.) *Stanford encyclopedia of philosophy*. Available at: http://plato.stanford.edu/entries/moral-animal/ (Accessed: 10 January 2024).

Jaworska, A. (2007) 'Caring and full moral standing', *Ethics*, 117, pp. 460–497.

Kant, I. (1797) *The metaphysics of morals*, translated by M. Mac Gregor. New York: Cambridge University, 1991.

Kennett, J. (2002) 'Empathy and moral agency', *Philosophical Quarterly*, 52, pp. 340–357.

Leopold, A. (1968) *A sand county almanac*, 2nd ed. New York: Oxford University Press.

Mercer, P. (1972) *Sympathy and ethics*. Oxford: Clarendon Press.

Metz, T. (2007) 'Toward an African moral theory', *Journal of Political Philosophy*, 15, pp. 321–341. Revised edition in I. Ukpokolo (ed.) *Themes, issues and problems in African philosophy*. London: Palgrave Macmillan, 2017, pp. 97–119.

Metz, T. (2010) 'Animal rights and the interpretation of the South African Constitution', *Southern African Public Law*, 25, pp. 301–311.

Metz, T. (2015) 'How the West was one: The Western as individualist, the African as communitarian', *Educational Philosophy and Theory*, 47, pp. 1175–1184.

Metz, T. (2024). 'Defending a relational account of animal moral status', in M.J. Tosam and E. Masitera (eds.) *African agrarian philosophy*. Cham: Springer, pp. 105–124.

Miller, H. (1988) 'Science, ethics, and moral status'. Available at: http://parismount.blogspot.co.za/2012/03/copyright-1988-harlan-b.html (Accessed 10 January 2024).

Noddings, N. (1984) *Caring*. Berkeley: University of California Press.

Nozick, R. (1989) *The examined life*. New York: Simon & Schuster Inc.

Regan, T. (2004) *The case for animal rights*, 2nd ed. Berkeley: University of California Press.

Singer, P. (1993) *Practical ethics*, 2nd ed. New York: Cambridge University Press.

Slote, M. (2007) *The ethics of care and empathy*. New York: Routledge.

Stone, C. (1972) 'Should trees have standing?', *Southern California Law Review*, 45, pp. 450–501.

Stone, C. (1985) 'Should trees have standing? Revisited', *Southern California Law Review*, 59, pp. 1–154.

Taylor, P. (1986) *Respect for nature*. Princeton: Princeton University Press.

Van De Veer, D. (1995) 'Interspecific justice and intrinsic value', *Electronic Journal of Analytic Philosophy*. Available at: http://ejap.louisiana.edu/EJAP/1995.spring/vandeveer.1995.spring.html (Accessed 10 January 2024).

Warren, M.A. (2003) 'Moral status', in R.G. Frey and C. Wellman (eds.) *A companion to applied ethics*. Malden, MA: Blackwell, pp. 439–450.

6 African Ecological Ethics and the Moral Status of Nonhuman Nature

Kai Horsthemke

Introduction

Perhaps the greatest challenge facing humankind today is the state of our planet and, intimately connected with this, our relationship with the natural environment. Most, if not all, other concerns – however significant – are necessarily secondary in this regard. The human impact on the environment has been, and continues to be, enormous. Human population increase, advances in technological ability and control, as well as human obsession with growth continue producing previously non-existent environmental problems. What is at stake here is nothing less than the survival of the earth in its present state, as being inhabitable, and therefore also human survival itself. At the very least, it is a matter of the quality and conditions of all life, present and future. If this is correct, then one of the greatest priorities – if not the greatest – of academic research, scientific, philosophical, educational, and other, may be how to arrest and possibly reverse the present decline. There are difficulties, of course. The problem is not only climate change denialism, i.e., doubts whether the diagnoses and prognoses are correct. Substantial philosophical and pedagogical questions also concern the articulation and definition of the approaches we human beings should adopt in response to the environmental predicament/s. Additional questions are raised about the metaphysical and ethical foundations of our concern for our planet and about the political, educational, and pedagogical implications. Should legislation, educational policy and practice, etc. be guided by a concern for nature and the environment for our (human) purposes? Or should we teach and learn, legislate and act for the natural environment in and for itself?[1] And what is the contribution African philosophy could make in this regard?

African Environmental Philosophy and the Hierarchy of Beings

Just as African philosophy was for the longest time ignored within mainstream academic philosophy, so nonhuman nature and the environment have until quite recently been ignored within African philosophy. It is only during the last 25 years that African philosophers have begun giving serious thought to ecological ethics and the moral status[2] of other-than-human organisms and the environment (see, for

DOI: 10.4324/9781003480914-7

example, Ramose 2005; Murove *passim*; Ojomo *passim*; Chimakonam, ed. 2018b; Ebert and Roba, eds. 2018; Chemhuru, ed. 2019, 2022).

It might be argued, however, that African philosophy has, to date, not yielded a compelling deep-ecological ethic. As Nigerian philosopher Ada Agada (2018: 190, 191) explains, "Deep ecology's emphasis on the intrinsic worth of all living things ... [and] romanticisation of the environment goes against the African privileging of the human". African moral theory and practice has tended to be anthropocentric, placing human beings at the centre of moral concern and above nonhuman nature in the ontological and ethical hierarchy. Like Agada (191), Nigerian scholar Otto Dennis (2022: 188) is quite clear about this: "For a traditional African, both corporeal and incorporeal forces exist and interact in an inseparable, constant, hierarchical mixture, where humankind is at the centre". An implication for ecological ethics is that the vast majority of African conceptions of ethical relations and of moral status can at best be seen to yield a shallow environmentalism.[3]In the great chain or hierarchy of beings (see, e.g., Imafidon 2014: 40–42), God (Nguni/Zulu[4]: *uNkulunkulu*or *uMveliqangi* – "the One who appeared/emerged first") is situated at the apex and then, at different levels below, the ancestors (*amadlozi*) or "living-dead" (*abaphansi*), then human beings (*bantu*; singular: *muntu*), and finally the rest of animate and inanimate creation, including animals (*isilwane*). The hierarchy of beings constitutes an interesting commonality in African religious and secular thought because it invariably places other-than-human animals and nature in an inferior position to humans, because of their lack of moral agency. In acting responsibly and morally, human beings are fulfilling their divinely (or traditionally) allocated role. They have moral responsibilities and duties primarily to God, the ancestors, and their fellow humans and also, albeit indirectly, to the rest of creation. Thus, although evil originates with human beings and not with God, the former are perceived to be moral creatures. Morality is understood to be a matter of human relationships, with one another as well as with God, the ancestors, and nonhuman creation. The ancestors or living-dead play a vital role in the lives of Africans in that they act as a link between God and living human beings, i.e., they bridge the gap that exists between *uMveliqangi* and *bantu*. They are consulted regularly, for example, during ceremonies at which animals (usually cows, bulls, goats, or sheep) are routinely slaughtered. Failure to engage in such sacrificial activities is believed to provoke the anger not only of the ancestors but also of God.

The hierarchy of beings remains intact in more secular conceptions, the cancellation of God notwithstanding. Here, the top rung is occupied by the ancestors. The human realm consists of the living, the living-dead, and those who are yet to be born. In fact, living human beings have strict duties of obedience not only to the dead but also, and especially, to the yet-to-be-born. A popular Kenyan proverb states that the "world was not given to you by your parents – it was lent to you by your children" (see Horsthemke 2015: 26, 42). Animals and the nonhuman environment generally are understood to have merely instrumental status. In this regard, human responsibilities and duties are essentially perceived to be *indirect*. In his discussion of intergenerational justice in relation to environmental issues, Dennis (2022: 189) refers primarily to human obligations to (human) posterity.

We humans are given temporary custodianship over the (natural) world, but we are required to "return" it in *at least* the condition in which we received it, not primarily for the sake of the environment but, rather, for the sake of our fellow (present and future) human beings. In more secular conceptions, then, our *direct* responsibilities and duties are essentially not to God, in respect of his creation, but to other human beings (future, present, *and* past, i.e., our ancestors).

Botho/ubuntu/hunhu

According to South African philosopher Mogobe Ramose (2005: 106),

> *Motho ke motho ka batho* is a Sotho proverb found in almost all the indigenous languages of Africa. It means that to be human is to affirm one's humanity by recognizing the humanity of others and, on that basis, establish humane relations with them. Accordingly, it is *botho* understood as being human (humanness/humanity) and a humane (respectful and polite) attitude towards other human beings which constitute the core or central meaning of the aphorism.

The equivalent proverb in the Nguni language group is *Muntu ngumuntu nga bantu*, while the Shona variant is *Munhu munhu navhanhu*. Usually translated as "A person is a person through other persons" or "I am because we are", *botho* and its cognates *ubuntu* (Nguni) and *hunhu* (Shona) are often cited as the basis of an African environmental ethic, insofar as the "principle of wholeness" alluded to here "applies also with regard to the relation between human beings and physical or objective nature" (Ramose 2005; see also Murove 2009; Le Grange 2012, 2015[5]). The implied difference relegates all of nonhuman nature to a realm beyond the nonphysical (psychological, spiritual, cognitive) and subjective. It follows not only that *botho* and its cognates are human-centred by definition but also that there is comparatively less to be taken into account when it comes to "physical or objective nature" – no subjective awareness, no primary perception, and ultimately no intrinsic value. This does not mean that defenders of *botho* deny that nonhuman nature has value – its value, however, is typically instrumental: not only plants, the ecosystem, etc., but nonhuman animals, too, are conceptualised as relational *objects*.

The ambivalent status of nonhuman nature – part of but at the margins of the relational web – is also rendered explicit by Kenyan academic Catherine Odora Hoppers. *Ubuntu* ethics, she writes (2005: 4–6),

> stresses ... the essential interrelatedness and interdependence of all phenomena – biological, physical, psychological, social and cultural. Indigenous cosmology centres on the co-evolution of the spiritual, natural and human worlds. ... Experiences from indigenous communities in other parts of the world emphasise the fact that knowledge is relationship, and relationship brings with it responsibilities and obligations and extends into ecological practice.

All the while, says Odora Hoppers, "Relationships between people *hold pride of place*, expressed in the various philosophies across Africa and best captured by the African concept of *ubuntu*" (2008: 30; see also Odora Hoppers 2005: 3). It is difficult, then, if not impossible, to extract a deep-ecological ethic from *botho* or *ubuntu.*

"African Modal Relationalism"

Since 2005, Thaddeus Metz has produced an impressive body of work on and within African philosophy, notably African ethics. His endeavours have not met with unanimous approbation and acclaim from African philosophers. Given the nature of philosophy and philosophical discourse, this is unsurprising. What is noteworthy, however, is that the reasons for the dismissal of Metz's work have tended to have little to do with philosophical substance, the critique more often than not amounting to thinly veiled *ad hominem* attacks on the US immigrant to South Africa and non-African-language speaker. Nonetheless, Metz's contribution must stand out as a systematic and far-reaching attempt to make African philosophy analytically compelling and normatively sound, by providing a sympathetic but never sycophantic (re)interpretation of some of its central concepts and concerns. Metz seeks at once to highlight African philosophy's distinct contribution to global, academic philosophy and to strengthen it in instances where it may appear wanting. His work on an African conception of moral status is a case in point.

Metz's "modal relationalism" (Metz 2022, 2023: 229) is offered as a viable and compelling African alternative to what has traditionally been a debate between individualist and holist accounts. He takes the view that an entity possesses moral status in accordance with its capacity to be a party to harmonious or communal relationships, i.e., relationships of identity and solidarity, to be a salient feature of African ethics. According to sub-Saharan moral views, the greater an entity's capacity to be part of a communal relationship with us, i.e., normal adult human beings, the greater its moral status (Metz 2022: 168–171, 2023: 109). In other words (Metz 2023: 105, 106), "a being has greater moral status the more it is capable of communing with us (as a subject)" or the more we can commune "with it (as an object)".

It is one thing, philosophically, to mine African philosophical thinking for a conception of moral status that may have existed only implicitly within relevant belief systems, writings, and discourses. It is quite another, however, to maintain that such a conception – "African" though it may be – is actually compelling and worth embracing. In other words, the pedigree of being "African" does not automatically yield justification or legitimation. This is especially significant for, or within, environmental ethics. Like *botho* or *ubuntu*, Metz's conception of moral status, although recognisably "African", is characterised at best by a shallow environmentalism. The reasons he enumerates for not destroying a tree or an ecosystem have nothing directly to do with the tree or the ecosystem but everything to do with the effect of this behaviour on humans and, possibly, some nonhuman animals (2022: 254). While Metz accepts that there may be pragmatic reasons "to think that trees and ecosystems have moral standing", e.g., in order to halt or reverse the present climate crisis, he denies that there is any "evidential reason to believe" (2022) that trees and ecosystems (and many nonhuman animals, for that matter) are directly morally considerable. Thus, he argues that:

> if there is reason not to destroy an ecosystem that includes rocks, shrubs, and mosquitos, it is not directly because these beings would be negatively affected, but rather indirectly because negatively affecting them would wrong other kinds of beings, roughly those with intention and a quality of life.
>
> (2023: 110)[6]

Although Metz's *ubuntu*-inspired account of relational ethics succeeds admirably in giving *ubuntu* and related principles and values normative teeth, establishing them as action-guiding is unlikely to render them useful from a deep-ecological perspective.

The Ethics of Nature-Relatedness and Eco-Bio-Communitarianism

Perhaps it is uncharitable to focus exclusively on *botho* or *ubuntu* – which may constitute an improvement on ethnocentrism[7] but is still decidedly anthropocentric – as characterising African ethical attitudes towards nonhuman nature. According to ethicists like Segun Ogungbemi (from Nigeria), Godfrey Tangwa (Cameroon), Bénézet Bujo (Democratic Republic of Congo), and Munyaradzi Felix Murove (Zimbabwe) (and, building on the work of some of these philosophers, South Africans Kevin Behrens and Lesley Le Grange), Africa offers additional conceptual resources that may assist in addressing questions around direct ethical responsibility regarding nonhuman nature (Ogungbemi 1997; Tangwa 2004; Bujo 2003, 2009; Murove 2008, 2009; Behrens 2014, 2018; Le Grange 2012), resources that involve an extension of the traditional concepts of "community", "holism", "relatedness" and "relationality".[8] Whereas Ogungbemi refers to his favoured perspective as "ethics of nature-relatedness", Tangwa calls his theory "eco-bio-communitarianism". Bujo emphasises African holism, "solidarity with creation as a whole" (Bujo 2009: 284) and "cosmic community" (Bujo 2009: 296), Murove draws on the notion of *ukama* originating in Zimbabwe (Murove 2009: 315–316). Le Grange builds on the ideas of Bujo and Murove (Le Grange 2012: 61–62), as does Behrens (with additional inspiration from Tangwa and Metz), who works within both African communitarianism and what he calls "African relational environmentalism" (Behrens 2014).

Ogungbemi has identified both the reasons for Africa's environmental crisis as well as the traditional conceptual and intellectual resources available for arresting (if not reversing) the decline. His contribution is significant insofar as he does not lay the blame squarely on colonialism and its consequences. He acknowledges poverty, the inequitable distribution of global wealth, as well as international economic pressure as principal factors contributing to Africa's ecological predicament. Thus, most of the people live in rural areas and lack many of the basic amenities, and there has been a noticeable trend on the part of governments to adopt Western-style economies – both of which have led to rapid increases in environmental degradation and exploitation. Yet, Ogungbemi also targets political conflict and military spending, misuse of science and technology, dramatic increases in population, and sheer ignorance as key determinants. Interestingly, he contends that poverty and ignorance do "not necessarily exonerate people from their contribution to environmental hazards" (Ogungbemi 1997: 204). Thus, the most common source of energy is fuelwood, and a fairly common hunting method involves bush burning. Ogungbemi believes that ordinary, traditional Africans can be held accountable in these regards. A similar argument must surely hold for poachers or those who collaborate in the trafficking of animal body parts. Finally, while Africa's contribution to the global ecological crisis is relatively minor when compared with that of industrial and technological giants like the United States and China, Ogungbemi points out that this may be due largely to an inability to exploit nature because of low levels of industrial, technological, and economic development.

Nonetheless, Ogungbemi identifies certain conceptual resources in traditional African thought and practice, a return to which would contribute substantially at least to halting the process of progressive destruction of African fauna and flora. He postulates an "ethics of care" (204; see also Ramose 2005: 106) that is manifested in the traditional African relationship with nature, where men and women have recognised the significance of "water, land and air management"':

> To our traditional communities the ethics of not taking more than you need from nature is a moral code. Perhaps this explains why earth, forests, rivers and wind and other natural objects are traditionally believed to be both natural divine. The philosophy behind this belief may not necessarily be religious, but a natural means by which the human environment can be preserved. The ethics of care is essential to traditional understanding of environmental protection and conservation.
>
> (Ogungbemi 1997: 204)

Ogungbemi acknowledges that the African ethics of care is different from the feminist ethic of care, in that it is essentially human-centred: humans are taken necessarily to rely on the nonhuman world for survival. It is precisely for the sake of present and future human well-being that we must treat the natural world in which we live with respect. This is why it is important to consider certain fundamental questions:

> How do we know how much we need, given the nature of human greed? Who judges whether we have been taking more or less than we need from our natural resources? If we have been taking more than we need, what are the penalties and how fair are they?
>
> (208)

Such an African ethics of care does not appear to be all that different from other anthropocentric preservationist ethics. Furthermore, if Africans actually

> practiced the ethics of care, forests would not be disappearing. Nor would deserts be advancing at the rate which they are.
>
> (Peterson 2013: 108)

One could also refer to the dwindling rhino and great ape populations in this regard. These considerations led Ogungbemi to formulate what he has called the "ethics of nature-relatedness" (Ogungbemi 1997: 208), which:

> asserts that our natural resources do not need man for their existence and functions. ... The ethics of nature-relatedness can be succinctly stated as an ethics that leads human beings to seek to co-exist peacefully with nature and treat it with some reasonable concern for its worth, survival and sustainability[9].
>
> (208, 209)

A noteworthy feature of this contemporary reformulation of the African ethics of care is that Ogungbemi's theory is completely divorced from religion. It implies neither that natural existents have a spiritual or "divine" nature nor that their existence can be attributed to a Supreme Being or creator. Instead, it has three key components: experience, reason, and the will. It thus retains the all-important experiential aspect of practical morality (for without any sense of relevance it would be that much more difficult to get people to reflect on their practices) while introducing two significant cognitive aspects of moral agency. Ogungbemi's ethics of nature-relatedness is therefore an improvement on its predecessor, although it comes at a cost that is not inconsiderable. Being divorced from "African cultural experience and ontology" and no longer being "grounded in ... African metaphysics" (Ojomo 2011: 110), one wonders in what ways (if any) it remains an "African" ethic.

In almost complete contrast, while Tangwa retains "the metaphysics of African cultural belief system[s] ... in his construction of a meaningful African environmental ethics" (Ojomo 2011), he fails to recognise "the manners and ways in which traditional Africans contributed to the degradation of the environment" (Ojomo 2011). Tangwa describes the "precolonial traditional African metaphysical outlook" as "eco-bio-communitarian" and as "implying recognition and acceptance of interdependence and peaceful coexistence between earth, plants, animals, and humans". He takes this to contrast with the "Western outlook, which might be described as anthropocentric and individualistic" (Tangwa 2004: 389). His account is a mixture of both invocation of the environmental friendliness of traditional African (in this instance Cameroonian) worldviews and philosophies, on the one hand (389–390), and the arguments employed by Western pro-animal and environmental activists against the beliefs and practices in their (Western) societies and cultures, on the other (390–391). In the latter regard, he focuses, amongst other things, on the condemnation of (Western) "cruelty to animals (including using them in experimentation and killing and eating them)", as well as on arguments for "animal liberation" and "animal rights" (391), for example, the dismissal of the "claim that humankind is the apex of biological existence" as "arrogant" and "speciesist" (388). In the former regard, he refers to the (African) worldview in which "the distinction between plants, animals, inanimate things, between the sacred and the profane, the communal and the individual, is a slim and flexible one" (389).

The focus on (and tacit endorsement of) the arguments by "some human militants for the rights of animals and/or plants" (388) is on the mark. Yet, while the "live and let live" attitude Tangwa attributes to eco-bio-communitarianism looks compelling on paper, one wonders to what extent it is actually manifest in traditional worldviews. He appears to be unaware of the contradictions and blind spots inherent in his exposition. Thus, although he refers to a tendency towards cosmic humility and cautiousness "in their attitude to plants, animals, and inanimate things" (389) on the part of ordinary Africans, and although the "traditional worldview ... does not suppose that human beings have any mandate or special privilege, God-given or otherwise, to subdue, dominate, and exploit the rest of creation", Tangwa mentions "frequent offerings of sacrifices to God ..." (389–390). What do these offerings of sacrifices comprise, if not "plants, animals, and inanimate things"? How is the traditional "live and let live" attitude to be squared with the anthropocentrism of ordinary Africans and their "contribution to environmental hazards", as identified by Ogungbemi (1997: 204)? Finally, what if Africa had the financial resources of the so-called "West"? It is not at all clear whether this would lead to people becoming even more "cosmically humble, respectful and cautious" in their dealings with one another and the natural world – quite the contrary, one might assume.

African Holism, **ukama,** *and "Relational Environmentalism"*

Tangwas's ideas are echoed by Bujo. Human well-being and self-realisation, according to Bujo, are crucially dependent on "peaceful coexistence with minerals, plants and animals" (Bujo 2009: 281; see also Tangwa 2004: 389). To make sense of the apparent absurdity of "peaceful coexistence with minerals" (and even with plants, arguably), perhaps "peaceful" should be understood to mean "non-exploitative". Yet, we learn that "effective healing" ("reconciliation with the cosmos"), as it is practised and modelled by the African traditional healer, requires not only embracing "the community of the living and the dead but also natural elements such as animal bones, teeth and hair, plants, pieces of wood and minerals" (Bujo 2009: 284). And indeed: "Animals, plants and minerals and other inanimates are the forces made available by God as vital energy *for use by human beings*" (Bujo 2009: 290; emphasis added). It would appear, then, that the notions of "cosmic community" and of the relationality of all life, and of holism *tout court*, remain perfectly compatible with a largely instrumental view of nonhuman nature.

What about *ukama*? Unlike *ubuntu*, this concept signals that a person can only be a person in, with, and through not just other (past, present, and future) human beings but also in, with, and through the natural environment. As Murove explains, *ukama* (meaning "relatedness"; Murove 2009: 302)

> is a Shona word implying relationship and an understanding of reality in terms of interdependence. Grammatically *ukama* is an adjective constructed *u-kama*. … *Kama* becomes a word meaning to milk a cow or goat. In Shona thought the idea of milking suggests closeness and affection.
>
> (Murove 2009: 316)

Tellingly, the "closeness" in question derives from an animal being used, first and foremost, for human ends and purposes. Indeed: "*Umuntu* [man] is always in need of others and these others, as suggested in *ukama*, also imply the natural environment" (Murove 2009: 324). Just insofar, one might add, as the "natural environment" (which may be taken to include animals) meets and satisfies *umuntu*'s needs – personal, cultural, or other. Murove continues:

> In its adjectival form, *ukama* means being related or belonging to the same family. However, in Shona, as in many other African languages, the meaning of *ukama* is not restricted to marital or blood ties. This culture tends to see all people as *hama* (relatives).
>
> (Murove 2009: 316; see also Nhemachena 2013: 118–123)

"All people" – which effectively excludes nonhuman animals and the environment. Therefore, any nonanthropocentric orientation that Murove perceives in African ethics appears to be illusory, i.e., not properly founded. Similarly, Le Grange's assertion that "the idea of oneness of humans with nature expressed in Arne Naess's … deep ecology, and the notion of wholeness of the earth expressed in James Lovelock's Gaia …, have resided among African peoples for many centuries through notions such as *ukama*" (Le Grange 2012: 62) is rather misleading in that it contains a false analogy. Neither Naess's nor Lovelock's concepts or ethics are human-centred – unlike *ukama*.

Even if this verdict appears too harsh, a further important question concerns *umuntu*'s actual responsibilities with regard to nonhuman nature. The mere moral injunction to "harmonise [humanity's] behaviour with the natural environment" (Murove 2009: 329) does not establish much about *umuntu*'s concrete, specific responsibilities and duties. In fact, such "harmonisation" could be – and indeed has been – considered compatible with, perhaps even to require, the bare-handed slaughtering of bulls for the sake of "good relations" among *umuntu*, between *umuntu* and *amadlozi* (ancestors), and even "future generations". Murove's account of visiting and dining with a friend's relatives in a rural area 40 km from Harare in Zimbabwe (an event he takes to illustrate cordial relations even between strangers on the African continent) provides further telling information:

> A live sheep was presented to us according to custom. After we clapped our hands in gratitude, the sheep was taken away for slaughter.
>
> (Murove 2008: 85)

There is no acknowledgement at all that there may be a moral issue here. Insofar as *ubuntu* and *ukama* have any action-guiding content at all, this is unlikely to have any primary, direct beneficiaries other than human beings. Working with the ideas of Tangwa, Bujo, Murove, and Metz, Behrens attempts to counter the prevalent assumption that African moral thought and practice is inherently anthropocentric and characterised by what I have referred to as shallow environmentalism (Behrens 2014: 64, 2018: 43–46). He argues that a promising African account of moral considerability can be derived from a communitarian foundation – especially, as Behrens conceptualises it, an African emphasis on the interrelatedness or interconnectedness between everything in nature (Behrens 2014: 65). The idea is that natural entities (present, past, and future human beings, animals, plants, communities like species and whole ecosystems, even inanimate natural objects) are all bound up in a "web or fabric of life" (76). It is neither the individuals nor the community that are given moral priority (66), but rather the relations that exist between them, the fact of their harmonious (co)existence (76). From these considerations, Behrens concludes that all living things, by virtue of their existing in harmonious relations with one another, are morally considerable. It follows, he contends, that human beings have a moral obligation to live in harmony with the rest of nature and to respect all natural entities. In the case of natural existents like rivers and inanimate objects, this obligation cannot be direct; nor can it be, says Behrens, in the case of metaphysical ideas like spirits and ancestors (77–81). But normal adult human beings are nonetheless bound up in relationships with all these entities that could be harmonious or disharmonious. Insofar as the former are preferable to the latter, this is sufficient for grounding not only moral considerability but also moral responsibility, whether direct or indirect.

Although Behrens sometimes maintains (Behrens 2018: 46) that it is the entities themselves that matter morally (be they individuals or groups/species), a thoroughgoing relational theory would have to give moral priority to the relations between entities, to the fact of their interconnectedness, of belonging to a web of life. This lack of clarity means that he may face a similar objection to that levelled against Metz's relational ethic: that moral standing is determined by something outside those who can reasonably be said to possess it. In other words, how can belonging to a "web or fabric of life" (Behrens 2018) also bestow moral considerability on the entities in question and not only on the relation(s) between them? Although he distances himself from Metz's conclusion that many animals, plants, and inanimate natural existents cannot be morally considerable, the question arises whether his African relational environmentalism is really egalitarian in a way Metz's theory is not, or whether it also allows for differing degrees of moral considerability. It is, quite unequivocally, the latter. While Behrens (Behrens 2018) maintains that "all living organisms are morally considerable because they are part of the web of life", he suggests that they "may have different degrees of moral status". From a deep-ecological perspective, however, the notion of gradational moral status renders human environmental obligations and responsibilities superficial, contrary to what Behrens intends.

What is remarkable about the theories advanced by both Metz and Behrens is that they draw attention to the practical implications of their theories of moral status and considerability, respectively, and seek to establish the content of moral obligations based on their respective African ethics. In stark contrast, as Nigerian philosopher Philomena Aku Ojomo argues, authors like Ogungbemi and Tangwa fail to raise, let alone discuss, some fundamental questions. One of these concerns "the nature of African obligation and the role of the relationship between Africans ([including] non-Africans in Africa…) towards … future generation[s] in Africa, sentient beings, nonhuman animals, [the] African environment in particular and nature in general". Another concerns "the political, cultural, economic, educational, legal, and moral imperatives to be taken into consideration in the construction of an African environmental ethics in order to salvage the African environment from further deterioration" (Ojomo 2011: 111).

"Earth-Eco-Socialism" and "Eco-Humanism"

Two promising orientations that have been formulated in recent years are Ojomo's "earth-eco-socialism" and fellow Nigerian Michael Onyebuchi Eze's "eco-humanism". Ojomo (2019) seeks to reconstruct Aldo Leopold's land ethic by borrowing from classical Marxian socialism the scathing critique of "exploitation, alienation, domination and subjugation" and applying these to the ravages humankind has visited upon the earth and the nonhuman "inhabitants of the biotic community". Normatively, earth-eco-socialism has elements of deontology, consequentialism, and an ethic of care (304). Unfortunately, Ojomo does not explain how these different normative-ethical strands are to be reconciled or ordered. An additional concern is that her favoured orientation does not appear to be distinctly "African". But is this a problem? Not obviously: given that Ojomo wishes to address "the peculiar environmental crises in Africa" (306), she establishes a distinctly African set of priorities that determine attitudes, policies, and concrete actions towards "maintaining, preserving and sustaining the earth" (304) and its inhabitants.

Similar observations might be made with regard to Eze's *eco-humanitatis* (Eze 2017). By contending that the focus of an "African view" of environmental ethics (630) is neither on duty nor on utility (i.e., not on the question "*what I must do as an ethical agent*") but rather on the question, "*Who am I?*" (i.e., "how do I become a good or responsible virtuous agent?"; Eze 2017), he neatly sidesteps the problem of reconciliation or ordering of normative principles. Instead, however, he faces the question of in what way, if any, his favoured view guides both policies and concrete actions. Eze is quite unequivocal about the second set of concerns raised above. Eco-humanism, with its "idea of living within nature as part of nature", he admits, "is not something unique to Africa" (Eze 2017). Yet, it remains possible, *contra* Eze, that "particular instances, situations, or dilemmas" (631) constitute distinct priorities for an African ecological ethic that require "*responsible (good, virtuous) individual*[s]" to come up with "practical solutions" (Eze 2017).

Conclusion – *I Am because We Are*

Traditional African ethics (whether religious or secular) has demonstrably tended to be anthropocentric. Indeed, most conceptions of ethical relations and of moral status within African ecological ethics can at best be seen to yield a shallow environmentalism. Nevertheless, anthropocentrism is not a necessary or essential feature of African moral thought. On the contrary, the latter contains certain normative resources for a substantial overhaul of the traditional worldviews and stances that inform it. In other words, African morality contains the conceptual resources for a more radically egalitarian environmentalism. Thus, the Igbo notion of *ohanife* unites "people, usually in a group, community of network" (*oha*) and nonhuman life (*ife*) (Chimakonam 2018a: 123), which arguably amounts to a rejection of anthropocentrism. "When things fall apart in the architecture of the environment, the only entity to be blamed is *oha*" (127). In fact, *ife* (nonhuman nature) manages rather well in the absence of *oha*. *Ife* does not need humans and would almost certainly thrive without *oha* (or *muntu*). While earthworms, ants, bees, butterflies, and dung beetles are of vital importance for the diversity and integrity of *ife*, *oha* could be removed from earth ecosystems without adverse effect. *Oha* (*muntu*) is the one species the earth could well do without.

Furthermore, "I am because we are" could quite plausibly be taken beyond the human realm. This realisation suggests the need for both sincerity and practical consistency on the part of African environmental ethicists. Given the brutal, dehumanising ravages of colonialism, "land dispossession" (Green 2013: 3), racism and apartheid that Africans have historically been subjected to, it seems reasonable to invite concerned and committed thinkers in sub-Saharan Africa, especially, to reflect on an even longer, more deeply entrenched historical process of exploitation, oppression, and discrimination – that of species apartheid. Human freedom is not expressed in terms of land repossession, sustainable development, and "environmental management" (Green 2013: 5), in terms of being free to develop and subjugate anything that can be developed and subjugated. Perhaps the minimal insight one could reasonably expect from African liberatory ethics is that true human liberation also involves the act of human beings freeing themselves from the (historical) role of ruthless developers and subjugators, from the oppressive and exploitative relationship they have with the rest of nature.

Notes

1 This introductory paragraph is an adaptation of my initial Call for Papers for the 17th Biennial Meeting of the International Network of Philosophers of Education (INPE) that was scheduled to take place from 28 to 31 July 2020 at the Universidad Nacional Autónoma de México, Mexico City, and for which I acted as conference chair. Following the outbreak and rapid global spread of a new coronavirus, the conference was cancelled and rescheduled to take place online later that year.

2 To some extent, therefore, I agree with Metz (2022: 228; all references here pertain to the e-reader version of his book) that "Much environmental ethical reflection about how to treat nonhuman parts of nature is crucially influenced by what is judged to have moral

status and to what degree". My disagreement with Metz concerns his gradational or quantitative conception (2022: 233–235; Metz 2023: 109, 110). My preferred egalitarian conception of moral status focuses on relevant (dis)similarities. Thus, men and women have equal moral status as moral agents, although this does not mean that men can have a right to pregnancy termination on demand. Humans, cetaceans, and the great apes have equal moral status as moral subjects, although this does not mean that whales, dolphins, chimpanzees, gorillas, etc., can have a right to vote or to run for public office. Finally, sentient animals and plants like trees have equal moral status as moral recipients, but this does not mean that plants can have moral subject status, with all this entails. A further implication is that organisms lacking subjective awareness or individuality and inanimate natural/environmental entities at the receiving end of agency (the actions and omissions of agents), for example, the soil, air, water, rocks, and minerals, have moral object status (see Horsthemke 2010, 2018: 323), rather than no status at all.

3 Very roughly, deep ecology – the position that nature is valuable in and of itself – is contrasted here with shallow environmentalism – the position that nature only has instrumental value, i.e., relative to the needs and interests of human beings.

4 In what follows, I use the Nguni/Zulu terms as these are the ones with which I am most familiar.

5 *Ubuntu*, says le Grange (2012: 63), implies that "humanness needs to be understood relationally". This "humanness is an expression of interconnectedness between people themselves, and between people and the biophysical world": in other words, between "the self, society and nature". This "sense of wholeness and interconnectedness of self with the social and natural by implication means that caring for others also involves a duty to care for nature". I hope to make clear in this section that this "interconnectedness" is usually conceptualised in anthropocentric terms and that the last-mentioned "duty" is at best indirect.

6 See also note 2 above.

7 There is, however, no consensus in this regard. Thus, Dennis's interpretation of the statement, "I am because we are, and since we are, therefore I am" (2022: 188), is essentially ethnocentric, in that it expresses a sense of solicitude for persons in "Negro African society". It pertains essentially to "Africans", i.e., "the black peoples of the continent. It is the Negroes that the description 'Africans' is properly so accorded" (187). In addition, *ubuntu* has certain disturbing androcentric features (see Horsthemke 2015: 159n 10).

8 Additional resources that have been cited (see, for example, Odora Hoppers *passim*; Green 2013; Garuba 2013; Nhemachena 2013; Masoga and Shokane 2019) include animism and traditional or indigenous African knowledge. For reasons why these may not be as compelling as their defenders make them out to be, see Horsthemke 2021.A traditional healer's grasp of "bush medicines" ("*bossiesmedisyne*"; Cohen 2013: 93), the medicinal properties of plants, constitutes an insight that may have a particular geographic origin and that may not be shared by many, but it has universal value and application. The same goes for the knowledge of Zimbabwean "water harvester" Zephaniah Phiri Maseko (see Mabeza 2013).

9 On sustainable development, sustainable use, and environmental sustainability, see also Ngcoza 2019; Masoga and Shokane 2019; Pesanayi *et al.* 2019. I have critically interrogated these ideas elsewhere (Horsthemke 2018: 135–143, 188–191), largely because of their anthropocentric orientation. They are also irreconcilable with a deep-ecological ethic, insofar as nonhuman organic life and the environment are considered replenishable resources of, at best, instrumental value. An especially glaring example of this kind of thinking is found in Shava and Masuku's (2019: 77) view of "livestock" as "indigenous [or traditional] living currency" that has "created and sustained multifaceted livelihood strategies", a "sustainable economy that enhances the socio-ecological resilience of indigenous rural communities".

References

Agada, A. 2018. Catalyzing climate change action in Nigeria: Moderate anthropocentrism and the African perspective of the cosmos. In J.O. Chimakonam (ed.), *African philosophy and environmental conservation*, pp. 177–195. London and New York: Routledge.

Behrens, K.G. 2014. An African relational environmentalism and moral considerability. *Environmental Ethics* 36(1): 63–82.

Behrens, K.G. 2018. An African account of the moral obligation to preserve biodiversity. In J.O. Chimakonam (ed.), *African philosophy and environmental conservation*, pp. 42–57. London and New York: Routledge.

Bujo, B. 2003. *Foundations of an African ethic: Beyond the universal claims of western morality*. Nairobi: Paulines Publications Africa.

Bujo, B. 2009. Ecology and ethical responsibility from an African perspective. In M.F. Murove (ed.), *African ethics: An anthology of comparative and applied ethics*, pp. 281–297. Scottsville: University of KwaZulu-Natal Press.

Chemhuru, M. (ed.) 2019. *African environmental ethics: A critical reader*. Cham: Springer.

Chemhuru, M. 2022. *Environmental justice in African philosophy*. London and New York: Routledge.

Chimakonam, J.O. 2018a. *Ohanife*: An account of the ecosystem based on the African notion of relationship. In J.O. Chimakonam (ed.), *African philosophy and environmental conservation*, pp. 120–134. London and New York: Routledge.

Chimakonam, J.O. (ed.) 2018b. *African philosophy and environmental conservation*. London and New York: Routledge.

Cohen, J.B. 2013. Cultivating *krag*, refreshing *gees*: Ecologies of wellbeing in Namaqualand. In L. Green (ed.), *Contested ecologies: Dialogues in the south on nature and knowledge*, pp. 90–109. Cape Town: HSRC Press.

Dennis, O. 2022. African environmental intuitionism and the obligation to posterity. In A. Graness, E. Etieyibo and F. Gmainer-Pranzl (eds.), *African philosophy in an intercultural perspective*, pp. 181–190. Berlin: J.B. Metzler.

Ebert, R. and Roba, A. (eds.) 2018. *Africa and her animals: Philosophical and practical perspectives*. Pretoria: Unisa Press.

Eze, M.O. 2017. *Humanitatis-eco (Eco-humanism)*: An African environmental theory. In A. Afolayan and T. Falola (eds.), *The Palgrave handbook of African philosophy*, pp. 621–632. New York: Palgrave Macmillan.

Garuba, H. 2013. On animism, modernity/colonialism, and the African order of knowledge: Provisional reflections. In L. Green (ed.), *Contested ecologies: Dialogues in the south on nature and knowledge*, pp. 42–51. Cape Town: HSRC Press.

Green, L. 2013. Contested ecologies: Nature and knowledge. In L. Green (ed.), *Contested ecologies: Dialogues in the south on nature and knowledge*, pp. 1–9. Cape Town: HSRC Press.

Horsthemke, K. 2010. *The moral status and rights of animals*. Johannesburg: Porcupine Press.

Horsthemke, K. 2015. *Animals and African ethics*. Basingstoke: Palgrave Macmillan.

Horsthemke, K. 2018. *Animal rights education*. Cham: Palgrave Macmillan.

Horsthemke, K. 2021. *Indigenous knowledge: Philosophical and educational considerations*. Lanham: Lexington Books.

Imafidon, E. 2014. On the ontological foundation of a social ethics in African tradition. In E. Imafidon and J.A.I. Bewaji (eds.), *Ontologized ethics: New essays in African meta-ethics*, pp. 37–54. Lanham: Lexington Books.

Le Grange, L. 2012. *Ubuntu*, *ukama* and the healing of nature, self and society. *Educational Philosophy and Theory* 44(S2): 56–67.

Le Grange, L. 2015. *Ubuntu/botho* as ecophilosophy and ecosophy. *Journal of Human Ecology* 49(3): 301–308.

Mabeza, C. 2013. Metaphors for climate adaptation from Zimbabwe: Zephaniah Phiri Masekoand the marriage of water and soil. In L. Green (ed.), *Contested ecologies: Dialogues in the south on nature and knowledge*, pp. 126–137. Cape Town: HSRC Press.

Masoga, M.A. and Shokane, A.L. 2019. Indigenous knowledge systems and environmental social work education: Towards environmental sustainability. *South African Journal of Environmental Education* 35: 195–206.

Metz, T. 2022. *A relational moral theory: African ethics in and beyond the continent*. Oxford: Oxford University Press.

Metz, T. 2023. Defending a relational account of moral status. In M.J. Tosam and E. Masitera (eds.), *African agrarian philosophy*, pp. 105–124. Cham: Springer.

Murove, M.F. 2008. On African ethics and the appropriation of Western capitalism. In R. Nicholson (ed.), *Persons in community: African ethics in a global culture*, pp. 85–110. Scottsville: University of KwaZulu-Natal Press.

Murove, M.F. 2009. An African environmental ethic based on the concepts of *ukama* and *ubuntu*. In M.F. Murove (ed), *African ethics: An anthology of comparative and applied ethics*, pp. 315–331. Scottsville: University of KwaZulu-Natal Press.

Ngcoza, K.M. 2019. Education for sustainable development at the problem-posing nexus of re-appropriated heritage practices and the science curriculum. *South African Journal of Environmental Education* 35: 135–144.

Nhemachena, A. 2013. Are petitioners makers of rain? Rains, worlds and survival in conflict-torn Buhera, Zimbabwe. In L. Green (ed.), *Contested ecologies: Dialogues in the south on nature and knowledge*, pp. 110–125. Cape Town: HSRC Press.

Odora Hoppers, C.A. 2005. *Culture, indigenous knowledge and development: The role of the university*. Johannesburg: Centre for Education Policy Development (CEPD), Occasional Paper No. 5: 1–50.

Odora Hoppers, C.A. 2008. Culture, language, indigenous knowledge and the role of universities in sustainable rural development. Proceedings of the Conference hosted by the Centre for Education Policy Development (CEPD) and held at the Kopanong Conference Centre, Johannesburg (South Africa), 6–7 March. Braamfontein: Centre for Education Policy Development (CEPD): 29–35.

Ogungbemi, S. 1997. An African perspective on the environmental crisis. In L. Pojman (ed.), *Environmental ethics: Readings in theory and application* (2nd edition), pp. 330–337. Belmont, CA: Wadsworth.

Ojomo, P.A. 2011. Environmental ethics: An African understanding. *The Journal of Pan African Studies* 4(3): 101–113.

Ojomo, P.A. 2019. Niger Delta environmental crises and the limitations of Aldo Leopold's Land Ethic: Towards an earth-eco-socialist model. In M. Chemhuru (ed.), *African environmental ethics: A critical reader*, pp. 291–308. Cham: Springer.

Pesanayi, T., O'Donoghue, R. and Shava, S. 2019. Situating education for sustainable development in southern African philosophy and contexts of social-ecological of change to enhance curriculum relevance and the common good. *South African Journal of Environmental Education* 35: 213–221.

Peterson, A.C. 2013. Articulating moral bases for regional responses to deforestation and climate change: Africa. *William & Mary Environmental Law and Policy Review* 38(1): 81–117.

Ramose, M.B. 2005. *African philosophy through* ubuntu (revised edition, 1st publ. 1999). Harare: Mond Books Publishers.

Shava, S. and Masuku, S. 2019. Living currency: The multiple roles of livestock in livelihood sustenance in the context of rural indigenous communities in southern Africa. *South African Journal of Environmental Education* 35: 77–90.

Tangwa, G. 2004. Some African reflections on biomedical and environmental ethics. In K. Wiredu (ed.), *A companion to African philosophy*, pp. 387–395. Oxford: Blackwell.

7 An African Relational Environmentalism and Moral Considerability

*Kevin Gary Behrens**

Introduction

Much of the work of Western environmental philosophers in recent decades has sought to challenge and displace a dominant anthropocentric, materialist, and ultimately environmentally damaging Western worldview. Indeed, the ongoing project of articulating a truly "environmental ethic" is grounded in the conviction that the roots of the ecological crisis facing the world are philosophical in nature: what is required to ensure more prudent and responsible treatment of nature is a change in how persons value the environment. In their search for an environmental ethic, philosophers have often turned to the beliefs, values, and practices of other traditions, cultures, and religions for ideas that might inform and enhance a sound environmental philosophy. It is notable that in this process African thought has largely been disregarded as a potential source of indigenous ecophilosophical values. Workineh Kelbessa writes that:

> Those who have studied non-Western religions and philosophies... have overlooked the contribution of Africa to environmental ethics. They have either kept quiet or what they said about Africa was rather thin compared to what they said about Native Americans, Asians and Australian Aborigines.
>
> (2005, pp. 19–20)

One important reason for this is a strong presumption that African thought is inherently anthropocentric. In his comprehensive survey of the ecological ethical traditions of communities across the globe, *Earth's Insights*, J. Baird Callicott claims that:

> African thought orbits, seemingly, around human interests. Hence one might expect to distil from it no more than a weak and indirect environmental ethic, similar to [a] type of ecologically enlightened utilitarianism, focused on long-range human welfare.
>
> (1994, p. 158)

The presumption that African thought is essentially strongly anthropocentric and thus unable to make any meaningful contribution to an environmental ethic is widespread.[1] My broad project, of which this article forms a part, is to challenge this presumption and demonstrate that African thought need not be construed as

DOI: 10.4324/9781003480914-8

hopelessly anthropocentric and incapable of offering anything of value to environmental ethical discourse. My purpose is not to claim that African thought necessarily offers unique perspectives but rather to counter the assumption that it offers nothing. In this context, I often contrast distinctively African themes and ideas to a previously pervasive human-chauvinist, environmentally exploitative Western worldview. To the extent that African thought is able to challenge the values inherent in that worldview, it can be seen to make a contribution to a truly environmental ethic. The African environmentalism I describe clearly shares much in common with other environmental ethical perspectives that have similarly sought to challenge the mainstream Western worldview.[2]

A central concern in environmental ethics relates to the question of what sorts of entities are morally considerable. In this article, I am concerned with African thought regarding the environment and with establishing a plausible account of how an African environmentalism might determine what is morally considerable. The scope of my enquiry is the philosophical and religious ideas of the indigenous people of sub-Saharan Africa. I use the phrase "African thought" as shorthand for this.[3] In doing so, I do not claim that a single worldview is shared by these peoples, but I believe, with most African philosophers, that there are common themes to be found. Some of these themes imply an attitude of respect toward nature, and my enquiry addresses these. African thought is often characterised as anthropocentric. Indeed, there can be little doubt that human interests and needs figure prominently in accounts of traditional African values and in the work of many African philosophers. However, such seeming anthropocentrism does not fully reflect African views of the world. Contrary to anthropocentric ideas, there is a strong emphasis on the inter-relatedness or interconnectedness of human beings and the rest of nature that is also evident in African thought, providing a basis for a promising African environmentalism.[4] A small representative selection of examples of this emphasis in the words of African writers will suffice to defend my claim. Munyaradzi Felix Murove, an ethicist who specialises in the thought of the Shona people of Zimbabwe, writes of "our dependence on and interdependence with all that exists, and particularly with the immediate environment on which all humanity depends" (Murove, 2004, p. 195). Godfrey Tangwa, a philosopher born in Cameroon, states: "The ... traditional African metaphysical outlook... impl[ies] recognition and acceptance of interdependence and peaceful coexistence between earth, plants, animals and humans" (2004, p. 389). Malawian theologian Harvey Sindima describes an African cosmology that stresses the bondedness, the interconnectedness, of all living beings" (1990, p. 137). The philosopher H. Odera Oruka and Calestus Juma, an African authority on sustainable development, both originally from Kenya, write of adopt[ing] a holistic outlook in which everything is related to everything else. This inter-relatedness requires a corresponding philosophical approach that looks at nature in its totality (1994, p. 117). Benezet Bujo, a Central African theologian, asserts: "The African is convinced that all things in the cosmos are interconnected. All natural forces depend on each other, so that human beings can live in harmony only in and with the whole of nature" (1998, pp. 22–23).

According to many African theorists, this belief in the interdependence of natural entities underlies a requirement that people should respect and live in harmony with the community of nature.[5] The robustly communitarian character of much African ethics informs this moral requirement to live in harmony. Individual members of the community of nature can be fulfilled only in and through their relationships with others. These relationships are often characterised in familial terms, emphasising the need for mutual support, solidarity, care, and nurturing. I have suggested elsewhere that this approach to the environment holds promise for environmental ethics because of its relational focus. I have called this perspective "African relational environmentalism," and I claim that on this relational approach, virtue is achieved through maintaining harmonious relationships that prioritise neither the individual nor the community, while respecting both. Since the interests and needs of both individuals and groups always count and always need to be kept in balance, this approach is able to avoid the extremes of both individualism and holism (Behrens, 2010, p. 482).

In developing this nonanthropocentric African Relational Environmentalism further, I now seek to establish what kinds of things might plausibly be considered part of the "community" of nature and capable of being included in these familial or communal relationships. Since, on this African approach, harmonious relationships with other natural entities ought to be nurtured, which entities are we talking about? Is it meaningful to talk about forming a harmonious relationship with an insect or a rock? Would an African relational environmentalism most plausibly take such things into account, morally, or would it limit its basic moral concern only to, for instance, animate entities? How would such an ethic compare to Western approaches such as biocentrism and ecocentrism?[6] Essentially, I ask what objects can be said to count morally, or be due moral consideration. My main claim in this article is that African thought extends moral considerability to include all things that are part of the interconnected web of life, that is, all individual living things, groups of living things such as families, species, and ecosystems, as well as inanimate natural objects such as rivers and mountains. Granting moral considerability to other natural entities entails a rejection of strong anthropocentrism, inasmuch as the interests of nonhuman nature are also considered to be of moral significance.

After briefly explaining the concept of moral considerability, I identify some of the entities different African conceptions imply are morally considerable. I then seek a single criterion that is able to account for how these entities might have moral standing, as well as undergird an approach to the environment in which harmonious relationships with nature are prized and nurtured. Next, I evaluate whether this approach to moral considerability is plausible. Finally, I compare this approach to other more conventional Western perspectives on moral standing and suggest that some of the distinguishing features of this African approach are worthy of serious consideration and represent a promising, comprehensive theoretically appealing perspective.

What Is Moral Considerability?

Before considering moral considerability from an African perspective, I need to clarify what is included in the concept. I use the terms *moral standing and moral considerability* synonymously in this article. For an entity to be morally considerable, it must be capable of being regarded as making some kind of morally significant claim against a moral agent or agents. In turn, moral agents must, in some sense, be said to have some kind of basic moral obligation toward such an entity. This view implies that we should have a direct moral obligation toward entities that are morally considerable. It is possible to have indirect moral obligations with respect to almost anything. Based on city bylaws, I have a legal and moral obligation not to damage the benches and other picnic facilities in my local park. But my obligation is not toward the benches or tables, but to the local authorities, and perhaps my fellow burghers. I have merely an indirect moral responsibility regarding the benches, in that the benches do not themselves have a moral claim on me. Direct moral considerability is accorded to entities toward which we are able to have a non-derivative moral obligation.

To claim that X is morally considerable is not the same as claiming that it has intrinsic value, final moral value, or should always be treated as an end in itself. Y. S. Lo distinguishes between these concepts as follows:

> In the literature of environmental philosophy, the notion moral considerability is quite often used interchangeably with the notion of moral value. But the two notions are not exactly the same. Moral value is something that ought to be protected and/or promoted. But to say that something has moral considerability is to say that its existence, well-being, interest, preference, and/or some other aspect of it ought to be directly (rather than derivatively) given positive weight in our moral deliberation about actions that are likely to affect it. Hence, while things of moral value are morally considerable, it is not necessarily the other way round.
>
> (Lo, 2001, p. 355)

I am concerned in this article with the kinds of things that are morally considerable, that ought, at bottom, to be taken into account morally. Note though, that to claim that X is morally considerable is not to claim that it deserves equal moral consideration as Y. As Kenneth Goodpaster expresses it:

> Whether a tree, say, deserves any moral consideration is a question that we must keep separate from the question of whether trees deserve more or less consideration than dogs, or dogs than human persons. We should not expect that the criterion for having moral standing" at all will be the same as the criterion for adjudicating competing claims to priority among beings that merit that standing.
>
> (Goodpaster, 2003, p. 184)

A single criterion might be able to establish that an entity is or is not morally considerable, but other criteria are required to weigh the relative claims of different morally considerable entities on a moral agent. For the purposes of this article, I seek only to identify a criterion that establishes that entities are morally considerable at all.

A number of different necessary and sufficient criteria have been proposed by various theorists to establish the moral considerability of things. Goodpaster claims that having life is that criterion and rejects the more limited criteria of moral agency, rationality, sentience, or having interests as candidates (Goodpaster, 2003, pp. 183–88). While the aforementioned criteria are all properties intrinsic to individual entities, Mary Anne Warren points out that there are also moral theories that base moral status on a single type of (social or ecosystemic) relational property between entities (2005, pp. 439–440). I aim, with this article, to derive a single criterion of moral considerability from African thought – one that is different from those proposed by mainstream environmental philosophers thus far. I seek a plausible account of moral considerability that is able to undergird and complement an African relational environmentalism that promotes harmonious, balanced relationships within the community of nature.

What Kinds of Things Are Morally Considerable in African Thought?

I begin by trying to establish what sorts of things different African conceptions might entail as morally considerable. African writers make a variety of claims about African worldviews that imply that not only human beings are morally considerable. Some of these claims are based in metaphysical or even overtly religious conceptions; others might be regarded as more philosophical. A number of themes recur and are repeated by different voices from vastly different geographical locations on the subcontinent. In my quest for a single criterion for moral considerability, I first consider some of these themes and their implications for what African thought, in general, might regard as falling within the ambit of moral considerability.

Before proceeding, I need to make the point that phrases such as "moral considerability," "moral standing," and even "moral agent" or "direct moral duties" are obviously not part of the vocabulary of moral discourse in traditional African thought. But, that does not imply that such ideas cannot be recognised in indigenous conceptions. It is not difficult to perceive that a Christian belief that human beings are created in the image of God and possess an eternal soul may be interpreted as a religious conviction that, when secularised, seems to be analogous within the idea that persons have a dignity that requires respect, or that persons are morally considerable. So, too, we find beliefs in African thought that are analogous to that of moral standing, and, regardless of their metaphysical origins, imply that certain entities are morally considerable. Certainly, it is appropriate to assume that any belief that entails that persons ought to treat another entity with respect is likely to be something akin to a conception that such an entity has moral standing.

Morally Considerable Entities Entailed by Interrelatedness

I have already identified a strong emphasis in some African thought on the interrelatedness or interconnectedness of nature, and I consider this theme first. This belief is often understood as providing the basis for ethics, both between persons and between persons and the rest of nature. It also provides the foundation for strong communitarian values. Furthermore, it appears to be fundamental to many claims that human beings have a moral obligation to live in harmony with the rest of the family of nature and to respect natural objects. One of the very strong claims made about this worldview is that it differs from that of a traditional Western exploitative and instrumental view of nature in that it does not separate humanity from the rest of nature. As part of an interrelated community of nature, humans do not stand outside of it, but are integrally a part of it. Oruka and Juma write:

> The Judaeo-Christian view of nature which has permeated Western philosophical thought has nurtured a form of possessive individualism which is disrupting the complex web of being in which humans are a part. The Judaeo-Christian ethic has placed humans apart from nature, a fact that has contributed to global environmental degradation. There is a need fora shift toward a new epistemological outlook in which humankind is viewed as part of a complex and systematic totality of nature ... an ecophilosophical approach which recognizes the totality of (spatial, temporal, and other) interlinkages in nature.
> (1994, p. 115)

Sindima contrasts African cosmology with that of the Western world:

> For some time the people of Africa have been influenced by a cosmology inherited from the West: the mechanistic perspective that views all things as lifeless commodities to be understood scientifically and to be used for human ends. Yet these people have an alternative way of looking at the world, an alternative cosmology, which can better serve their needs for cultural development and social justice in an ecologically responsible context. This alternative way might be called a life-centered way, since it stresses the bondedness, the interconnectedness, of all living beings.
> (1990, p. 137)

Richard B. Peterson, who has studied Central African thought and the environment, makes similar claims:

> African dialectical thought reminds us that we are nature, we cannot get ourselves out of it. Thus, an "African ecology"' if you will, can contribute a corrective to our Western dualistic ecology, and help to amend our disconnection with nature by emphasizing that we must manage for a whole system, humans included, Africans certainly recognize the important differences between humans and animals (as one major non-human part of nature), but

> they have less of a tendency than we in the West to set the two up in dualistic opposition. The two are parts of one whole.
>
> (Peterson, 2004, p. 173)

The claim being made here is that an important implication of understanding all of nature as interconnected is that it implies a worldview in which humans cannot be perceived to be separate from the rest of nature. Such alienation from nature is a foreign idea, as is thinking of other parts of nature as having no value of their own but existing purely for their instrumental value to people. This perspective on nature was inconceivable to the African worldview in pre-colonial times. A belief that all natural things are interconnected and that humans are part of, rather than set apart from, nature provides a sound foundation for treating other entities with respect and for valuing other natural objects as morally considerable entities. In trying to understand this conception, we must remember that this same belief in interconnectedness and bondedness is often proffered as the foundation of the unshakable commitment to communitarianism in African ethics. Tangwa describes this outlook as:

> eco-bio-communitarian, implying recognition and acceptance of interdependence and peaceful co-existence between earth, plants, animals, and humans. This contrasts with the Western outlook which might be described as anthropocentric and individualistic. Within the African traditional outlook, human beings tend to be more cosmically humble and therefore not only more respectful of other people but also more cautious in their attitude to plants, animals, and inanimate things, and to the various invisible forces of the world.
>
> (2004, p. 389)

As much as being part of nature implies that other individual natural entities are morally considerable, the community, the tribe, the clan, and the family have moral standing, too. In fact, it is only through co-participation with others in promoting harmony and welfare within the group that the individual is able to reach moral maturity. The suggestion of the Nigerian political scientist, Claude Ake, that African thought requires a concept of group rights is just one example of how African communitarianism is often seen to entail moral considerability for social groups:

> It is necessary to extend the idea of human rights to include collective human rights for corporate social groups such as the family, the lineage, the ethnic group. Our people still think largely in terms of collective rights and express their commitment to it constantly in their behavior.
>
> (1988)

While some Western theorists of a more individualist ilk might have difficulty conceptualising a direct moral duty to a group, such as a community or a species, or understanding in what sense a moral agent could harm or benefit such groups (in any way other than through the combined benefit or harm done to the collection of individuals comprising the group), African communitarianism would find these concepts intuitively understandable and even appealing.

Morally Considerable Entities Entailed by Life Force

Another recurrent theme in African thought is the conception that all beings possess some kind of life force or vital energy. Augustine Shutte, a South African philosopher of religion, writes:

> The African conception of life includes both the physical and the spiritual. And it applies to everything. Stones are alive as well as animals. The difference is that animals have more life-force than stones, and we have more than animals. . . . So the universe can be seen as a graded system of life-force, emanating from the source of all force, God, and then going from the strongest, the ancestors who have died and the heads of clans and families, to the weakest, animals and material objects.
>
> (1993, p. 22)

Bujo approvingly quotes the famous theologian Placide Tempels: "All beings in the universe possess vital force of their own: human, animal, vegetable, or inanimate" (1998, p. 209). Bujo asserts: "All beings, organic and inorganic, living and inanimate, personal and impersonal, visible and invisible, act together to manifest the universal solidarity of creation" (1998, p. 210). and goes on to establish a link between life force and health:

> it is the task of the human person to study the cosmos to identify plants, animals and minerals possessing that force that can liberate one from physical and psychic suffering. One can actually remain healthy in a holistic sense, only by living in harmony with the whole of creation.
>
> (1998, p. 211)

Some of these concepts (especially the belief that ancestors who have already died continue to possess a "life force") are difficult for the non-African mind to accept. But, it is clear that a belief in a shared life force inhabiting living persons, human ancestors, animals, plants, stones, minerals, rivers, etc. suggests respect for more than just human beings, and a conception akin to moral considerability which embraces not just persons and other living things, but even inanimate and "spiritual" entities.

Morally Considerable Entities Entailed by Totemism, Inhabiting Spirits, and Folklore

Another recurring idea in African thought or religious belief that has implications for an African understanding of moral considerability is the widespread practice of totemism. In some African groups, specific animals are associated with the tribe or its people symbolically. It is forbidden to harm these totemic animals.

Murove writes of the observations of Phillip Junod (from the 1930s), who studied the baPedi people, that:

> he observed that in totemic ancestorhood, "there are feelings of affection and interdependence, of participation in one way of life... Junod went on to say that "Totemism shows well one characteristic of the bantu mind: the strong tendency to give a human soul to animals, to plants, to nature as such, a tendency which is at the very root of the most beautiful blossoms of poetry, a feeling that there is community of substance between the various forms of life.
>
> (2004, pp. 201–202)

In a similar way, rivers, rocks, trees, and forests may be venerated and understood, alternately, as being inhabited by some sort of spirit or an ancestor. Ancestor spirits may inhabit the bodies of other living things or take up residence in inanimate objects. Tangwa writes:

> Within [the African] world-view the distinction between plants, animals, and inanimate things, between the sacred and the profane, matter and spirit, the communal and the individual, is a slim and flexible one. For instance, the belief that humans, in certain circumstances can transform into animals and plants, or into plants and forces, such as the wind, is very prevalent within this system and has very significant implications for the way nature is approached.
>
> (2004, p. 389)

Murove points out that the moral teaching of Africa is often contained in folklore and claims that nature frequently plays a central role in tales or fables. Other animals come to the aid of humans, intervene in human affairs, and often exhibiting remarkable wisdom. "Human existence was only meaningful when seen in a continuum with all that exists. Such a widened sense of Ukama [relationality] enriched human experience through the inclusion of animals as co-partners in the moral palaver" (2004, pp. 206–207). Once again, the ideas imply a respect for and valuing of other natural objects, living things, inanimate things, and even ancestors who have died, in a way that confers on them at least some moral status or considerability. Certainly, when contrasted with the Judaeo-Christian belief that only human beings are spiritual or possess a soul, this worldview appears to be better able to include nonhumans within the ambit of the morally considerable.

Thus far, I have claimed that African thought on the environment is not absolutely anthropocentric and have shown that an emphasis on the interconnectedness of all nature provides an apparent foundation for a requirement to treat other natural entities with respect. It is also at the root of communitarian conceptions that are so central to African thought, implying that not only are natural individuals able to be morally considerable, so are groups such as clans, families, ecosystems, and species.[7]

I have considered a number of themes that suggest that African worldviews regard a wide variety of objects as requiring respect or having moral standing. These include other living humans, animals and plants, the dead (in the form of ancestors), and inanimate objects such as rocks, forests, rivers, mountains, and even minerals. To the non-African, the claim that inanimate objects or dead ancestors may have a life force or be inhabited by spirits may appear strange and difficult to accept. It ought, however, to be acknowledged that the consequence of believing that all of these objects have some "spiritual" value grants them a status akin to moral considerability. While some mindsets have difficulty with a belief in spirits, intangible life forces, and the continued participation of dead ancestors in the affairs of the living, it can still be appreciated that these African views do require a certain respect for the environment which was largely absent from the historically dominant anthropocentric and exploitative worldview of the West.

Interconnectedness in the Web of Life As the Basis for Moral Considerability

Having identified entities to which African conceptions typically grant moral considerability, I now search for a single unifying criterion that is plausibly able to ground the moral standing of these identified entities. Of the themes considered in the previous section, the one I find most promising as providing a plausible, single conceptual framework able to embrace the implications of the others is the notion of the inter-relatedness of nature. Implicit in this notion is a sense that all natural objects "share" something. This might turn out to be a relational property, but could equally turn out to be a single, intrinsic, common characteristic. The claim of interconnectedness would remain equally valid, however, whether the single shared characteristic turned out to be an intrinsic property, such as possessing a life force, a spirit, or an ancestor, or a relational quality, such as mutual dependence. Natural entities would still be understood as being interrelated by virtue of sharing this one attribute. Interrelatedness, then, on the face of it, entails the other candidates. Furthermore, there is reason to believe that the idea of inter-relatedness is more likely to be understood and embraced by non-Africans than the other options, which require much more contested metaphysical beliefs. It is also clear that the notion of interconnectedness is equally emphasised by those Africans who ground it in life force, inhabiting spirits, or the ancestors. For these reasons, I now explore how the conception that all nature is interconnected might ground the moral considerability of the entities I have identified.

Believing in an interconnectedness of everything in nature might well imply a healthier perspective on humanity's position vis-à-vis the rest of nature than that entailed by traditional Western anthropocentric views. But, how does an averment that everything in nature is interconnected provide a theoretical basis for a claim that all of these things are morally considerable? Sindima uses a construct more familiar to non-Africans to explain this worldview. He claims:

> the African understanding of the world is life-centered. For the African, life is the primary category for self-understanding and provides the basic framework for any interpretation of the world, persons, nature or divinity. ... Part of the very process of life involves a tendency toward self-transcendence, which itself aims for *umunthu*, or the fullness of life, In the human sphere the process of lite achieves fullness when humans are richly connected to other people, other *creatures* and to the earth itself. Humans realize their own fullness by realizing the bondedness of life.
>
> (1990, p. 143)

What are we to make of the claim that this view is life-centered? As I have shown above, for many Africans this worldview includes within the compass of the morally considerable not just living human beings, but also other living animals and plants, spirits and the recently dead (ancestors), and "inanimate" objects such as plants, rocks, rivers, etc. It also claims that fullness of life is achieved only in connection, community, and the bondedness of all life. Individual lives are almost inconceivable without strong communal bonds. So, groups such as species and families and communities must also be morally considerable. But, in what sense can all of these morally considerable entities be entailed by a "life-centeredness"? How exactly can the term life itself be understood in a way that could conceivably encompass all of these things? From a Western biocentrist perspective, life is a biological thing, and each individual organism possesses life, and has a life of its own, until it dies, physically. Such an understanding of a life-centered ethic could certainly not account for regarding as morally considerable: inanimate objects, ancestors, or spirits (although it might be able to make a case for the moral consideration of groups of living organisms). If most of these entities can be morally considerable on an African conception of life-centeredness, the very concept of life must have to be different. Sindima might well suggest this when he explains the "African vision of interrelatedness and bondedness with nature" (1990) as follows: "persons are one, woven by creation into one texture or fabric of life, a fabric or web nature and characterized by an interdependence between all creatures" (1990) Similarly, Oruka and Juma:

> postulate pan-organism as the basic truth underlying all nature. ... [This] means that all aspects of nature are interconnected, so that the ecological activities are a network. So a break or imbalance in one aspect has serious consequences in other aspects of the domain.
>
> (1994, p. 122)

Life is not just something possessed by individual organisms, it is also a "woven ... texture or fabric of life, a fabric or web characterized by interdependence" (1990, p. 143). Life, on this account, is not merely something each individual living organism possesses; it is a complex network of which any individual living thing is only a part. Furthermore, living things only fully have life because of their interconnectedness with other things. It is as though life is something one "plugs into," rather than just possesses. We share life, rather than own it. A river may not have a good of its own, or interests in its own right, but its health and integrity provide the medium that can sustain so much other life. So, it forms part of the web of life, even though it is not, itself, alive. An ancestor may be physically dead, but his or her moral teachings, his or her example, and perhaps his or her wisdom, still influence his or her human descendants, perhaps even affect their choices. So, in this sense, an ancestor can be part of this web of life. Clearly all animals, organisms, and plants are part of life's fabric and groups such as species or families or communities or ecosystems are subsets of the web of life. If one understands life as existing in individual things as well as in an interconnected web, then this African view may well be described as life-centered. To distinguish it from more conventional biocentric or life-centered approaches, I refer to this view as "web of life centered." What grants moral considerability to an entity is not that it has an individual life or telos of its own, but that it is part of the web or fabric of life. And then on this view, all things that can be part of this web of life, that themselves share or enable this life, can be morally considerable. Since everything that forms part of the web is in some way interconnected or interdependent with other parts of the web, all of these entities need to be taken into account morally. It seems to me, then, that the most plausible interpretation of the various overlapping conceptions of moral considerability I have identified in African thought is that they ground moral considerability in being part of the interconnected web of life.[8]

My project in this article is to identify a criterion of moral considerability based on African ideas that is able to complement what I have called "African relational environmentalism." That is to say, I seek to describe a plausible conception of moral considerability that is consistent with a general ethical outlook in which what is most prized is neither individuals or their capacities, nor communities or groups, but harmonious relationships. This web-of-life-centered approach to moral considerability is able to do that. Life, writ large, in this sense of an interconnected web or fabric shared by individual living things and other natural, and even supernatural, entities is central. Because of this, it is imperative that harmonious relationships between natural objects be maintained. If what is morally considerable is anything that is part of the interconnected web of life, then it makes sense that moral agents should promote and preserve that which holds the web of life together. Harmonious, balanced relationships within the web of life should be maintained and prized. As Peterson puts it:

> [African] lifecenteredness is oriented less toward individual entities (rocks vs. trees vs. animals vs. people) and more toward the relations between them. More attention is paid to processes and the flow of forces between entities than the entities themselves. Emphasis falls on relating rather than existing since it is the quality and nature of the relationship that determines whether the whole will sink or swim. The relationship between any two entities affects all of the rest of life since all of life is bonded.
>
> (2004, p. 172)

No doubt, Western philosophers might be quick to object, fundamentally, to the notion that the "is" of interconnectedness can imply the "ought" of a moral requirement to maintain harmonious relationships in nature. I acknowledge that this is a logical problem that cannot be ignored. However, on many African accounts, it is clear that the fact of interrelatedness is understood as almost obviously providing grounds for an obligation to maintain harmonious relationships. While this could simply be another example of how easily people fall into the trap of assuming that an "ought" is implied by an "is," I propose that there is a deeper sense of connection here. The assuredness with which so many African voices claim that the fact that we are interconnected entails that we are morally responsible for one another suggests to me that something else may be at work here. Holmes Rolston III provides insight that is helpful.

> Ecological description [the "is"] finds unity, harmony, interdependence, stability, etc, and these are valuationally endorsed [the "ought"]. ... Still, the ecological description does not merely confirm these values, it informs them; and we find that the character, the empirical content of order. harmony and stability is drawn from, no less than brought to nature. ...What is ethically puzzling, and exciting, in the marriage and mutual transformation of ecological description and evaluation here is that an "ought" is not so much derived from an "is" as is discovered simultaneously with it. ... it is difficult to say where the natural facts leave off and where the natural values appear. For some observers, at least, the sharp is/ought dichotomy is gone; the values seem to be there as soon as the facts are fully in, and both alike are properties of the system.
>
> (Rolston, 1989, pp. 19–20)

Clearly, Rolston's claims are predicated on the findings of ecological science, which imply a naturalistic sense of interdependence in nature, ecosystems, and the biosphere. But, such views are not all that far removed from the more metaphysical claims of interconnectedness in African thought. In much the same way as Rolston suggests, the moral requirement to promote harmonious relationships and the recognition that all things in the web of life are interconnected seem to originate, or be discovered, together in African thought.

The Plausibility of This Approach

So far, I have mainly been concerned with simply describing a theory of moral considerability based in African thought. I now turn to evaluating the plausibility of such an approach. A web of life-centred theory of moral considerability might well be able to undergird an African relational environmentalism that locates life (conceived as a web of life) at the centre of moral concern, and consequently, prizes harmonious relationships in nature. It might also be able to account for how African thought can regard humans, other living things, animals and plants, individuals and communities (including species and ecosystems) as morally considerable.

While there is controversy even about some of the aforementioned entities, other respected philosophical approaches have extended moral considerability to all of these things, and it is within the mainstream of environmental philosophy to be able to conceive of such entities as having moral standing. So, I will be bold enough to defend the inclusion of all living organisms as well as groups of organisms as morally considerable entities, without further argument. Far more controversial are the claims that inanimate objects such as rocks, rivers, or mountains might be morally considerable. Most controversial, obviously, is the idea of including ancestors and spirits within the scope of the morally considerable. I turn now to considering these two more controversial claims.

The reason that the idea that inanimate objects can be said to be morally considerable is so controversial is primarily because it is difficult to understand how it is possible to regard as morally considerable entities that cannot be directly harmed. How can an agent have a moral duty to something that has no good or even interest of its own? Nonetheless, there is something philosophically attractive about the notion that inanimate natural objects that play a very important role in holding the web of life together ought to be understood as being morally considerable. If we consider a river, again, it is true that it is prima facie problematic to claim that we owe a direct moral duty to a river itself, and for itself. It does not matter to a river whether it is clear and pristine or polluted and toxic. But, it does matter to many other things that rely on the river for survival. Since a river is such a life-supporting thing (or, in African terms, part of the web of life), surely we must take the river into account in our ethical thinking? Perhaps even more pertinently, in these days of global warming and climate change, although we may not ordinarily think of an ocean current as something to which we can owe a direct moral duty, we must take into account the devastating repercussions on global climate of significant changes in the flow of, say, the Gulf Stream. Thus, it seems that conceiving of such entities as morally considerable might be attractive on the pragmatic grounds that if they were treated as being morally considerable, it might turn out to be for the good of many other organisms, including humans.

Conceiving of inanimate natural objects as morally considerable appears to fly in the face of the standard definition of moral considerability that requires that we have direct and non-derivative moral duties toward entities. But, despite this, we often use language that suggests that we do, in fact, fairly routinely think about at least some inanimate natural things as though they are things that can themselves be harmed or benefited. It is not unusual for people to claim that effluent will "harm" a river, that a mining operation will "scar" a landscape, or that pollution might "damage" the atmosphere. Of course, this could be put down to little more than imprecise or possibly allegorical use of language. But, I surmise that there may be more to it than just that. It seems that, implicit in such expressions, is some sense in which we are inclined to regard certain inanimate natural things as quasi-organisms. We are inclined to talk about particularly those natural inanimate entities that seem to play a very significant role in sustaining the life of many other entities, such as natural habitats, for example, watercourses or forests or mountains, as though they have a good of their own. Despite the theoretical concerns about this issue, there does seem to be something appealing about thinking in this way, at least from the perspective of environmental ethics. Since a severely polluted river is capable of doing enormous damage to all kinds of organisms and species, conceiving of a river as having a good to be protected is pragmatically attractive, if we hope to protect the natural environment. In what follows, I hope to show that conceiving of such inanimate natural entities as morally considerable can be theoretically defended.

The kind of entities I am focusing on in this section are natural things that play a systemically important role in sustaining the flourishing of other living things. Essentially, they are instrumentally important to life. I am concerned with trying to defend the idea that such entities can be morally considerable. In a challenging article on moral value, Shelly Kagan makes the claim that it is possible for things to be valued as ends, or to be "worth having 'for their own sake'" (Kagan, 1998, pp. 277–297). based on relational properties, and even on instrumentally valuable properties. So, he suggests that the pen used by Abraham Lincoln when signing the Emancipation Proclamation might be construed to be valuable for its own sake exactly because of its historically deeply significant instrumental value. Similarly, he claims that rare or uniquely beautiful objects might be valued as ends even though their value is grounded in a relational property, namely, their rarity (Kagan, 1998, pp. 282–287). If this view is at all plausible, then it is no less conceivable that inanimate natural objects could be thought of as being morally considerable because of their vital systemic contribution to the web of life. Their important instrumentality to other life is a relational property that allows us to consider them (directly) worthy of being taken into account morally. Their moral considerability is grounded in their instrumental and relational value. Whether one accepts Kagan's claims or not, if there is a case to be made that a river could be thought to have final moral value because of its relational qualities, it is surely conceivable that the less radical claim that a river can be morally considerable might be plausible.

To press this claim further, I return to Lo's distinction between moral considerability and moral value that I cited earlier in this article. On her account, in contrast to moral value, moral considerability is not something that we seek to promote or secure. It is used rather to denote a category of entities whose "existence, well-being, interest, preference, and/or some other aspect of it ought to be directly (rather than derivatively) given positive weight in our moral deliberation about actions that are likely to affect it" (Lo, 2001, p. 355). It is true that rivers and the like might not really have a well-being of their own, preferences, or even interests. But, they can be said to have an existence which we would want to positively take into account in our moral considerations. Now, if all that is required for moral considerability is that an entity can be said to have an existence, then almost anything would be morally considerable, and there would appear to be little point to the category at all. It is clear that Lo does not intend for us to make this inference. So, she must be suggesting that there are at least some kinds of entities whose mere existence is something we ought to take into account morally. Just as Kagan thinks that very rare or historically significant things could be thought to have value as ends in themselves, perhaps systemically crucial inanimate natural objects can be said to be morally considerable simply for the sake of their continued existence. One reason for this is precisely because of their vital instrumental systemic role in facilitating the thriving of other living things. But taken on its own, this criterion of instrumentally enabling the lives of other organisms could equally entail that a human-made artificial fertiliser is morally considerable – an idea that seems neither theoretically attractive nor something entailed by African conceptions. What is still needed, then, is some way of accounting for why the property of naturalness of an entity that contributes to the flourishing of many other living things is what makes it morally considerable. Political scientist Robert E. Goodin provides just such an account. In writing about the values endorsed by the so-called "green" political movement, he describes what he calls the "green theory of value." He claims that on this theory:

> What it is that makes natural resources valuable is their very naturalness. That is to say, what imparts value to them is not any physical attributes or properties that they might display. Rather it is the history and process of their creation. What is crucial in making things valuable, on the green theory of value, is the fact that they have a history of having been created by natural processes rather than by artificial ones.
>
> (1992, p. 27)

This should not be taken as a claim that only natural things can have moral value. But, it does go some way toward suggesting what it is about natural things, even inanimate natural things, that makes their mere existence something that we ought to take into consideration. This idea resonates with the "web of life" criterion for moral considerability I have suggested. Even geological formations are part of the web of life on the basis of their history and the natural process of their coming into existence. They are morally considerable because they have been forged by the natural processes of the Earth alongside and inextricably interconnected with the natural processes that continue to create and sustain life. They may not have a good or interests of their own, but their existence is important enough for us to take it into consideration. It seems to me, therefore, that it is not implausible to include inanimate natural entities within the scope of what is morally considerable.

Even more problematic than granting inanimate objects moral status are the more metaphysical ideas that ancestors and spirits are morally considerable because many people simply do not share a belief in the possible existence of such spiritual entities. Nonetheless, I suggest that there may be lessons to learn from such conceptions. First, the belief in ancestors implies a continuity and interdependence between generations. However much Western philosophers may have agonised and debated over whether we can have moral obligations to future generations, it is almost a non-question in African thought. Intergenerational moral accountability is intrinsic to African thought. Kwasi Wiredu writes:

> Of all the duties Owed to the ancestors none is more imperious than that of husbanding the resources of the land so as to leave it in good shape for

> posterity. In this moral scheme the rights of the unborn play such a cardinal role that any traditional African would be nonplussed by the debate in Western philosophy as to the existence of such rights. In upshot there is a two-sided concept of stewardship in the management of the environment involving obligations to both ancestors and descendents which motivates, environmental carefulness, all things being equal.
>
> (1994, p. 46)

There is no possibility of imagining that we do not owe it to future generations of living things to ensure they inherit a world capable of sustaining their lives. Second, this should serve to remind us that in ethical decision making, one cannot simply ignore the spiritual or religious and even cultural ideas that people hold, and that are often very important to them. The deeply held spiritual beliefs of people are an important part of their sense of identity and solidarity with others, as well as their well-being, and this is something we ought to take into account when making ethical decisions.

The Promise of This View

I now seek to compare this African perspective on moral considerability to more conventional Western approaches and consider its promise. How does a web of life theory of moral considerability compare with other approaches? Clearly one would expect to find some commonality with other life-centered or biocentric views, since they also entail that all living things (at least) are morally considerable. However, this strand of African thought differs from biocentric positions in one very important respect. Biocentrism is usually characterised as an individualist position, locating moral value in particular properties of individual organisms or in maximising utility for such individuals. The African relational environmentalism I have been describing is strongly grounded in a communitarian value system. It prizes fostering harmony in the community of all living things, stressing their interrelatedness and bondedness. I have claimed that the most promising aspect of this approach is its refusal to prioritise individuals over the community or family of nature, or vice versa (Behrens, 2010, pp. 477–478). As much as some African thought apparently accords moral considerability to all individual living organisms, it would seem to fly in the face of a fundamental communitarianism in African thought to imagine that it would not also accord moral considerability to wholes such as species, ecosystems, and the biosphere.

Now, just as this view is distinct from biocentrism, it is also distinct from ecocentrism, or holist[9] conceptions in which the biosphere, species, and/or ecosystems are accorded final moral value. One of the persistent challenges facing holists is that their consequences appear to be unavoidably misanthropic. If the good of the biosphere always trumps that of individual people or organisms, it is hard to see why we should not then cull people to save the Earth, for instance. This has led critics, such as Tom Regan, to accuse holism of "environmental fascism" (Palmer, 2003, p. 25). Can African thought, with its focus on communitarianism, escape these problems? As I have said before, one of the greatest strengths of this African perspective is that it does not prioritise individuals over groups or communities, or vice versa. Grounding moral considerability in belonging to a web of life entails promoting harmony and balance within the web. Such a position is essentially communitarian, and yet individuals matter, too. African relational environmentalism is neither biocentric nor ecocentric, neither individualist nor holist. It is relational, and it must therefore take into account the relationships and connectedness of everything in nature. It seeks to promote harmony, solidarity, respect, and mutual cooperation between natural things.

Conclusion

An African emphasis on the interrelatedness or interconnectedness of everything in nature is the foundation for a morality that prizes relationships that are harmonious within the community of all natural things. If what is central to morality is life, as in the web or fabric of life, then anything that forms part of this complex web of life can be morally considerable. Such a view enables us to include within the scope of our moral consideration criteria proposed by holists, ecocentrists, sentientists, and biocentrist individualists. In fact, it is able to expand this sphere even further by taking into account inanimate objects. Perhaps the most apt description of this view is Tangwa's claim that it is "eco-bio-communitarian," (2004, p. 389) essentially communitarian and relational, and thus able to embrace the most useful aspects of both individualist biocentrism and holist eco-centrism, while avoiding their extremes.

Notes

1 For some representative examples of thís claim, see Benezet Bujo, *The Ethical Dimension of Community* (Nairobi: Paulines Publications, 1998), p. 25; Michael Eze, "What is African Communitarianism?" South African Journal of Philosophy 27 (2008): 387: John Mbiti, Introduction to African Religion (London: Heinemann, 1975). p. 38.

2 There are commonalities between the African view I describe and some of the thoughts of Native American communities. I acknowledge this point, as well as that a comparative study would be very valuable. However, since my aim is to redress a lack of attention to African eco-philosophy, I do not attempt such comparisons in this article.

3 Commentators have raised two concerns about my project that I need to address. The first is that I possibly nullify the claim that the ideas I express are any longer truly African since I often "secularize" metaphysical beliefs. It should not be assumed that only traditional conceptions count as African thought, something professional African philosophers themselves reject. Recent African philosophy has deliberately moved beyond ethno-philosophy, which has been criticised as being primarily descriptive. In recent decades, trained African philosophers have sought to apply the usual argumentative and critical tools and methods of philosophy to many of the questions typically considered by philosophers to produce their own theoretical work. As might be expected, much of this work is done with reference to indigenous African thought, values, or context, but this need not necessarily be the case. See Didier Kaphagawani, "African Conceptions of Personhood and Intellectual Identities," in The African Philosophy Reader, ed. P. Coetzee and A. Roux (London: Routledge, 1998), pp. 97–98. I see my project in a similar light. A second concern that has been expressed is that in using African ideas to develop a theory, I might be guilty of a new kind of "colonialism" by

Westernizing African thought. My intention is rather to contribute to what seems to be a growing interest in environmental ethics by African philosophers, to stimulate further philosophical discourse, and to show that environmental ethics can be enriched by African thought.

4 My project is intended to be more analytical than simply descriptive. I ultimately hope to articulate an environmental ethic inspired by African thought. I do not seek to "discover" an African ecophilosophy and describe it, but rather to systematically develop an environmental ethic based on some pertinent and distinctive themes or conceptions in African thought.

5 Kofi Opoku, "African Traditional Religion: An Enduring Heritage," in Religious Plurality in Africa ed. Jakob Olupona and Sulayman Nyang (Berlin: Mouton de Gruyter, 1993), p. 77: Workineh Kelbessa, "The Rehabilitation of Indigenous Environmental Ethics in Africa," Diogenes 53 (2005): 25.

6 The terms biocentricism and ecocentrism are not used consistently in the literature. For the sake of clarity, I use biocentrism to denote the "view that any living thing has moral standing." Donald vanDe Veer and Christine Pierce, eds. The Environmental Ethics and Policy Book (Belmont: Wadsworth/Thomson Learning, 2003), D. 650. On this definition, biocentrism is an individualist view. By contrast, I use ecocentrism to denote holist views in which species, ecosystems, and the entire biosphere are regarded as morally considerable.

7 I have only suggested that based on its essentially communitarian nature, African thought can plausibly be understood to regard groups such as species and ecosystems as morally considerable. Far more argument is required to defend this claim robustly. However, ecosystems are merely subsets of the web of life, being part of which, I argue below, is the criterion that establishes moral considerability. There are also good reasons to think that African thought would regard species qua species as morally considerable, too. Kelbessa, for instance, claims that the Oromo people "pay attention to other species as valuable in and of themselves" and that "it is morally wrong to totally destroy a species." Kelbessa, "The Rehabilitation of Indigenous Environmental Ethics," p. 24.

8 This web of life approach is reminiscent of Albert Schweitzer's biocentrism. There are also parallels with some of the fundamental characteristics of deep ecology.

9 I am aware that I am comparing this view to very broad categories of alternatives, and that there are far more nuanced and sophisticated approaches within these broad categories that warrant consideration.

References

Ake, C. "The African Context of Human Rights," in *Applied Ethics: A Multicultural Approach*, ed. Larry May et al. Upper Saddle, NJ: Prentice Hall, 1988.

Behrens, K. "Exploring African Holism with Respect to the Environment," *Environmental Values*, 2010, 19(4), 465–484.

Bujo, B. *The Ethical Dimension of Community*. Nairobi: Paulines Publications, 1998.

Callicott, J.B. *Earth's Insights: A Survey of Ecological Ethics from the Mediterranean Basin to the Australian Outback*. Berkeley: University of California Press, 1994.

Eze, M. "What is African Communitarianism?" *South African Journal of Philosophy*, 2008, 27, 386–399.

Goodin, R.E. *Green Political Theory*. Cambridge: Polity Press, 1992.

Goodpaster, K. "On Being Morally Considerable," in *The Environmental Ethics and Policy Book*, eds. Donald Van De Veer and Christine Pierce. Belmont: Wadsworth/Thomson Learning, 2003, 308–325.

Kagan, S. "Rethinking Intrinsic Value," *Journal of Ethics*, 1998,, 2, 277–297.

Kelbessa, W. "The Rehabilitation of Indigenous Environmental Ethics in Africa," *Diogenes*, 2005, 52(3), 17–34.

Lo, Y.S. "The Land Ethic and Callicott's Ethical System (1980–2001): An Overview and Critique," *Inquiry: An Interdisciplinary Journal of Philosophy*, 2001, 44(3), 331–358.
Mbiti, J. *Introduction to African Religion*. London: Heinemann, 1975.
Murove, M.F. "An African Commitment to Ecological Conservation: The Shona Concepts of Ukama and Ubúntu," *Mankind Quarterly*, 2004, 45(2), 195–215.
Opoku, K. "African Traditional Religion: An Enduring Heritage," in *Religious Plurality in Africa* ed. Jakob Olupona and Sulayman Nyang. Berlin: Mouton de Gruyter, 1993, 67–82.
Oruka, H.O. and Juma, C. "Ecophilosophy and Parental Earth Ethics," in *Philosophy, Humanityy and Ecology*, ed. H. Odera Oruka. Nairobi: ACTS Press, 1994, 115–129.
Palmer, C. "An Overview of Environmental Ethics," in *Environmental Ethics*, ed. Andrew Light and Holmes Rolston, II. Malden: Blackwell, 2003, 15–37.
Peterson, R.B. "Central-African Voices on the Human-Environment Relationship," in *This Sacred Earth: Religion, Nature, Environment*, ed. Roger Gottlieb. New York: Routledge, 2004, 151–157.
Rolston, H. II, *Philosophy Gone Wild*. New York: Prometheus Books, 1989.
Shutte, A. *Philosophy for Africa*. Rondebosch: UCT Press, 1993.
Sindima, H. "Community of Life," in *Liberating Life*, ed. Charles Birch et al. Maryknoll: Orbis, 1990, 537–551.
Tangwa, G. "Some African Reflections on Biomedical and Environmental Ethics," in *A Companion to African Philosophy*, ed. Kwasi Wiredu et al. Malden: Blackwell, 2004, 387–395.
Warren, M.A. "Moral Status," in *A Companion to Applied Ethics*, ed. R.G. Frey and Christopher Wellman. Malden: Blackwell, 2005.
Wiredu, K. "Philosophy, Humankind and the Environment," in *Philosophy, Humanity and Ecology*, ed. H. Odera Oruka. Nairobi: ACTS Press, 1994, 30–48.

8 Deep Ecology, Ontic Relationality, and Positionality

Exploring the Ontology of Anthropogenic Waste

Elvis Imafidon

Introduction

Deep ecology is first and foremost an ontological point of view. As an ontological point of view, it has at its crux the claim that all beings, human and non-human, have as their most fundamental mode of being, the being-with-others. Perhaps a more profound way of stating this, which I will show subsequently to aptly represent an African understanding of such relationality or participation in relationships among beings, is being-together. The ontological point of view of deep ecology expands the Heideggerian idea of Dasein's mode of being since not only Dasein but all beings have the potentiality and actuality of participation in relationships. This ontically necessitated mode of being with other beings leads to several conclusions: all beings are interconnected and interwoven in a web of relationships; all beings contribute to, or have the potentiality to contribute to, building the relationships or communion; a being occupies an ontic position in the community of beings in relation to other beings such that its being-in-place, ontologically speaking, sustains balance and ontological equilibrium in the web of beings; and flowing from these, a being thus brings inherent and intrinsic ontological value to the locale of communion. Therefore, it is not surprising that interconnectedness, intrinsic value, inclusion, equity for all beings, and sustainability are at the heart of deep ecology discourses in such spheres as ethics, politics, policy and lifestyle. Considered in this way as an ontological point of view, while the concept of deep ecology may be quite recent, the ontological claims associated with it are certainly not new. Many philosophical and religious traditions of the world have chanted for ages the interconnectedness and the intrinsic worth of beings. In fact, eco-relationality or relational ecology is a fitting term to describe many of such traditions and their commitment to the well-being of the environment, and these traditions include the sub-Saharan African ontology of the ecosystem (Behrens 2014). But as I will show subsequently, such a term only accounts for the ontology of relationality and not the ontology of positionality, which is also crucial for the sort of ontological point of view of deep ecology that I hope to develop through the pages of this chapter.

Let me turn quickly to the idea of anthropogenic waste and how it connects to the relational and positional modes of being that I am going to elaborate on shortly. Anthropogenic waste, as I have it in mind here, is intrinsically tied up with the concept of ontic relationality or being-with-others and ontic positionality

DOI: 10.4324/9781003480914-9

or being-in-place because it emerges from human activities that fundamentally disrupt relationships and displace beings. Much has been said and written about the key causes of contemporary environmental crises and threats to deep ecological ways of being. They include industrialisation, urbanisation, rapidly growing energy consumption, overpopulation, overexploitation of natural resources, deforestation, climate change, and pollution. At the heart of these key factors is the ugly reality of anthropogenic waste. Perhaps, anthropogenic waste or human-produced waste is the most poignant consequence of advances in science, technology, industrialisation, capitalist models of modern societies, and the quest for human material, aesthetic, and ecstatic security. The waste of natural resources by humans in our time is unprecedented. To put this in some perspective,

> yearly, we directly waste or mismanage around 78% of the total water withdrawn, 49% of the food produced, 31% of the energy produced, 85% of ores and 26% of non-metallic minerals extracted, respectively. As a consequence, natural resources are getting depleted and ecosystems polluted, leading to irreversible environmental changes, biological loss and social conflicts.
>
> (Marín-Beltrán et al. 2022: 1)

There is substantial awareness in existing literature of how anthropogenic waste has disrupted the interconnectedness of things, led to human domination of other beings, and destroyed the relational mode of being as being-with-others or being-together. To be sure, there is much to be learnt about this mode of being, particularly from non-Western and indigenous forms of knowledge, ancient wisdom, and philosophies that have not been taken seriously enough in terms of how such a mode of being ought to be and how the interconnectedness and relationships that have been disrupted and destroyed can be revived. As Bill Devall (1985: ix) puts it,

> we may not need something new, but need to reawaken something very old, to reawaken our understanding of Earth wisdom. In the broadest sense, we need to accept the invitation to the dance – the dance of unity of humans, plants, animals, the Earth. We need to cultivate an ecological consciousness. And we believe that a way out of our present predicament maybe simpler than many people realize.

I am therefore in this chapter drawing heavily on the sub-Saharan African philosophy of relationality in exploring what it means to be together with other beings. However, a fundamental aspect of the impact of anthropogenic waste is the corresponding disruption of being-in-place, the wanton and colossal displacement of being, and the ontological imbalance and disequilibrium it leads to in the ecosystem. Therefore, I begin in the section that follows by exploring how anthropogenic waste creates being-out-of-place due to the very nature of its ontology, one defined by a jouissancial, aesthetic, ecstatic, and instrumental mode of being. Key features of this mode of coming-to-be of anthropogenic waste include putting beings out of ontic positions and feeding the jouissance of dissecting, disembodying, and storing up of beings. I then proceed to explore two fundamental modes of being from sub-Saharan African perspectives that are crucial for consideration for deep ecological commitments: being-together – ontic relationality and being-in-place – ontic positionality. Focussing on African indigenous healthcare systems, I show the value of such a positional and relational ontology by analysing the moral, aesthetic, and hermeneutic value of being-in-place and being-together in community and how such understanding shapes reciprocal use, value, and relationality of all beings, human and non-human. The chapter concludes by highlighting the need to rethink our capitalist and industrial structures if the goal to radically reinterpret the value of the non-human is to be taken seriously.

The Ontology of Anthropogenic Waste

Waste, in a sense, is simply ontic, factual, and existential. It is an existential fact of being to produce waste in the right amount and form. All beings are ontically waste-producing beings. Of course, the use of 'waste' here to represent this existential experience of being is itself problematic since, etymologically and definitionally speaking, it connotes uselessness and emptiness, for ontically produced wastes are not in themselves useless and empty but significantly contribute to the collective well-being of the ecosystem. Ontically released oxygen as waste from plants is fundamental for animal life; ontically produced carbon dioxide as waste from animals, oceanic and volcanic life is crucial for the survival of plants; animal faeces are significant sources of nourishment for the soil and energy production; and pineapple peels usually regarded as waste are now being used to produce Pinatex, a biodegradable and environmentally friendly leather-like material in place of the high demand on animal leather. Thus, rethinking the concept of ontically produced waste is important, although such is beyond the purview of the current paper. But this understanding of ontically produced waste in the sense just described is crucial for making sense of anthropogenic or human-produced waste specifically as used here. I use anthropogenic waste here to mean the waste generated by the wide range of human activities. Such activities are therefore not simply ontic but also consist of political, economic, technological, industrial, and aesthetically inclined activities. Thus, anthropogenic waste, as human-produced waste, comes into being through a wide range of human activities, both ontic-driven and norm-driven. An ontology of anthropogenic waste comprises an analysis of specific modes of being human that fundamentally lead to warranted and unwarranted waste of the nonhuman in ways that threaten ontological equilibrium and collective and relational coexistence in the ecosystem. I am particularly concerned here with two modes of being human that have become defining modes of being in modern industrial and capitalist societies and how they result in anthropogenic waste: disruptive ontic positionality or the disruption of being-in-place (being-out-of-place) and being-toward-jouissance.

Let us begin with an analysis of the first: disruption of being-in-place. The spatio-temporality of being is ubiquitous in the study of ontology. A quick look at the more recent history of Western ontology would reveal a detailed focus on the temporality and finitude of being in such works as Heidegger's *Being and Time* (1985), Husserl's *The Phenomenology of Internal Time Consciousness* (1964), Sartre's *Being and Nothingness* (1956), and Merleau-Ponty's *Phenomenology of Perception* (1962). But this is often heavily an anthropocentric discourse of temporality and finitude, not an ecocentric one. What would constitute a robust ecocentric discourse of being and time with a focus that goes beyond the human is a project worth exploring but not here and now. I am more concerned at this time with being and place or the ontic positionality of being that also requires a much-needed attention in the study of ontology. The hermeneutics of place already shows how significant it is to take place seriously in our quest to understand and interpret adequately our being and lived experiences (Janz 2017). An ontology of place further enriches the significance of the study of place in the search for meaning and a better understanding of being.

A being, human or non-human, ontically occupies a here, a place of dwelling. Such a place of dwelling is fundamental for the meaningful and authentic existence of such a being and determines the potentiality and possibility of contributing meaningfully to the existence of other beings. Being-in-place, therefore, consists of an entity occupying and dwelling in an ontically defined, strategically positioned, and appropriately situated place of dwelling, the right spot, in a web of being. By implication, being-out-of-place diminishes meaningful existence, reduces effective positionality, disrupts ontological balance and equilibrium in the ecosystem, and diminishes being in general. The ontic place of one's dwelling is not, in this sense, fixed or static but rather of a kind, a form, and a being is ontically and existentially conditioned to dwell in a topophilic place that contributes to defining its ontic, existential, and communal identity. There are thus places that are topophobic to an entity, such that when an entity is displaced from its ontically defined place and forced to dwell in a space of ontic incompatibility, it produces existential meaninglessness for such a being and disrupts the being's ability to contribute meaningfully to the well-being of the community to which it belongs.

Our modern world is essentially defined by, and profusely saturated with, human activities of taking beings out of place and moving beings from points of topophilic ontic positionality to places that produce imbalance and disequilibrium. Much of the human activities that are known to continually destroy the ecosystem, disrupt the coexistence and relationality of beings, and threaten the interwovenness of being can be linked to the disruption of ontic positionality. For example, we see this clearly in deforestation, which happens with such aims as urbanisation, increased production for consumption, and mining. Deforestation is a glaring example of the disruption and dispositioning of a community of beings, an ecosystem. The justification for deforestation often consists of the *need* for urbanisation and large-scale agricultural and animal farming due to a rapidly increasing population and the *need* for sustained forest economics in order to harness the natural resources of the forest such as timber, oil, and mining products (Oral 2020). I have deliberately italicised *need* as I question in the analysis of the second human mode of being as being-toward-jouissance that follows shortly the extent to which these activities emerge from need or from excessive want of pleasure, enjoyment, and intense ecstasy. The forest itself is a place of being adequately situated in the larger ecosystem to harbour several beings-in-place. Thus, deforestation does not only destroy and disrupt an ontic locale, but it also dispositions several beings resulting in beings-out-of-place and the consequent imbalance in the community of beings. The primary aims of deforestation such as urbanisation, large-scale agricultural production, and mining further perpetuate such ontic displacement as these activities are not simply pursued for the purpose of satisfying human needs but for human jouissance. Thus, the argument that such human activities are ontically necessitated and unavoidable should be questioned as it is difficult to, for example, reconcile human consumption needs with the colossal waste of produced consumable products globally. We harness and produce more than we need and still fail woefully to fairly and equitably distribute what we produce.

Anyone familiar with the psychoanalytic thought of Jacques Lacan will be familiar with his idea of jouissance, which I find interesting and apt in describing the unavoidable coming-together of pleasure and pain in the human quest for satisfaction and enjoyment. Dylan Evans (2006: 117) explains how the concept develops in the works of Lacan thus:

> In the seminars of 1953–4 and 1954–5, Lacan uses the term occasionally, usually in the context of the Hegelian dialectic of the MASTER and the slave: the slave is forced to work to provide objects for the master's enjoyment (*jouissance*). . . Soon after, the sexual connotations become more apparent; in 1957, Lacan uses the term to refer to the enjoyment of a sexual object. . .and to the pleasures of masturbation. . ., and in 1958 he makes explicit the sense of jouissance as orgasm. . . It is only in 1960 that Lacan develops his classic opposition between jouissance and pleasure. . . painful pleasure' is what Lacan calls jouissance; 'jouissance is suffering'. . . The term jouissance thus nicely expresses the paradoxical satisfaction that the subject derives from his symptom, or, to put it another way, the suffering that he derives from his own satisfaction.

Concerning the second mode of being, being-toward-jouissance, I am particularly interested in the Lacanian idea of jouissance as painful pleasure, pleasure so intense that its boundlessness and the inability to reach a perfect point of satisfaction results in pain – a painful pleasure. To be sure, jouissance in this sense is not disconnected from the more sexually oriented sense in which Lacan uses it at some point or even from its original etymological meaning for that matter. The boundless pursuit of sexual pleasure results in an unquenchable thirst for such pleasure and thus in pain. But the pain of sexual pleasure is only one form of such jouissance, others include the quest for material pleasure and aesthetic ones, and in many cases, a fine blend of these and more pleasures. Thus, as a being-toward-jouissance, the human being has been socially constructed and lured, particularly in modern times, to pursue an insatiable course of pleasure relentlessly, a circle of pursuit that ends in less pleasure and more pain. We are, in a sense, consumed by our quest for pleasure, and suffering is at the heart of every glimpse of satisfaction we get. Moreover, the painful pleasure, particularly material and aesthetic pleasures, sought by humans today, is essentially sought out from the non-human and since by its very nature jouissance is a boundless, painful pursuit of pleasure, the pain is indeed not only felt by the actor (the human) but also by those acted upon (the non-human and the community of beings). It brings to mind one of the initial senses in which Lacan uses jouissance, the Hegelian dialectical sense of the pain of a slave to provide objects for the master's enjoyment. In this sense, the slave feels pain for the pleasure of the master, but the master also experiences pain as satisfaction is always elusive; the pleasure is never enough and there is always a need for more exploitation and domination of the slave. Similarly, in the context of our discourse, human painful pleasure, particularly in the quest for material and aesthetic satisfaction, is dependent essentially on the pain of the non-human through an unending exploitation and domination by the human.

When we think, for example, about large-scale industrial animal farming and the unquenchable thirst and demand for meat by humans, this jouissance becomes glaring. The human quest for different forms of meat in terms of types, parts, and features is unending and boundless. It is both a quest for a material and an aesthetic jouissance and no longer the simple need to eat meat, which itself is contestable and has been highly debated in human history. The demand for all kinds of meat such as poultry, beef, pork, lamb and mutton, goat, seafood, game meat, and exotic meat along with different meat qualities, such as lean meat, tender meat, fatty meat, processed meat, and dark meat, as well as specific parts like ribs, drumsticks, steaks, thighs, and belly. The jouissance, ecstasy, and enjoyment humans derive from disembodying, dismembering, and aesthetically packaging and arranging animals and their body parts have consumed modern society. This has led to unwarranted and colossal waste of animal products, as well as questionable and unethical practices in animal research and production. The unquenchable painful pleasure that humans seek from meat alone can also be seen in the huge interest in the mode of rearing and preparing such meat and how such practices define price, class, and status as well as the pleasure that could be derived from the consumption of such meat. There is a long and growing list of descriptions of such meat forms that could heighten the pleasure humans derive from meat. It includes the organic, raw, intramuscular fat (marbling), and packaging nature of the meat, as well as the age of killing and strict rearing conditions. For example, one of the most prized steaks in the world, the Japanese A5 Kobe beef steak, is so expensive due to its intense marbling, tenderness, rich flavours, and rearing under very strict guidance with certification. The contents of this jouissance list result in ethical challenges concerning how the animals producing such meat are reared, treated, and objectified. Marbling or practices aimed at increasing intramuscular fat in animals could result in animal welfare issues caused by feeding practices, gene alteration, and restricted movement. Beyond this effect of the painful pursuit of pleasure in relation to meat is the dialectical pain that results for humans as well. There is perhaps too much meat produced and made available today, meat that is not equitably distributed globally, resulting in too much meat for consumption by a few. The implication beyond the colossal waste this results in is the growing health issues for humans who consume more meat than they need, such as weight gain, kidney issues, cardiovascular disease, and cancer. To be sure, the fundamental point of such unquenchable pursuit of pleasure from meat in this instance and other forms of jouissance in relation to the non-human is that it displaces and dispositions the non-human on a large scale and causes imbalance and disequilibrium in the community of beings.

Two African Modes of Being

The differences and diversity of African places today make it important to begin this section with a few words about what sort of locales in sub-Saharan Africa I have in mind here. To be sure, I have in mind here the sort of locales I lived in for decades, locales that have not been made overly urban or that are not vigorously chasing the mirage of the urban such that they lose touch with the local (which often follows a Western or colonially defined model of what is urban and industrially developed). These are locales that still retain cultural and indigenous values, ontological systems, and localised ways of being, knowing, and coexisting. I am, of course, struggling here to avoid using the terms 'modern' and 'traditional' to describe these locales, as has often been done conventionally (cf. Gyekye 1997; Lauer 2003; Kumar 2006). For example, modernity in Africa has often been defined by what Rekhar A. Kumar (2006) calls a European ideology where European ideas of such coordinates as knowledge, education, technology, individual rights, and politics dominate, and this is often contrasted with tradition in Africa, which then becomes the idea that such a European ideology has yet to take deep roots in such places, and indigenous ways of being still dominate. But this is a simplistic and challenging way of defining these locales in Africa, where you will easily find both the so-called modern and traditional coexisting in one locale and where you find individuals moving fluidly and flexibly between the 'modern' and the 'traditional' based on need and topophilic affinities. Thus, the sort of African places I have in mind here are perhaps those places where indigenous ways of being still thrive, where daily existence and approaches to issues such as healthcare, upbringing, farming, and social relations still hinge largely on local knowledge and ontologies. It is in such indigenous locales that I explore these two interrelated modes of being: being-together and being-in-place.

There already exists a rich literature on community and relationality as the highest good or most prized value in African indigenous places or communities (e.g., Gyekye 1996; Ramose 1999; Tutu 1999; Ikuenobe 2006; Metz 2021). The defining ontological principle of communing and relationship in these African communities is being-together. Being-together as a fundamental mode of being is in some way different from the idea of being-with-others. While the latter emphatically states the onticness of being with others and not being alone and is often used in the context of human relationships, the former essentially includes not only an ontic reality but also a moral responsibility between the human and the non-human. Being-together acknowledges that the fundamental nature of our being is that we do not simply exist with other beings but must establish and sustain active relationships with these other beings to maintain a balanced and sustainable ecosystem. In African places, these beings who not only coexist but are in relationship include the human and a broad category of the non-human such as ancestors, deities, the Supreme Being, plants, animals, forests, and water bodies. Polycarp Ikuenobe aptly captures this point thus:

> In the traditional African view, reality. . . is a continuum and a harmonious composite of various elements and forces. Human beings are a harmonious part of this composite reality, which is fundamentally a set of mobile life forces. Natural objects and reality are interlocking forces. Reality always seeks to maintain an equilibrium among the network of elements and life forces.... Because reality or nature is a continuum, there is no conceptual or interactive gap between the human self, community, the dead, spiritual or metaphysical entities and the phenomenal world; they are interrelated, they interact, and in some sense, one is an extension of the other.
>
> (Ikuenobe 2006: 63–64)

Beings in an African community recognise that one's well-being, survival, and existence depend on the existence, survival, and well-being of other beings. The African relational ontology holds that a being is a being through other beings, which means that to be is to be together, to coexist with and depend on other beings, and this necessitates the moral responsibility of promoting and caring for the well-being of other beings in the community. Thus, sub-Saharan African ethics have often been theorised as a theory of the good founded on the extent to which an action promotes or impedes communal harmony and collective wellbeing (Metz 2007; Imafidon 2014). This ontologized ethics of promoting being-together can be seen, for example, in the relationship that exists between the healthcare provider and the forest. The indigenous healthcare giver or healer has a topophilic affinity with the forest in particular and with gardens and plants in general. They are significant sources of knowledge and resources for healing. The forest is not merely an objectified and lifeless site of resources but a lively locale of being that a healer shares a close relationship with. There is a clear acknowledgement of the shared essence and unifying force of being that both the healer and the forest as a source of healing have in common. A typical instantiation of this understanding that permeates everyday healing processes in sub-Saharan African communities is the conversation that a healer would usually have with beings in the forests – trees, plants, water bodies, etc. – before taking only what is needed from such beings. In such conversations, the healer affirms that both he or she and the being in the forest are beings together, sharing a common essence of life and a common locale and requiring mutual support and respect. The words of a Senegalese healer, Malik Sow, during an interview conducted by the renowned Sudanese-British journalist, Zeinab Badawi, aptly capture this:

> My work improves people's lives and provides them with protection. . . I inherited the ability to heal from my father and grandfather. In one of the methods, we use the roots of a tree. I have to request what I need from the

> tree and I must give it something in return. We reach a mutual agreement about both my needs [such as the bark of the tree, a root stem or some leaves] and what the tree wants in return. Only then can I cut a bit from the tree and take those trimmings for the purpose of healing. I repeat a series of chants and speak to the tree in the same way I would to a human being. I bring something for the tree and offer it.
>
> (Sow as interviewed by Badawi 2017)

These chants to, and talks with, the non-human, such as trees, plants, ancestors, deities, and water bodies, that punctuate daily existence in African communities spring up interesting ideas and questions about communicative action with the non-human in African places that I cannot explore further in this paper but it also implies that being-together, as in this particular case of the healer (Malik Sow) and the tree, requires fundamentally the second mode of being, being-in-place, to which I now turn.

In an African community, being together fundamentally requires mutual recognition and respect for ontic positioning in being to promote ontological balance and equilibrium that the dispositioning of being would create. To be together in a community, therefore, implies to know ones place in the ecosystem or ontological community and to respect, value and preserve the place of the others in the community. This relational and positional ontology prevents unwarranted dispositioning of beings for the sake or benefit of one or a few beings in the community. As seen in the case of a healer speaking to a tree, there is clearly respect for the place of the tree in the community and the need to sustain the status quo. Hence, only what is needed at a given time is taken from it and nothing more. There is no justification for taking out of place the entire tree, dissecting it into pieces, and packaging it into individual parts in order to effectively fulfil one's role as a healer. Taking only what is needed from the community of beings and acknowledging and respecting other beings as co-existents in the community minimises anthropogenic waste. This, perhaps, explains the interpretation in African places of numerous natural resources as the embodiment or place of dwelling of spiritual forces, and this interpretation serves as an effective ontological construct to prevent human being from taking for granted the non-human and from exploiting it beyond necessary need.

An objection to the following line of reasoning about sustaining beings in place to promote being together is the argument of population differences between the sort of communities I am describing here and large industrial societies. The argument would be that for the sake of resource security, large industrial places would need to take more from other beings than in smaller communities. There are two key flaws with this reasoning. First, the need for more from the non-human community of beings in capitalist industrial societies for the sake of resource security does not justify the pursuit of jouissance that results in much waste as discussed above, nor is it reconcilable with the wanton waste of natural resources that currently exist in industrial societies globally. More so, the uncontrolled pursuit of jouissance and resource security happens within the framework of the objectification of non-human beings as mere objects for human exploitation. This is certainly different from the indigenous African ontological framework where all beings have a collective stake in the sustenance of the ecosystem. Second, as seen in the narrative of the Senegalese healer, to take what is needed from a being requires forming a mutual relationship with that being, acknowledging and protecting the positioning of that being, and giving back to that being its needs. Taking on the basis of an objectified framework, the dispositioning of being, and without giving back leads to ontological imbalance and would entail moral irresponsibility to the community. Thus, the argument on the basis of a growing world population is not sufficient to justify the unwarranted anthropogenic waste of our ecosystem today.

Concluding Thought

Ontic relationality or being-together, and ontic positionality or being-in-place, as explored above from indigenous African perspectives, invite us to radically rethink three key aspects of the human–non-human encounter in our discourse of deep ecology today: the interpretation of the non-human by the human, the basis and justification for taking resources from the non-human, and the nature of giving back to the non-human as reciprocation for taking from it to sustain eco-justice. Concerning the first aspect, African relational ontology provides an inclusive ontology where all beings, human and non-human, coexist and collectively encounter one another. All beings contribute fundamentally to the well-being and sustenance of the community and its individual members. Thus, it radically shifts away from the objectified view of the non-human in modern ways of being fuelled by scientific, technological, calculative, instrumental, and economically productive ways of thinking. Concerning the second, African positional ontology defends the preservation of being in place to sustain balance and equilibrium in the ecosystem and sees as justifiable the use of only what is needed from a being in the community, which checks against anthropogenic waste. But as we can see from the foregoing, modern ways of being fundamentally shape dispositions and continually extract more than what is necessary from the ecosystem in the bid not simply to satisfy needs but also jouissancial wants. Concerning the third, replenishing or giving back to beings that we get our needs from requires in African tradition a communicative action, where we sense and understand what the other being is telling us it needs in return and we provide such, leading to an equitable and reciprocal use of resources. While in modern times there are growing attempts in the forms of laws, policies, and ethical practices to give back to the non-human in our ecosystem by, for example, planting trees and reducing paper printing, it is not clear to what extent these efforts have emerged from deliberately listening to, feeling, hearing, seeing, and understanding our non-human cohabitants, particularly as acts of communication are often conventionally restricted to the human. Therefore, while the materialist, anthropocentric, realist, and instrumentalist ontologies of modern life are very deeply entrenched in modern ways of being and may thus militate against a relational and positional ontology, we must find ways to dislocate thriving ontologies of modern life and revive old ways of understanding our being and our relationship with the environment. One way to begin the journey of doing so is to ensure the representation of indigenous voices on the ecosystem in global spaces of deep ecological discourses and practices.

References

Badawi, Z. and Sow, M. The Senegalese Healer who talks to Trees. *History of Africa. British Broadcasting Corporation*, August 18, 2017. https://www.bbc.co.uk/news/av/world-africa-40982735.

Behrens, K. G. Toward an African Relational Environmentalism. In Elvis Imafidon and John I. A. Bewaji (eds.) *Ontologized Ethics: New Essays in African Meta-ethics*. Lexington Books, 2014: 55–72.

Deval, B. *Deep Ecology*. Peregrine Smith Book, 1985.

Evans, D. *An Introductory Dictionary of Lacanian Psychoanalysis*. Routledge, 2006.

Gyekye, K. *African Cultural Values: An Introduction*. Sankafa Publishers, 1996.

Gyekye, K. *Tradition and Modernity: Philosophical Reflections on the African Experience*. Oxford University Press, 1997.

Heidegger, M. *Being and Time*. Translated by J. Macquarrie and E. Robinson. Basil Blackwell, 1985.

Husserl, E. *The Phenomenology of Internal Time-Consciousness*. Edited by Martin Heidegger and Translated by James S. Churchill. Indiana University Press, 1964.

Ikuenobe, P. *Philosophical Perspective on Communalism and Morality in African Traditions*. Lexington Books, 2006.

Imafidon, E. On the Ontological Foundation of a Social Ethic in African Traditions. In Elvis Imafidon and John I. A. Bewaji (eds.) *Ontologized Ethics: New Essays in African Meta-ethics*. Lexington Books, 2014: 37–54.

Janz, B.B. *Place, Space, and Hermeneutics*. Switzerland: Springer, 2017.

Kumar, R. A. *Tradition and Modernity: A Reflection on Influence of European Ideology*. Boleswa Occasional Papers in Theology and Religion, 2006.

Lauer, H. *Tradition versus Modernity: Reappraising a False Dichotomy*. Hope Publications, 2003.

Marin-Beltran, I., et al. Scientists' Warning Against the Society of Waste. *Science of the Total Environment*, 811, 2022: 1–14.

Merleau-Ponty, M. *Phenomenology of Perception*. Translated by Colin Smith. Routledge and Kegan Paul, 1962.

Metz, T. *A Relational Moral Theory: African Ethics in and Beyond the Continent*. Oxford University Press, 2021.

Metz, T. Toward an African Moral Theory. *The Journal of Political Philosophy*, 15(3), 2007: 321–341.

Oral, H. V. Deforestation. In S. Romaniuk, et al. (eds.), *The Palgrave Encyclopaedia of Global Security Studies*. Springer, 2020.

Ramose, M. U. *African Philosophy through Ubuntu*. Mond Books, 1999.

Sartre, J-P. *Being and Nothingness: An Essay on Phenomenological Ontology*. Translated by Hazel E. Barnes. Washington Square Press, 1956.

Tutu, D. *No Future without Forgiveness*. Random House, 1999.

9 Peripherality, Non-philosophy, and Ecology in African Philosophy

Bruce Janz

Introduction

A discussion of ecology, and in particular non-human life, has been a fairly recent focus within contemporary African philosophy. The focus in the mid to late 20th century was to answer back to the depictions in Western philosophy about the nature and capabilities of Africans, to decolonise philosophy, and to set out some fundamentals of thought that would enable thinking in an African place. These fundamentals included the reconstitution and recovery of African intellectual traditions, the conceptualisation of personhood, community, philosophical method, and a host of other necessary elements of a robust philosophical tradition.

What did not receive much attention was the natural world beyond the human. This fact led some to see African philosophy as irredeemably anthropocentric, a charge that Edwin Etieyibo has argued against (Etieyibo, 2019). If we take the tables of contents of significant collections of and introduction to African philosophy up to about 2004, we see very little attention paid to environmental ethics. Coetzee and Roux's *African Philosophy Reader, 2nd Ed*, does not have an entry in the area (Coetzee & Roux, 2003), nor does Emmanuel Eze's (Eze, 1998). Kwasi Wiredu's has one chapter on biomedical and environmental ethics, out of 47 (Wiredu, 2004). Richard Bell's introduction does not mention nonhuman environmental or ecological issues, (Bell, 2002), nor does Lee Brown's (Brown, 2004) or Tsenay Serequeberhan's (Serequeberhan, 1991).

In more recent years, collections of essays have paid more attention to questions of ecology in African philosophy. In the recent *Handbook of African Philosophy* (Imafidon et al. eds., 2023), there is a section on philosophy and the nonhuman, with two essays. *The Palgrave Handbook of African Philosophy* (Afolayan & Falola, eds., 2017) likewise has two entries on environmental questions. We have also had whole volumes devoted to environmental questions (Chimakonam, ed., 2018; Chemhuru, ed., 2019). An excellent overview of the history of environmental concerns within African philosophy was done by Workineh Kelbessa (Kelbessa, 2018).

My point here is not to castigate any of the editors or authors from times past. Nor do I mean to suggest that there have been no writings about environmental matters in African philosophy before the last few years – there clearly have been. The empirical point is simply that when it comes to the narratives we have about

DOI: 10.4324/9781003480914-10

African philosophy as represented in major overviews and text collections, questions related to the environment have not been at the forefront of the development of African philosophy in the modern era, and that only in relatively recent times do we see greater attention being paid to the question of the environment and/or the nonhuman. This says to me not only that there were other questions that demanded greater attention during these times, but also that the environment or ecology were not "activated concepts" until recently, that is, concepts that had much currency in the conversations about the nature of African philosophy, despite the efforts of those who wrote about environmental issues earlier.

My goal for this chapter will be to start from this observation and survey some possible ways that African philosophy has or could engage with the environment and the nonhuman. The goal here will be to move toward a way of thinking about the environment using the resources of African philosophy. In particular, I am interested in how we might think about the other while being in the middle of it.

Thinking the Other in African Philosophy: Peripherality and Non-philosophy

In a recent chapter (Janz, 2018), I argued that African philosophy must take seriously its own peripherality and its own non-philosophy. Peripherality refers to the contextualizing, framing, and scaffolding elements that make our rational gaze possible. As with vision, we do not simply look at a thing and ignore everything else. We focus on something at a moment in time, having already focused on other things, and recognising that any act of focusing is possible only through the cognitive scaffolding of what has come before, what potential projects are in place, and what kinds of connections exist for the thing our intentional consciousness is engaging. Our ability to act using sight, as opposed to simply collecting sense data through vision, is embedded in temporal experience, particularly in a memory of what is not there, what is "out of frame," and what has been. Every act of seeing is already an act of comparing and relating, and our peripherality is the present absence of sight. This is true for our philosophical focus as much as it is for our sensory focus.

Non-philosophy is more than just that which is the raw material for philosophy.

> Some assume that non-philosophy is the material source of philosophy; some assume that it is the motivation; some of these are simply alternative forms of disciplinary knowledge; and some might be, for lack of a better term, the antidote to philosophy.
>
> (Janz, 2018, 15)

In any of these cases, non-philosophy is something that philosophy must make disappear, either by subjecting it to its own rational gaze or by rendering inert any challenge to philosophical knowledge. It is what philosophy tames by making it rational. And it is why the environment is inherently difficult for philosophers to address – it will not be tamed, or rather, the environment tamed is no longer the environment but something else.

These concepts become relevant as we think about the environmental philosophical thinking that has emerged in African philosophy. As has already been noted, it emerged relatively recently. And, as with patterns of thought in the West, it has emerged as a problem to be solved. We have accounts in the West of, for instance, moral agency and standing for humans. We have accounts of human existence and knowledge which place us at the centre of our world. And so, when we come to think about the environment, the task is to understand how it fits into already established ways of thinking.

The emergence of environmental thinking in Africa potentially faces a similar pattern. While we might think differently in African philosophy about, for instance, the nature of human existence or the social world, the environment remains a problem, one that is often solved by the extension of some aspect of that philosophy into what we might see as non-philosophical space. These extensions include:

1. **Extending communalism beyond the human world**. One way of including the environment within our metaphysical, epistemological, and ethical space would be to see it as part of an African understanding of community. This is, in a sense not much different from rationales used in the West to justify concern for the environment based on the necessity of individual and societal human thriving. Our concern for the environment, in this line of reasoning, is necessitated by our concern for ourselves and our own survival.
2. **Leveraging the metaphor of family**. H. Odera Oruka (Oruka & Juma, 1994) was one of the earliest African philosophers to explicitly address questions of the environment from an African point of view. His "parental earth ethics" made the case that an ethics which arose from African experience would have to take two principles into account: the "parental-debt" principle and the "individual luck" principle. The first of these started from the idea that the survival of the family is central to ethics and must be preserved. The second provides for individuals to benefit from their own fortune, but it is overruled if it comes into conflict with the first principle. Oruka and Juma then argue that "the earth or the world is a kind of a family unit in which the members have kith and kin relationship with one another" (125–126). Using this logic, they rebut territorial claims, either on the part of colonialists or original inhabitants, if issues of the survival of the world family are at stake.

   Oruka and Juma's position extends a known social structure, that of the family, into the nonphilosophical territory of the environment, and in so doing, provides a set of principles that are mainly in place to deal with conflicts between other claims.
3. **Expanding the circle**. This is an approach I outlined in Janz 2018. The idea is to expand the scope of moral agency and moral standing past the human context. If one starts from some version of humanism, that suggests that those with moral standing have it because of some aspect of their humanity. An extension of the regard we have for humans could be made either by seeing

nonhumans as having some characteristic of humans which would give them moral standing (e.g., the ability to feel pain and the ability to have emotions), or by seeing their value as essential to human survival. This strategy is most often connected to Peter Singer's work.

4. **Seeing relationality beyond the human**. Some versions of Ubuntu are seen as already being anti-anthropocentric, in that the relations that animate human community are really meant to be seen in all things, not just within the human world (see, for instance, Museka & Madondo, 2012; Chibvongodze, 2016). These versions of Ubuntu tend to take the exemplar of relations as ones we are familiar with, and extend these models to the living world.
5. **Arguing for a version of neo-romanticism**. We might imagine that the natural world has value because it has a spiritual or divine energy that animates everything. Some versions of vitalism could come close to this approach, and one can see the rationale for not damaging nature if it is, in fact, part of the emergent nature of the divine. Others, such as Mangena (Mangena, 2015), see the necessity of including traditional African beliefs, such as those in spirits and ancestors, to ground an African version of deep ecology.

What we see in these options is, in a sense, the same kind of move we see in Western philosophy. There is a metaphysically real starting point that is taken for granted. In Western philosophy, that tends to be the individual, and in African philosophy, it can be some combination of the individual and the community. In either case, the environment becomes a problem in need of a solution. That problem might be epistemological (how can we know nature in a non-reductive sense?), it might be metaphysical (what is the status of nature in relation to humans?), or it might be ethical (how can we see nature as having moral standing if we start from a conception of ethics based on either the individual or the community?). We can remain committed to anthropomorphism in either a Western or an African version of environmentalism and see the rationale for caring about the environment as indexed to our level of care for ourselves or for some collection of humans – the family, the nation, or all of humanity.

If the environment is presented as a problem to be solved, the solution will always come out of the resources available to us. It will always be an extension of what we already think we know, in other words, as solutions always are. A solution is always based on the question we ask that produced the problem in the first place.

This means that in both African philosophy and in other traditions, it is important to find a way to reframe the question of ecology. If it is only a problem to be solved, it will be solved with the intellectual tools we have available. These tools will inevitably render it comprehensible to us, not just in scientific terms but in philosophical ones. In so doing, we use what we know to shape and frame what we do not. This brings with it a certain amount of comfort – after all, problems are there to be solved, and if we can find a way of thinking about nature that gives a rationale for preservation, what could be wrong with this? And yet, as we have seen many times in the past, our conception of nature is often limited, and our idea of adequate ethical solutions can usher in more significant long-term problems.

The concepts of peripherality and nonphilosophy, already mentioned, give us a way to understand how to ask the kind of questions that does not just turn nature into something we already recognize. We are simultaneously in the middle of nature and are an other to it. This does not mean that we are not also thoroughly natural as well, or that we stand as an exception to anything we say about it. It does mean, though, that our impulse to reduce nature to a function of our world will be a partial, limited snapshot of complex systems with emergent properties.

The African Milieu

I have elsewhere (Janz, 2009b, 2011) argued for an ethics of place, which would include a way of thinking about ecological questions. This has involved seeing ourselves, both as individuals and communities, as always "in the middle of," that is, in the milieu. This is in contrast to a common philosophical positionality, which has us above or beneath the world, before or after it, but not all of these at once. This formulation may be recognisable as Heidegger's Fourfold (das Geviert) – the earth, the sky, the gods, and the mortals – which he uses to place us within a space of thought and action. We could see these four as coming in pairs, the first two pointing to the particular and the universal and the second two pointing to the eternal and the temporal. For Heidegger, we are not either before or after, nor are we either above or below – we are, in fact, all of these at once, and this was Heidegger's form of resistance to the reductive or transcendental philosophy that came before him. We are in the middle, both spatially and temporally. Any one of these by itself places us apart from the world, whereas all together place us in the middle of it. We are at home in the world, we in-habit it. The point is that privileging any one of these over the others leads us to misunderstand our place in the universe and leads us to engage in a space of thought not adequate to reality, nor for that matter to ourselves.

We do not need Heidegger, though, to think about Africa from within its midst, even if he does give a useful vocabulary to think about the tensions and anxieties, but also the richness, of thinking within the milieu. Thinking in the milieu means giving up the Western Enlightenment conceit that we can, at least in principle, know everything. Once we give that up, we realise that we exist with peripherality. We face the world with our intentional consciousness and know it, while always recognising that what we know exists in the context of things that we are not currently facing, and perhaps (in some cases) could not face. And we think with the recognition that there is non-philosophy, that is, that not everything is available to us as an object of philosophical thought. We therefore need to negotiate while we think. We need to question, and every act of questioning opens up a space of thought and also closes down or limits other spaces. The search for an ethics based in principlism, in other words, is at best going to be provisional, open to continual revision as our knowledge of the world changes and as the questions that motivate those principles evolve (for a good example of how this can happen, see Clarke, 2019).

Thinking from within an African milieu starts from thinking in place, that is, recognising the conceptual space made available by lived experience within a place, broadly understood (Janz, 2009a). This is a different move from the abstractive one that often happens within African philosophy, in which principles are derived from some putative, culturally specific ideas. It is also different from some versions of anti-colonialism and anti-racism, which focus most of their energy on ridding African thought of the imprint of non-African questions and modes of thinking but spend little time exploring the kinds of concepts that might be adequate for becoming-African. While abstraction and anti-colonial thought are important, in themselves, they do not ask us to think in a place that is being created by our actions, and which changes us as it is being created.

This becomes important when we think of the environment not as an object of reflection or as a problem to be solved, but as the space in which thought happens. I have elsewhere argued (Janz, 2009b) that places know. Not in some quasi-religious or metaphysical sense that there are genius loci or spirits, but in the same sense that we say the body knows. To put it in these terms means seeing our human knowledge not as something that stands over and apart from nature, but rather as something that is always in a state of encounter with the knowledge of other systems within the natural world, systems foreign to us and not primarily for us, but which interface with us.

Michael Onyebuchi Eze gives a version of this approach to environmental thought (Eze, 2017). Eze too finds many versions of environmentalism to be limited, suffering from similar problems of being animated and framed by Western thought. He reframes the human relationship to nature as "dialectical subjectivism" (627). We are in constant symbiotic dialogue with other things. As he puts it,

> One's subjectivity is fully expressed only in dialogue with other beings within the eco-community. The environment thus becomes a critical part of human society and vice versa. The earth, seas, rivers, forests, trees, animals, humans, spirits, forces, animate and inanimate things, everything together constitutes this eco-community. It is no wonder that in many African societies animate and inanimate beings are very often anthropomorphized.
>
> (627)

I agree with his approach here, and indeed I want to take this a step further. The dialogue of which he speaks is a complex one. As we know from human dialogue, if engaged in the right spirit, we can find our own limitations and misunderstandings in the process. We can realise that we have been asking the wrong question of the other. The pattern of much environmental thought has been to interrogate nature, to ask about its true structure and to search for principles of behaviour that would maximise desirable outcomes or live up to whatever sense of duty we might have. We do not, on this model, see the dialogue as going in the other direction. We as humans are not interrogated, except to ask whether we are living up to the ethical commitments we have formulated. But this is not a real dialogue. The environment remains an object of knowledge, and we remain the ones who know and strive for a complete description of the environment.

The dialogue is complex in another sense as well, that is, it is best understood through complexity theory. The interaction of forms of knowledge produces emergence, that is, results that cannot be deduced from the original systems themselves. "Dialogue" perhaps does not do justice to the range and types of encounters between systems within nature, including our own systems of knowledge made possible by the questions we ask. To even think of ourselves as separate from nature ignores the fact that a great deal of our DNA bears witness to the history of our encounters with bacteria, viruses, fungi, and other life processes. The same is true for other living beings. Our "dialogue" is sometimes confrontational, often cooperative, and almost inevitably autopoietic.

Eze knows this, and he wants to take thinking about the environment in Africa to the same place I would – away from questions of duties and towards more fundamental questions of who I am, or who we are. (630) He advocates what he calls "Humanitatis-eco," which is "a way of life, a culture, a religious experience, a way of being as part of the eco-community" (631).

Eze has opened the door to something that looks very much like what Arne Naess would have seen as deep ecology, or in his terms, "Ecosophy-T" (Naess, 1989). We can walk further through this door if we look at another African philosopher: Mogobe Ramose.

Deep Ecology through Ubuntu

Deep ecology has made a recent appearance in African philosophy (Mangena, 2015; Ibanga, 2017). A number of thinkers have appealed to Ubuntu or equivalent concepts in other languages to ground environmentalism in African concepts (e.g., Etieyibo, 2017, 2019; Mangena, 2018). At least one (Le Grange, 2005) has looked to Felix Guattari as a way of augmenting Ubuntu theory. However, deep ecology and Ubuntu have not been seen by most theorists as mutually supporting, except by Mogobe Ramose. Ramose's work is in some cases a significant source for this grounding, even though he does not say a great deal about ecology. What he docs say, though, is suggestive, and I would argue, fits into the picture I have been trying to give of a way to question within a milieu. In what follows, I will take these suggestions that he makes and try to tease out some of the implications for an environmental philosophy based in African thought that does not simply look for easily applied normative content.

In *African Philosophy Through Ubuntu*, he has a short chapter on "Ecology Through Ubuntu" (Ramose, 1999, 154–159). While Ramose does not directly reference Arne Naess's (Naess, 1989) environmental thought, his approach bears much in common, even to the similarity of language in using "ecosophy" (in Ramose's case, he speaks of the "ecosophical dimensions" of a proverb (157), and in Naess's case, he calls his own approach "Ecosophy-T" (Naess, 1989, p. 163ff)). And the parallels do not end with the use of a single word. Both use David Bohm's work in *Wholeness and the Implicate Order* to make a case for wholeness as an approach to nature. Both emphasise the importance of potentiality and becoming, and (although neither uses this term) emergence. And neither foregrounds the project of finding ethical norms or guidelines as the reason for engaging the environment. While ethics are important, they will not be found in maxims, principles, or guidelines that can be applied to the human or natural world. Naess and Ramose are not interested in just finding an ethics of the environment, nor are they interested in extending our familiar metaphors and frameworks for ethics into unknown territory. They are interested in asking new questions of nature, and of ourselves.

Ramose's chapter on ecology comes late in his book, after he has already established his approach to Ubuntu and explored a range of ways that Ubuntu is a space of thought and not just an approach to ethics or politics or some other aspect of human life. His term for this space of thought is "wholeness":

> Wholeness speaks to the processual character of reality or being. It speaks to the primary observation of both philosophy and the natural sciences, namely, that underlying all reality is the principle of motion. Without motion, being as enfoldment cannot unfold. Secondly, we prefer the term wholeness because as experience, wholeness cannot yield easily to absolutism and dogmatism in order to establish its authority.
>
> (Ramose, 1999, p. 155)

We see in Ramose's formulation here no attempt to extend familiar human forms to include the nonhuman. We have something else, though – a recognition of the generative nature of all reality. It is tempting to think of wholeness as a unitary or singular thing, but that is not what is meant here. Ramose follows Bohm's sense of wholeness, which is contrasted to fragmentation. Fragmentation has led to compartmentalization, our tendency to separate ourselves off from the environment. This can be instrumentally useful, but:

> when this mode of thought is applied more broadly to man's notion of himself and the whole world in which he lives (i.e., to his self-world view), then man ceases to regard the resulting divisions as merely useful or convenient and begins to see and experience himself and his world as actually constituted of separately existent fragments.
>
> (Bohm, 1980, p. 3)

Bohm's concern was scientific theory and its tendency to isolate parts of nature in order to come to clearer causal explanations for phenomena. Bohm does not want to recombine these once the analytic work is done, the way Descartes might advise in his *Discourse on Method.* For Bohm, wholeness is in fact reality, and our tendency to isolate and fragment is a move that we make in order to accomplish an epistemological task. Reality, though, always has been interconnected. What ought to be recognised is what he calls "Undivided Wholeness in Flowing Movement" (Bohm, p. 14).

This is what I take Ramose to be doing when he uses the term "wholeness." It means that the attempts to understand nature by extending some aspect of the human world into the natural world start from the wrong position. That move starts from the givenness of fragmentation, rather than the givenness of wholeness. It starts from the idea that we have some epistemological grasp on our own relations and lack a similar grasp on relations in the natural world, and so the best thing to do is to extend our existing knowledge into new areas and use it until it is shown to be inadequate. This has an intuitive appeal to it, but what it actually does is to replicate human ways of thinking about nature onto nature itself. Even with the best of intentions, this not only has unintended consequences, but it does something more – it tames nature, making it something recognisable by us and, in so doing, releases us from the responsibility of asking new questions.

Ramose's brief comments about ecology suggest the centrality of two concepts I have already mentioned: peripherality and nonphilosophy. Taming nature makes it into philosophy. It makes it into something that can function within an ontology we can understand. Nonphilosophy, though, stands as a question. How do we think that which is not part of an ontology we recognize? We can make that ontology very abstract, but that then undermines the possibility of the other concept, peripherality. Thinking in place, in the middle of a place, always entails contextualising concepts and functions and phenomena. This is what wholeness means, in the sense that Ramose and Bohm have it. It is not a monolithic totalising machine, as they both make clear. Ramose talks about wholes within wholeness:

> There is never a final immutable whole but only enduring and transient wholes always governed by the principle of motion responsible for change. Because of this, the various wholes are, properly, aspects of the universe as a wholeness, that is to say, a perpetual process of generation, death and regeneration; of the construction of an intrinsic order from the ruins of apparent destruction and chaos.
>
> (Ramose, p. 157)

For Ramose, the perennial barrier to realising this reality in Africa is colonialism, but a more immediate problem is the "somewhat disconsolate comfort and 'easy life' brought about by technological advancement" (Ramose, p. 158). Technology is, for Ramose, a cause of the acceleration of the threat of collapse, specifically ecological disaster. While we might want to think further about the agency of technology (we might ask, for instance, is it really technology or is it the capitalist system in which technology provides short-term advantage at the cost of long-term sustainability?), the effect of technology here is to further fragment reality and lead us to see our engagement with it as instrumental. And like fragmentation, technology tames the world, or at least aims to. Whether it succeeds in this or not, it makes us believe that it is possible if only we have the right discovery, invention, or technique.

Ramose ends his brief chapter on ecology through Ubuntu by arguing for peace, to both address ecological issues and the threat of nuclear annihilation. In practical terms, this seems easier said than done, but his point is that both environmental and nuclear disasters have the same basic form and cause – both are the result of fragmentary thinking and an inability to see past our own interests and our own projections on the rest of the world. Understanding what the problem is must come before we can imagine a lasting solution.

Peripherality and Nonphilosophy in Deep Ecology

What we see in Mogobe Ramose, David Bohm, and Michael Onyebuchi Eze, we also see in a very similar manner in Arne Naess's Ecosophy-T (Naess, 1989, pp. 163–212). While we cannot go through his thought in any great detail, we can pick out at least one significant idea that resonates with the version of Ubuntu we have seen already: the concept of "friluftsliv":

> [I]n Norwegian, there is a clearer, more value-laden word that refers to the type of outdoor recreation that seeks to come to nature on its own terms: to *touch the Earth lightly*. Literally, 'friluftsliv' means 'free air life', but it has been translated as 'open air life' and 'nature life.' … The satisfaction of the need for outdoor life and the need for machine-oriented technical unfolding cannot take place *simultaneously*. At present, the socio-economic forces in the industrial countries are lobbying in favour of priority for the capital-intensive apparatus: the apparatus-poor life is a hindrance to 'progress'. We should see true *friluftsliv* as a route to paradigm change.
>
> (Naess, 1989, p. 178)

We see several things in this passage. We see Naess expressing a similar scepticism towards technology as Ramose does. For Ramose, it accelerates disaster by giving us a comfortable and easy life. For Naess, it stands in the way of encountering the Earth itself.

More than just diagnosing a problem, though, Naess suggests a solution, or at least the context in which a solution might be made available. *Friluftsliv* can bring about paradigm change, something that Ramose also talks about in similar terms. How is this possible? I believe that it is with these concepts of nonphilosophy and peripherality. We encounter nature. Living with it, facing it, will sooner or later make us recognise that our categories of understanding it are inadequate. It cannot be fully rationalized, even if it can be known at some level. Our scientific knowledge is essential, but only the kind of science that is acutely aware of its own provisionality and dependence on wholeness over fragmentation. Neither science nor any other part of our Enlightenment toolkit can give us an adequate framework for understanding, not just because of our lack of knowledge, but because of the autopoietic nature of the systems of nature, including our own cognitive systems. And so, the necessity of nonphilosophy remains. Nature must remain wild (and so must we, but that is another discussion).

And if this is true, peripherality comes along with it as well. *Friluftsliv* is a kind of play, very different from the structured and commodified encounters with nature that have become so common. Play is a context of learning, of asking new questions. Ramose speaks of the principle of motion, a constant becoming that exists within wholeness. The play that Naess has in mind is a kind of engagement with motion. That which we look at directly becomes fixed, while that which is in the periphery moves. Our rational gaze attempts to fix the world, because that is how we think we can know it. But in fact, what is giving the space and context to our gaze is that which constantly must question and amend that knowledge, and the only way we can access it is through play, through *friluftsliv*.

This is the significance of Ubuntu, beyond (or before) its moral code in a traditional sense. It is not simply an artefact of thought but the space of thought, the potentiality of thought, in the context of the nonphilosophy that the world presents and with the tools that play affords.

What Difference Does It Make? Ethics and Deep Ecology

If one's goal for environmental philosophy is to generate policy positions or optimise the ethical outcomes of decisions involving ecological entities, the sort of deep ecology we see in this paper might seem abstract, theoretical, or even useless. Ramose, Eze, and Naess do not foreground ethical principles, if by that we mean seeing our ethical duty as discharged through a set of intellectual maxims that can be applied and even universalized. There are few payoffs in this approach that can be turned into policies at a state or corporate level. So, is this just all empty philosophising within an African frame of reference?

The problem here, I think, lies more with our assumptions about what constitutes ethics than it does with deep ecology itself. In the West at least, and perhaps also in some approaches to African philosophy, we think about ethics as its own separate domain, perhaps even coming after metaphysics or epistemology. We might imagine that we should first get straight what is, and only then can we think about what ought to be. And, what ought to be often amounts to avoiding what ought not to be, that is, it stands as sets of restrictions on a given state of affairs, already established in our metaphysics, that optimise reality according to some metaethical standard.

This scenario assumes that philosophy's primary task is descriptive of reality, and the evaluative task (decisions and implementations of "what ought to be") comes after description, as a result of understanding human needs and optimising them. In the examples given earlier of the extension of human thought into the non-philosophical world, the pattern remains the same. Ethics comes at the end, even in the case where we broaden the definition of what might hold ethical status.

Deep ecology of the sort we have seen here does not come to offer another ethical option to the existing ones. It overturns this view of the place of ethics entirely. If we look to deep ecology to come up with a better decision as to how to interact with the natural world, we are already too late. Ethics is not what comes after the description of reality. It is already embedded in the questions we ask of reality, and in our willingness to revise those questions as they become evidently inadequate. It is in our ability to maintain the nonphilosophical status of the natural world as nonphilosophical, and not rationalise it into something that we can fit into a system of values we recognize. Putting it this way does not assume an implicit dualism between the known self and the unknown nature, though, because our "known" self is also unknown nature, and the unknown nature is not unknowable, at least not in some ways, for a particular amount of time, until the emergent results of the ongoing interaction between all the parts of the natural world, including ourselves, change the conditions of existence, necessitating new questions.

We see something like this in Ramose's essay "The Ethics of Ubuntu" (Ramose, 2003). He argues there that:

> The logic of *ubuntu* is towards -ness. This logic stands in opposition to the dogmatism of fragmentative reasoning. One of the first principles of *ubuntu* ethics is the freedom from dogmatism. It is flexibility oriented towards balance and harmony in the relationship between human beings and between the latter and the broader be-ing or nature.
>
> (326)

It is tempting to turn this statement into a metaethical one and see him as arguing for an objective thing called "balance" which can guide and govern individual ethical decisions. But this would be to misunderstand the "-ness" nature of ubuntu. It is humanness, but not only that. To work from this familiar context, though, to be human in the world means to question what it means to be human, and more than that – what it means to be embedded as an emergent feature of a world where we cannot extricate ourselves from the myriad other processes in the world. Our ability to question is our ability to change our questions as the need arises. And we see this in ecosystems on a regular basis – all things continually ask what it means to exist as they are, and they change or adapt as circumstances change. In some cases, they change so much that they evolve into a new species. In other cases, there is no answer to their questioning, and they go extinct. The point is that this questioning is not done once and then enshrined as if it were the only question that could be asked. Ramose's idea of balance is earned every day, and the ethics of ubuntu is living through ubuntu by living in a state of learning and questioning.

Conclusion

The importance of Ramose's approach to Ubuntu (along with versions we see in Michael Eze and others) is to establish a way of thinking and being within the natural world. Without this, we merely have principles which, however well-intentioned, will never address the reality of the world because they attempt to render it static. The natural must remain as nonphilosophy, and we must learn to play in the space that peripherality affords. This will not leave us with traditional ethical principles, but it will leave us with the ability to think with and in nature, and it will help to heal the damage we have done by thinking in reductionist and commodified terms.

References

Afolayan, A., Falola, T. (eds.) (2017) *The Palgrave Handbook of African Philosophy*. Palgrave Macmillan.

Bell, R. (2002) *Understanding African Philosophy: A Cross-Cultural Approach to Classical and Contemporary Issues*. Routledge.

Bohm, D. 1980. *Wholeness and the Implicate Order*. United Kingdom: Routledge.

Brown, L. (ed.) (2004) *African Philosophy: New and Traditional Perspectives*. Oxford University Press.

Chemhuru, M. (ed.) (2019) *African Environmental Ethics: A Critical Reader*. Springer.

Chibvongodze, D.T. 2016. Ubuntu is not only about the Human! An Analysis of the role of African Philosophy and Ethics in Environment Management. *Journal of Human Ecology*, 53(2): 157-166.

Chimakonam, J.(ed.) (2018) *African Philosophy and Environmental Conservation*. Routledge.

Clarke, M. (2019) "New Waves: African Environmental Ethics and Ocean Ecosystems," in Chemhuru, M. (ed.) *African Environmental Ethics: A Critical Reader*. Springer, pp. 153–172.

Coetzee, P., Roux, A. (eds.) (2003) *The African Philosophy Reader*, 2nd ed. Routledge.

Etieyibo, E. (2017) "Ubuntu and the Environment," in Afolayan, A., Falola, T. (eds.) *The Palgrave Handbook of African Philosophy*. Palgrave Macmillan, pp. 633–657.

Etieyibo, E. (2019) "African Philosophy and Nonhuman Nature," in Hull, G. (ed.) *Debating African Philosophy: Perspectives on Identity, Decolonial Ethics and Comparative Philosophy*. Routledge, pp. 164–181.

Eze, E. (ed.) (1998) *African Philosophy: An Anthology*. Blackwell.

Eze, M. (2017) "*Humanitatis-Eco (Eco-Humanism)*: An African Environmental Theory," in Afolayan, A., Falola, T. (eds.) *The Palgrave Handbook of African Philosophy*. Palgrave Macmillan, pp. 621–632.

Ibanga, D. (2017) "Is Deep Ecology Inapplicable in African Context: A Conversation with Fainos Mangena," *Filosofia Theoretica: Journal of African Philosophy, Culture and Religions*, 6(2), pp. 101–119.

Imafidon, E., Tshivhase, M. and Freter, B. 2023. *Handbook of African Philosophy*. E. Imafidon, M. Tshivhase, and B. Freter (Eds.) Switzerland: Springer International Publishing.

Janz, B. (2009a) *Philosophy in an African Place*. Lexington Books.

Janz, B. (2009b) "Thinking Like a Mountain: Ethics and Place as Travelling Concepts," in Drenthen, M., Keulartz, J., Proctor, J. (eds.) *New Visions of Nature: Complexity and Authenticity*. New York: Springer, pp. 181–195.

Janz, B. (2011) "Watsuji Tetsuro and Fu-do," *Journal of Global Ethics*, 7(2), pp. 173–184.

Janz, B. (2018) "Peripherality and Non-Philosophy in African Philosophy," in Chimakonam, J. (ed.) *African Philosophy and Environmental Conservation*. Routledge, pp. 9–23.

Kelbessa, W. (2018) "Environmental Philosophy in African Traditions of Thought," *Environmental Ethics*, 40, pp. 309–323.

Le Grange, L. (2005) "Guattari's Philosophy of Environment and its Implications for Environmental Education in (Post) Colonial Africa," *Southern African Journal of Environmental Education*, 22, pp. 35–45.

Mangena, F. (2015) "How Applicable is the Idea of Deep Ecology in the African Context?" *Filosofia Theoretica: Journal of African Philosophy, Culture and Religions*, 4(1): 1-16.

Mangena, F. (2018) "Zimbabwe's Environmental Crisis: Questioning Ubuntu?" in Chimakonam, J. (ed.), *African Philosophy and Environmental Conservation*. Routledge, pp. 196–208.

Museka, G. and Madondo, M.M. 2012. The Quest for a Relevant Environmental Pedagogy in the African Context: Insigts from Unhu/Ubuntu Philosophy. *Journal of Ecology and the Natural Environment*, 4 (10): 248-265.

Naess, A. (1989) *Ecology, Community and Lifestyle*. Translated and Edited by David Rothenberg. Cambridge University Press.

Oruka, O., Calestous, J. (1994) "Ecophilosophy and the Parental Earth Ethics," in Oruka, O. (ed.) *Philosophy, Humanity and Ecology: Philosophy of Nature and Environmental Ethics*. ACTS Press, pp. 115–129.
Ramose, M. (1999) *African Philosophy Through Ubuntu*. Mond Books.
Ramose, M. (2003) "The Ethics of Ubuntu," in Coetzee, P., Roux, A. (eds.) *The African Philosophy Reader*, 2nd ed. Routledge, pp. 324–330.
Serequeberhan, T. (ed.) (1991) *African Philosophy: The Essential Readings*. Paragon House.
Wiredu, K. (ed.) (2004) *A Companion to African Philosophy*. Blackwell Publishing.

10 Selective Subordination for Harmony and Holistic Balance from an African Perspective

Emmanuel Ofuasia

Introduction

Consider the world as a sophisticated automobile. No matter how glamorous it appears, a bad water pump or radiator will frustrate the journey. Analogously, the African considers the world as a cosmic and enduring relationship between the physical and spiritual, where the lack of balance in one necessarily affects the other negatively. This duality is central to the flourishing of life (see Ofuasia 2024b). Whereas the present discussion is not about cosmology, it needs to be stated that it has implications for cosmology. My principal agenda is to show, from the African parlance, how the doctrine of "selective subordination for harmony" in human–animal relations may birth a holistic balance on personal, communal, and ecological grounds. While making a case for my "selective subordination for harmony" thesis, I evidence the inherent problems with the egalitarian and "subordination without cruelty" theses concerning the nature of the "relationship" between humans and non-humans, not just animals. I disclose how the latter is limited in addressing how the human and non-human animal relations can serve as a basis for ecological balance as they seem to overlook the role of non-physical forces and how, in their analyses, the outlook that nature is inherently anti egalitarian is downplayed.

Limitations of the Egalitarian Thesis and the Emergence of Other Frameworks

The concept of egalitarianism within the context of human and non-human animal discourse is usually based on the principle of sentience. If both species share this feature, then the latter possesses as much worth as the former. This, roughly put, is the central claim of egalitarianism and has been affirmed by some scholars. For instance, Donaldson and Kymlicka (2011) maintain, from the perspective of the egalitarian thesis, that humans have no status higher than non-human animals so long as the latter are sentient beings. This is why for them both human and non-human animals have equal rights (see Ebert *et al.* 2022).

After Donaldson and Kymlicka (2011), Akande and Ofuasia (2017) have also offered an egalitarian ground on the basis of evolution following Singer's (1985) position that sentience serves as the basis for moral worth. Akande and Ofuasia (2017, p. 7) argue that "in addition to sentience, good neighborliness is a factor in

DOI: 10.4324/9781003480914-11

determining the moral worth of non-human animals." Akande and Ofuasia (2017) add that cruelty done to these non-human animals in the name of being subordinates to humans "may produce violent and wild species of these animals' kind in a future evolution, thereby endangering the lives of future human generation, through negative alteration of genes." Much as they are concerned with how cruelty breeds future violent species through evolution, it may not be wished away that the central contention which is unique to their position is egalitarianism.

Before them, Singer (1985, p. 479), talks about "equal consideration of interests, and it is this principle that must extend to any non-humans who have interests." Singer (1985, pp. 479–480) goes on to maintain that "consciousness, or the subject capacity for subjective experience is both a necessary and sufficient condition for having an interest." Whereas the debates concerning the real essence of consciousness remain unsettled, it is helpful to relay that in 2012, *The Cambridge Declaration on Consciousness,* written by Philip Low (2012, p. 2), has established and argued that the idea of consciousness operates both among human and non-human animals:

> [c]onvergent evidence indicates that non-human animals have the neuroanatomical, neurochemical, and neurophysiological substrates of conscious states along with the capacity to exhibit behaviors. Consequently, the weight of evidence indicates that humans are not unique in possessing the neurological substrates that generate consciousness. Non-human animals, including all mammals and birds, and many other creatures, including octopuses, also possess these neurological substrates.

Some studies have questioned the egalitarian thesis. Specifically, McMahan (2002) shows dissatisfaction with the egalitarian view by insisting that the egalitarian must be willing to grant that the killing of a fish is just the same as killing a human. On this note, even when McMahan (2002) is concerned about the status of non-human animals, he offers a different framework – one which works for humans and another for non-human animals. McMahan (2002) implies that egalitarianism works for humans, but consequentialism or utilitarianism works for non-human animals.

From yet another perspective, in his attack against egalitarianism, Kagan (2019) argues that egalitarians cannot overlook the absurdity in the position that it is morally wrong to kill non-human animals for self-preservation or self-defense. Even the most consistent egalitarian will, when the circumstance demands, preserve themselves before asking if these non-human animals have rights or moral worth. The deduction is that, for Kagan, this is enough reason to uphold the view that all non-human animals do not possess rights in the same way that humans do.

In their 2020 publication entitled *Just hierarchy*, Bell and Pei offer an argument concerning how humans are justified in their treatment of non-human animals on the basis of hierarchy. What this means is that they do not admit the central argument of egalitarianism. Their argument is captured in the phrase "subordination without cruelty" (SWC, hereafter), which among other things, establishes that humans are justified to treat non-human animals as subordinates who are to serve human ends so long as cruelty is not involved (Bell and Pei 2020). This outlook is clearly a counter-argument against the egalitarians. The basic argument of egalitarians is that insofar as non-human animals are sentient, then it is pointless to deny them rights or moral worth.

Before showcasing their own SWC thesis, as an improvement, Bell and Pei (2020) offer some arguments against the thesis of egalitarianism and how moral worth extends to non-human animals. They explain that there are some implications tied to the arguments of the egalitarians, such that when followed to their logical conclusion, even the egalitarians will not accept them.

The first implication deduced by Bell and Pei (2020) in the argument of the egalitarians is that if humans and non-human animals are equal, there ought to be semblances in moral outrage on the part of the former when the rights of the latter are transgressed or violated. Bell and Pei (2020) mention that humans rarely respond to such a status quo and this goes to show that humans and non-human animals are not on the same pedestal of rights. There are, however, a few instances such as when the soccer player of West Ham Football Club, Kurt Zouma, was called out and convicted on two count charges for inflicting needless pain on a cat, which almost cost him his career (see Weaver 2022). However, such instances have not gained traction as egalitarians would have wanted. Nevertheless, it does not mean that there are no moral outrages when the interests of non-human animals are frustrated as Bell and Pei (2020) argue.

The second argument deployed by Bell and Pei (2020) against the egalitarians is one which is connected to the proportionality factor between crime and punishment. They reflect that were humans and non-human animals to be on the same level of rights, then advocates of animal rights ought to extend the same proportion of punishment to those that violate the rights of non-human animals to those that violate the rights of humans as well. Bell and Pei (2020, p. 150) attest that what this implies is the "call for life-long imprisonment of animal meat eaters or people who kill insects." On this showing, Bell and Pei (2020, p. 151) move to conclude that "it definitely does not make sense to say that animals and humans have equal rights." In another passage, they tender their conviction, and I agree with them that: "if it is a question of life and death, humans should have priority" (Bell and Pei 2020, p. 154).

In addition to the foregoing, I am suspect if humans and non-human animals are equal on the basis of sentience. Moral vegetarianism, for instance, contends that humans should focus on a plant-based diet alone and shun animal consumption. Whereas I agree with them, it is important to add that sentience may also be extended to the plants following the various findings underscored by some botanists (Mallatt *et al.* 2023; Segundo-Ortin and Calvo 2023). These studies have even gone beyond studies on plant sentience to admit that there are reasons to acknowledge *putative* cognition among plants. This is derived from the capacity to relate with their biotic local environment (see Arimura and Pearse 2017), make flexible decisions from available or multiple choices (Ofuasia 2024a; Karban and Orrock 2018) and even learn from previous experiences (see Baluška *et al.* 2018).

The lesson here is that if sentience is something that may serve as a reason to confer worth on non-human animals and to go as far as desisting from using them for food, then it is also true that we must also desist from using the plants for food. In a recent paper, I have already indicated the metaphysical basis that endorses such an outlook and I will not pursue the matter any further here (see Ofuasia 2024b).

One way to emerge from the dilemma generated by the extension of sentience to plants and the need to desist from using them to serve human and even animals' palates (for instance) is to allude to the principle of hierarchy and the nature of the alimentary canal of humans and non-human animals. Whereas I will return to emphasise this in a later portion of this chapter, for the meantime, I return to offer my grouses with Bell and Pei's assessment of the egalitarian thesis, having already hinted at the places of agreement with them in the preceding paragraphs.

First, Bell and Pei took Donaldson and Kymlicka (2011) as the exemplar for their criticism of egalitarianism. It is worth mentioning that a Strawman may be noticed. Donaldson and Kymlicka (2011, p. 133) do not foreclose the possibility of punishing those who violate non-human animals just as we do to humans who violate human rights. They are, in fact, looking for a status quo, where in future such can be possible. This is different from the total foreclosure deduced by Bell and Pei (2020) which in turn makes their criticism of Donaldson and Kymlicka (2011) a Strawman. Nevertheless, it is worth stating that this is the main argument of Bell and Pei (2020) which in turn provides a reason for them to jettison the egalitarian view in favour of their hierarchical perspective.

For Bell and Pei, even if these non-human animals are subordinate to humans, they must still not be treated cruelly. This is summed up in their words that as humans, we "have a duty to avoid cruelty to animals, both because animals can suffer and because those who are cruel to them are more likely to be cruel to humans" (Bell and Pei 2020, p. 153). The central focus then is: what does cruelty mean for Bell and Pei (2020)? Does cruelty occur when non-human animals are used for painful experiments with benefits for both human and non-human?

Bell and Pei argue that it is morally permissible to subject non-humans to cruel experiences for the sake of medical research so long as "we would try to use animals who have less in common with humans such as rats" (Bell and Pei 2020, p. 154). What this means is that non-human animals may experience cruelty from humans in the name of medical research, physically punishing them, confining them, and even eating them. This deduction does not seem to be in consonance with their outlook that non-human animals "are ends in themselves who have their own goals and can suffer, and we have a duty to minimize their suffering" (Bell and Pei 2020, p. 165).

In their reaction to the SWC of Bell and Pei, Ebert *et al.* (2022, p. 37) affirm that "what makes these actions permissible on Bell and Pei's view, is precisely that we are permitted to treat animals as mere means to human ends." While there is nothing inherently wrong with non-humans serving human ends, it needs to be stressed that the ends must be properly channelled toward the promotion of ecological balance. These medical researches using non-human animals may be relevant to future species of not only humans but non-humans too. However, the contention is that Bell and Pei did not succeed in making this point very clear given the contradiction with which SWC is confronted. In other words, it is clear that the position of Bell and Pei concerning cruelty to non-human animals is not robust enough. Is the lesser resemblance of non-human animals to humans justifiable, for Bell and Pei, to subject them to cruel scientific/medical experiments? Or are these scientific/medical experiments pleasurable to these non-human animals with a lesser semblance to humans? These are tough questions that disclose a contradiction in Bell and Pie (2022). On the one hand, they maintain that humans "have a duty to avoid cruelty to animals, both because animals can suffer and because those who are cruel to them are more likely to be cruel to humans" (Bell and Pei 2020, p. 153). On the other hand, they argue that non-human animals may be used for things that advance human goals such as medical research so long as "we would try to use animals who have less in common with humans such as rats" (Bell and Pei 2020, p. 154). It is clear that one cannot hold both positions consistently.

Even when the works of McMahan (2002) and Kagan (2019) indicate the absurdity in the arguments of the egalitarians, they are unlike Bell and Pei (2020) unhappy with the ways that non-human animals are treated for human ends. Nevertheless, the ways in which they respond to the relation between humans and non-human animals have been mainly from the physical or materialistic dimension, as they have failed to take cognisance of the non-physical approach to the subject or discourse. This is the case when the hierarchical structure is examined from the African perspective.

The important point which may be noticed from the various arguments that have been offered by those who oppose egalitarianism vis-à-vis those who endorse it is that neither may be held consistently. On the one hand, the egalitarians cannot treat the killing of a bear that tries to attack a human baby as morally commendable. On the other hand, the outlook that non-human animals ought to be killed indiscriminately for human ends cannot also be morally commendable because humans, evolutionarily speaking, need non-human animals for their environmental well-being (see Akande and Ofuasia 2017). The lesson to deduce, then, is that the SWC offered by Bell and Pei (2020), coupled with the dilemma highlighted in a previous paragraph, is limited in understanding the nature of the relationship between humans and non-human animals.

The African perspective, it needs to be said, has not been very vocal in matters of animal rights, especially when it is recalled that African religions usually demand the use of non-human animals for sacrifice. Are humans justified in using these non-human animals for sacrifices? Does their use for sacrifice not clearly validate the locus that these non-human animals are lower in hierarchy compared to humans? These are two questions to which I now turn to in the next section as I argue for what I term the "selective subordination for harmony" (SSH, hereafter) thesis. This thesis, among other things, not only upholds the hierarchical view but goes on to establish how the killing of non-human animals is pertinent not for humans but for the sake of holistic balance. This, at least, is the position of Yorùbá philosophic sage, *Ọ̀rúnmìlà,* who upholds the hierarchical view with implications for holistic balance.

"Selective Subordination for Harmony": An African Perspective to Human–Animal Hierarchy

Perhaps it is important to commence this section with the evidence of hierarchy in nature. On the National Geographic Channel, there are horrid scenes where lionesses attack the newborn of a deer for food, even when the latter has barely taken ten steps in the new environment outside its mother's womb. The remains are either devoured by hyenas or the vultures that are usually lurking close by to feast on whatever may be left of what was a newborn deer. In Ota, a popular town in South-West Nigeria, an eagle dives low and picks one of the newly hatched chicks of a hen. The eagle, obviously, will feed her eaglets with the intestines of the chick. Regardless of how far or high the eagle flies, it must surely return to the face of the earth to pick food, which is obviously a non-human animal. There are numerous illustrations in nature that validate the outlook that, ontologically, all entities are unequal and there is an order or hierarchy that must be maintained for the sake of balance. Humans, for instance, are omnivorous – they eat flesh and plants, same as some other non-human animals. Cattle, on the other hand, are herbivorous. Their elementary canal is suited for grass/plants. Hyenas, tigers, and some others are carnivorous animals. They are natural flesh eaters. So, it is not considered an abnormality for a lioness to chase down and kill a calf of a bison for food, for instance. This shows there is an established order in nature – a hierarchy that is maintained for ecological balance to thrive.

One crucial point to understand is that even among non-human animals, hierarchies reinforced by nature persist. It is not possible for a deer to feed her young ones with the cubs of the lioness, just as the hen cannot feed her chicks with the eaglets. Surely, their alimentary canals are dissimilar. There is a hierarchy in the natural instinct for survival, irrespective of how egalitarians in the discourse on animal rights may wish to argue. When this understanding is applied to Yorùbá thought, a new dimension may be discerned – one which takes cognizance of both the physical and non-physical dimensions of reality. Whereas the egalitarians and those who endorse the SWC thesis have considered the ways that humans and non-humans can cohabit without one cavalierly causing injury or harm to the other, the Yorùbá have since moved beyond this locus by focusing on how such cohabitation may signify both holistic and ecological balance.

Specifically, when automobile accidents are noticed too frequently, especially as it pertains to a dog being run over by automobiles, the Yorùbá are likely to infer that the deity, Ògún, who is said to be the "patron saint" of blacksmiths, hunters, and people whose livelihood depends on iron works, may need to be propitiated. Unless this is done, the Yorùbá are of the view that the community will be in chaos due to the occurrence of future violence and misfortune. This outlook, which is common among the traditional Yorùbá, is well-captured by Rene Girard (2018) on the interplay between communal violence and sacrifice, but more on this later.

For the moment, it is worth adding that there are other instances wherein the blood of non-human animals is needed to appease some forces or deities so that there can be spiritual balance, which will in turn affect the environment. In some other instances, some of these deities or "forces" demand that the blood of non-human animals be spilled in order for a human whose sickness has defied remedy or healing from orthodox or western medicine to receive healing (see Akin-Otiko 2018). Some of these illnesses are properly referred to as somatoform disorders in western medicine, but this is beyond the scope of this inquiry, whose focus is not on medicine.

The fundamental question which comes to the fore at this juncture then is: To what extent may practices such as non-human animal sacrifice in traditional African religion inform holistic balance? In response to this question, an *Ifá* priest, Áwo Fáladé Òsúntólá (2013, par. 5) reveals:

> Animal sacrifice…nourishes us with the meat thereby completing the cycle and affirming our connection to our earthly and heavenly *ẹgbẹ* (community). Sacrificing an animal nourishes the spirits (with blood) and the community (with meat) thereby indexing the powerful link between humanity and the *òrìṣà* as they are both nourished by a single ritual process. Offerings of fruit, *amidu*, and even drinks all recharge our *òrìṣà*'s ase but they perform this task slowly and with coolness. Blood sacrifice, on the other hand, recharges an *òrìṣà* quickly and with heat.

What the foregoing implies is that animal sacrifice in *Ifá* and similar religious cultures in Africa is not seen mainly from the perspective of killing these non-human animals *arbitrarily*. It goes beyond the physical since there are spiritual dimensions to it which breed harmony and balance between the physical and non-physical spheres. As a way of avoiding the use of non-human animals for sacrifice, Ofuasia (2018) offers a critique of the ancient Yorùbá philosophic sage, Ọ̀rúnmìlà, who, in *Ògúndá Otura*, an *Ifá* chapter, has validated that for eternity, non-human animals will serve as "buy-back" for human lives. Ofuasia ends his analysis with the suggestion that fruits be used. However, an important point which Ofuasia overlooks is that Ọ̀rúnmìlà has not endorsed the inadvertent harm of non-human animals. Ọ̀rúnmìlà's declaration in *Ògúndá Otura* is similar to the contention of Bell and Pei (2020, p. 154): "if it is a question of life and death, humans should have priority." This is a position that the most enduring and consistent egalitarian cannot deny.

As a way of establishing that the harmony between the physical and non-physical sometimes involves animal sacrifice, there are instances where a human may be sick and such ailments command the blood of animals. The importance of using non-human animals for sacrifice and as a "buy-back" for human lives among traditional Africans also speaks to the need for harmony, even in matters of health. In an interactive session with an *Ifá* priest, Semiu Akinbode Shokunbi, Ofuasia *et al.* (2022) assess the importance of the use of non-human animals for sacrifice and their blood for harmony with the non-physical world was upheld as a practice that goes as far back as Ọ̀rúnmìlà. The *Ifá* priest admits that the blood of some of these non-human animals is important for reversing the lot of humans. Semiu Akinbode Shokunbi explains that it is not them, the *Ifá* priests, that make demands for these animals but the spiritual forces or deities in charge of the spiritual problems of humans (see Ofuasia *et al.* 2022). In this way, the priest is absolved of his involvement in blood-spilling and cavalier killing of animals. What, then, is sacrifice, and how is the concept employed among Africans?

Specifically, sacrifice is derived from the Latin etymology, *sacrificium*, meaning "sacred action" (Wepener 2021). Generally, sacrifices are "offered to propitiate, appease or send messages to the gods, ward off sickness and provide workers for elite citizens in the afterlife" (Ojo 2005, p, 379). When talking about sacrifice, Rene Girard is usually cited since he has done copious research on the topic.

Girard (2018) considers the outlook that societies are inherently violent and goes on to discuss what he calls the "mimetic nature of violence." By being mimetic, Girard (2018) argues when violence occurs in a persistent manner in a community, it will not stop or halt by itself but will serve as a push for further instances of violence. Girard (1996) therefore argues that sacrifice is usually employed as a means to halt violence. In the words of Wepener (2021, p. 4) "it is at this point that sacrifice, as well as a scapegoat, becomes operative, in the sense that the violence is redirected in the act of sacrificing something or someone who is innocent." As Girard (2018, pp. 1–2) himself attests: "Because the victim is sacred, it is criminal to kill him – but the victim is sacred only because he is to be killed… when unappeased, violence seeks and always finds a surrogate victim."

This is why a victim who is usually not a member of the community and could be a human or non-human animal (depending on the severity of the violence) is usually used as a sacrifice to "protect the community from *its own* violence" (Girard 2018, p. 105 *italics in original*). Girard (1996) agrees that the more serious the violence and catastrophe are, the more precious and unique the sacrifice victim must be. What needs to be learned outside the scope of Girard's approach to the understanding of the "sacrificial object" is that it is usually chosen through a non-physical selective mechanism.

The interactive session with the aforementioned *Ifá* priest has left an important impression that informed the thesis of Selective Subordination for Harmony (SSH). This is because among the Yorùbá, just like in some other religious cultures across Africa, the aimless killing of non-human animals for human pleasure is usually frowned upon. However, when it comes to holistic balance, there are no compromises. The forces are selective about what kind of being is needed to restore balance not only in the environment but also in the community and even within the human body.

SSH is able to reconcile the dilemma of SWC given its thesis: Compared to non-human animals, humans have more worth and their sustenance overrides that of non-humans even when non-human animals must not be treated with disregard. Given its root in African metaphysics, SSH contends that "The vitalist conception of reality as force has allowed Africans to develop reverence for nature since it is through it that they are able to relate on a multi-sensory plane with reality" (Ofuasia 2024a, p. 163). The reason why humans have more worth and rights is because they are higher on the hierarchy of being than non-human animals: "Nature deals in hierarchies and our ancient Africans understood since it has never occurred to them that non-human animals should have rights like humans and plant life, which is even crucial to both humans and non-human animals is excused" (Ofuasia 2024a, p. 164).

Although SSH admits the locus that non-human animals are lower compared to humans in the ontological hierarchy, its selective character is based on the fact that it is the wish of "non-physical forces" that determines the kind of non-human animals that will be used for establishing harmony for the society to experience balance as well as peaceful cohesion. During an interactive session with the Lagos-based *Ifá* priest, Semiu Akinbode Shokunbi, this is disclosed:

> Once we cast *Ifá* for divination and we are instructed to use a dog for sacrifice, we cannot go after a cat. If we are instructed to pick a goat, we have no business with the ram. If divination tells us to use a single kolanut, we do not go after a goat. It is what we are instructed via divination that we do…It is *Ifá* that will say use this and not that for the sacrifice.
>
> (see Ofuasia *et al.* 2022)

Again, what the foregoing excerpt reveals is that there is a "non-physical selective mechanism" which indicates what is needed for restoring balance in nature and society. This non-physical approach not only overcomes some of the hurdles and contradictions noticed in SWC and the negative implications inherent in the arguments of some egalitarian scholars, but goes on to provide a ground for why sacrifice is pertinent for communal and ecological well-being.

On first showing, SSH may be interpreted to mean that it places no value on non-human animals because they may be easily targeted and killed for human ends. This also means that SSH finds the hierarchy of humans over non-human animals to be just. On deeper reflection however, it needs to be said that SSH provides two ways of revising the human and non-human animal interaction: (1) Non-human animals are helping human animals reach the goal of harmony and holistic balance, which takes cognizance of both the physical and non-physical aspects of life. This outlook allows one to understand that (2) human animals *need* non-human animals; otherwise, life would be unbearable, both in the physical and non-physical spheres. It is based on the recognition of these two crucial points that SSH is able to overcome the contradiction noticed in SWC. Unlike SWC, which says in one place that humans "have a duty to avoid cruelty to animals" (Bell and Pie 2022, p. 153), and in another context says that, for the sake of medical research, "we would try to use animals who have less in common with humans such as rats" (Bell and Pei 2020, p. 154), SSH is bolder. It relies on the principle of hierarchy to articulate that non-human animals are on a lower pedestal vis-à-vis humans – although they may be used to attain ends that will be beneficial to themselves, to humans, and to the quest for ontological and ecological balance, they must be shielded from cruelty outside these benefits. No non-human animals should be killed for the fun of killing – using catapults to shoot down pigeons, crushing snails with boots, or subjecting non-human animals to conditions that do not promote well-being and holistic balance.

At this juncture, a critic may object: what happens if the "non-physical selective mechanism" demands a human and not a non-human animal for harmony? The answer is simple: there are such cases of human sacrifice in ancient Africa where the forces demand a human animal in place of non-human animals. In the histories of some cultures in ancient Africa, there are records of human sacrifice that were deemed necessary to guarantee ecological and social harmony. For instance, there is adequate evidence that "Ghana's Asante Kingdom's very origins are associated with human sacrifice. On the eve of war against Denkyira, individuals volunteered themselves to be sacrificed in order to guarantee victory" (Mbogoni 2013, p. i).

The Acoli of modern-day Uganda slaughter animals for the purpose of reconciliation. As correctly noted by Wepener (2021) the first blood spilt from the animal is placed on some objects of the shrine and then on the person seeking reconciliation. Once this has been done, the parties that sought reconciliation eat bread together and drink wine as a symbol of renewed friendship.

There are instances among the Yorùbá and the Igbo of modern-day Nigeria, where the blood of a human of high ranking such as a prince or princess was demanded by "non-physical selective mechanisms" for ecological and social balance or harmony (see Law 1985). The research of Law (1985) states how places like Dahomey, Asante, Old Calabar, and the riverine Igbo engaged in human sacrifice as a means to revive spiritual, social, and ecological balance when things go wrong in the community. None of these comments are calculated to encourage the use of humans for sacrifice. The point being aimed at is that even the strict egalitarian, at this juncture, will understand that when it is possible for humans to be exposed to sacrifice for the sake of order or balance, that non-human animals undergo, this does not vitiate human hierarchy over that of non-human animals.

Secondly, it is pertinent to add that non-human animals used for restoring social and ecological balance are not selected by humans but by non-physical "mechanisms," usually via divination. Once the necessary sacrifice has been conducted, harmony is restored both socially and ecologically. There are relevant *Ifá* verses that allude to this urgency in sacrifice for harmony. For instance, a verse in *Odù Ògúndá Méjì* makes the point clearer:

> *...When he had finished drinking the water, he saw two people quarrelling over a fish they had caught*
> *Ògún told them that they should be patient*
> *And that they should divide the fish and go home but they refused.*
> *The first person said he came from the West.*
> *The second person said she came from the East*
> *After hearing what they both had to say concerning the matter,*
> *Ògún took the sword which he had in his hand and cut the fish into two equal parts for them*
> *The first person thanked him and said that he wanted him to open a way for him so that he might reach is town*

He said that Ògún would become prosperous
He said that Ògún would acquire an abundance of good things.
And that he would suffer no hardship before reaching his town.
The second person also thanked Ògún
And she asked that Ògún not neglect to open the way for her to reach her town
Ògún told them that he understood
Therefore, since the day that Ògún divided the fish into two for the people who were quarrelling
He has been called Ògúndáméjì, Ògún who divides into two equal parts (see Karenga 1999, pp. 60–61).

Ògún, in the foregoing verse, performed sacrifice. He was instructed on what to do and what to retain. Eventually, this paid off as he was able to solve a problem and avoid violence that would have ensued between two persons from two different towns. Another interesting verse on sacrifice for safeguarding social harmony is present in *Odù Ọ̀sá Méjì* thus:

Ọ̀sá, the brightly shining one, Babalawo of the world
Divined for the physical world
They said that the physical world should stop making sacrifices for wealth
And instead make sacrifices that would protect the physical world from its enemies
In this way, we will live
And so we plead that as long as we live in the physical world,
The physical world shall not be destroyed (see Karenga 1999, p. 87).

Whereas the foregoing verses hermeneutically remind humans of the role that they ought to play as "protectors" of the universe, it is important to stress that sacrifices in most cases do not involve killing or spilling of blood. This is well-stated in another verse from *Odù Ìrẹtẹ̀-yẹ̀kú:*

...For your sacrifice is in vain, if your character is deficient.
Therefore, you should also take the sacrificial pigeons and chickens home.
You should release them
You should not kill them
You should give them food, if they come to eat at your house
And you must not kill anything whatsoever from this day on
For one who does not want death to kill him should not kill anything whatsoever either (see Karenga 1999, pp. 210–211)

The above verse indicates that there are also instances where the non-physical mechanism for SSH goes as far as barring an agent from killing but caring for sacrificial non-human animals. This is the case irrespective of the value adduced to the non-human animal. What *Odù Ìrẹtẹ̀-yẹ̀kú* also reveals is that it is possible for non-human animals to be cared for. It is also possible for them to be allowed to live without any effort on the part of humans to frustrate or bring about their demise. All of these go into the programme of seeing that for psychological, social, and ecological balance to be realized, humans must learn to live in harmony with other non-human beings that are also part of the ecological constitution of the world.

Conclusion

The essence of this paper has been to provide another means of observing the nature of the human–animal relationship from the purview of African traditional religion. Most times, conversations on whether or not non-human animals have rights or are subjects of moral worth have been synonymous with Western scholarship. This has birthed several theories, such as the SWC and the egalitarian theses, that have been considered early in this research. Having established that we cannot run away from the hierarchy where human animals are deemed higher than non-human animals, this research goes further to propose its thesis of selective subordination for harmony (i.e., SSH), which takes cognisance of how non-physical mechanisms are instrumental in dictating sacrifice and utilities for maintaining balance on personal, social and ecological grounds.

References

Akande, M.A. and Ofuasia, E. (2017). 'The chicken fallacy and the ethics of cruelty to non-human animals.' *Bangladesh Journal of Bioethics* 8(1), pp. 7–20 Available at: https://doi.org/10.3329/bioethics.v8i1.31079 (Accessed: February 12, 2024)

Akin-Otiko, A. (2018). *Uncommon methods of diagnosis and treatment in African traditional medicine (ATM).* Ibadan: Gold Press Limited

Arimura, G. and Pearse, I.S. (2017) 'From the lab bench to the forest: Ecology and defence mechanisms of volatile-mediated "talking trees". How plants communicate with their biotic environment.' *Advances in Botanical Research* 82, pp. 3–17

Baluška, F., Gagliano, M. and Witzany, G. (Eds.) (2018). *Memory and learning in plants.* Cham: Springer International Publishing

Bell, D.A. and Pei, W. (2022). *Just hierarchy: Why social hierarchies matter in China and the rest of the world.* Princeton: Princeton University Press

Donaldson, S. and Kymlicka, W. (2011). *Zoopolis: A political theory of animal rights.* New York: Oxford University Press

Ebert, R., Giroux, V., Pepper, A. and Voigt, K (2022). 'Is Daniel a monster? Reflections on Daniel A. Bell and Wang Pei's "subordination without cruelty" thesis.' *The Ethics Forum* 17(1), pp. 31–45

Girard, R. (1996). 'Violence and the sacred: Sacrifice.' In R.L. Grimes (Ed.), *Readings in ritual studies.* New Jersey: Prentice-Hall, pp. 239–256

Girard, R. (2018). *Violence and the sacred.* London: Bloomsbury Academic

Kagan, S. (2019). *How to count animals, more or less.* New York: Oxford University Press

Karban, R. and Orrock, J.L. (2018). 'A judgment and decision-making model for plant behavior.' *Ecology* 99(9), pp. 1909–1919

Karenga, M. (1999). *Odù ifá: The ethical teachings.* Los Angeles: University of Sankore Press

Law, R.R. (1985). 'Human sacrifice in pre-colonial west Africa.' *African Affairs: The Journal of the Royal African Society* 84(334), pp. 53–87

Low, P. (2012). 'The Cambridge declaration on consciousness.' Proceedings of the Francis Crick Memorial Conference, Churchill College, Cambridge University. Available at: http://fcmconference.org/img/CambridgeDeclarationOnConsciousness.pdf (Accessed: April 10, 2023)

Mallatt, J., Robinson, D.G., Blatt, M.R., Draguhn, A. and Taiz, L. (2023). 'Plant sentience: The burden of proof.' *An Interdisciplinary Journal on Animal Feeling* 33(15), pp. 1–10. DOI: 10.51291/2377–7478.1802

Mbogoni, L.E.Y. (2013). *Human sacrifice and the supernatural in African history*. Dar es salaam: Mkuki na Nyota Publishers

McMahan, J. (2002). *The ethics of killing*. New York: Oxford University Press

Ofuasia, E. (2018). 'On the interests of non-human animals in *yorùbá* culture: A critique of *ọ̀rúnmìlà*.' *Bangladesh Journal of Bioethics* 9(2), pp. 6–21. Available at: https://doi.org/10.3329/bioethics.v9i2.41186 (Accessed: February 6, 2034)

Ofuasia, E. (2024a). *Ìwà: The process-relational dimension to African metaphysics*. Cham: Springer International Publishing.

Ofuasia, E. (2024b). 'Ìwà as ontological duality: A prolegomena to an African version of event metaphysics.' In. A. Agada, E. Ofuasia and B.Y. Ikuli (Eds.), *Contemporary African metaphysical thought*. London: Palgrave Macmillan, pp. 61–84. Available at: https://link.springer.com/chapter/10.1007/978-3-031-73963-7_4

Ofuasia, E., Shokunbi, S.A. and Ibiyemi, S.O. (2022). 'Animal sacrifice in traditional African religion.' *YouTube*. Available at: https://www.youtube.com/watch?v=nY3wYOl-rOc (Accessed: February 11, 2024)

Ojo, O. (2005). 'Slavery and human sacrifice in yoruba land: Ondo, c. 1870–94.' *Journal of African History* 46(3), pp. 379–404

Òsúntólá, F. (2013). 'Is animal sacrifice important?' Available at: http://iyaifagbemi.blogspot.com.ng/2013/03/is-animal-sacrifice-important.html?m=1 (Accessed: February 11, 2024)

Segundo-Ortin, M. and Calvo, P. (2023). 'Plant sentience? Between romanticism and denial: Science.' *Animal Sentience: An Interdisciplinary Journal on Animal Feeling* 33(1), 1–33. DOI: 10.51291/2377–7478.1772

Singer, P. (1985). *Ethics: Thought and practice*. J: New Jersey: Prentice-Hall Inc.

Weaver, M. (2022). 'West ham footballer Kurt Zouma ordered to do community service over cat abuse.' *The Guardian*, June 1. Available at: https://www.theguardian.com/football/2022/jun/01/kurt-zouma-cat-abuse-videos-west-ham-footballer (Accessed: February 12, 2024)

Wepener, C. (2021). 'The practice of ritual sacrifice and the role of religion in development: A south African practical theological exploration.' *Stellenbosch Theological Journal* 7(2), pp. 1–18

11 Biodiversity Conservation in Nigeria

Contrasting an Anthropocentric and a Deep Ecology Perspective

Olajide Akinleye-Martins, Naziru Zakari Muhammad, Helen Kopnina, and Mike Russell

Introduction

Currently, only around 15 per cent of the Earth's land surface and 5.3 per cent of the global ocean are protected (MPAtlas, 2021), and thus unlikely to prevent mass extinction (Noss *et al.*, 2012; Dinerstein *et al.*, 2017; Taylor *et al.*, 2020). Nigeria is one of the most populous countries in Africa, with rapidly declining biodiversity and a high rate of plant and animal extinction (Sobere and Ihua-Madunenyi, 2019; Kopnina *et al.*, 2022; Oko *et al.*, 2022; Olaoti-Laaro *et al.*, 2024). Due to the expansion of extractive industries, agriculture, urban development, population growth, and the destruction of habitats, the number of threatened species is increasing (Holechek *et al.*, 2017; IUCN, 2020). Here we speak of "biodiversity", as a "mass term" referring to the existence of so many individual living beings, "teleological centres of life" in the words of Paul Taylor. More broadly, there are biological scientists, including cell and molecular biologists, who recognise a "new paradigm" that should finally put an end to Cartesianism ethics about biodiversity – for example as expressed in the book *How Life Works: A User's Guide to the New Biology*, by Philip Ball (2023), which, from the perspective of cell biology, emphatically recognises the "agency" of all life forms. The Nigerian biodiversity crisis (Akindele *et al.*, 2021) also triggers changes in the way the "agency" is perceived, especially due to the socio-economic turmoil, resulting in a "decline in human socio-economic activities with a resultant amplification in unemployment, poverty and crime rate" (Zungum *et al.*, 2019).

Traditional conservation projects seem insufficient in countering biodiversity loss because they rarely involve local communities as active stakeholders. On the one hand, traditional forms of philanthropy involved in conservation, especially in Africa, have been criticised for being top-down, western-orientated, and inefficient in garnering local communities' support. On the other hand, local involvement can be seen from an ethical perception, involving the discussion of colonialism as well as environmental philosophy and differences in beliefs ranging from anthropocentric (human-centred) to ecocentric (ecology or ecosystem-centred). Ecocentrism is commonly understood as emerging from the "land ethics" of Aldo Leopold (1949) and the concept of "deep ecology" of Arne Naess (1973). Additionally, intermediate positions, such as an "ecocentric turn" have been identified, which Ordóñez-Vargas *et al.* (2023) identify as a transition which is not free from contradiction,

DOI: 10.4324/9781003480914-12

ambivalence, or great challenges. They define "ecocentric turn" as an "epistemological movement from anthropocentrism to ecocentrism, with biocentrism as the middle stage between the two ends" (Ordóñez-Vargas *et al.*, 2023, p. 107). Other intermediate positions include concepts of geocentrism (Frigo *et al.*, 2024). The authors argue that the central concepts involved in geometrical thought represent examples of environmental virtue ethics and defend moral weak anthropocentric positions, such as "ethical", "responsible", or "enlightened" anthropocentrism (Frigo *et al.*, 2024, p. 1).

In conservation, ethical debates have recently centred around the so-called "fortress conservation" and "community-based conservation" (e.g., Adams and Hulme, 2001; Jones and Murphree, 2004; Washington *et al.*, 2018, 2024). The former, stricter protection conservation is based on the idea of protection of biodiversity for its own sake (ecocentrism), with scientific evidence that even supposedly sustainable use of biodiversity by human communities, results in its depletion and thus the exploitation of biodiversity is the end and the fulfilment of human needs (Jones and Murphree, 2004). The latter is defined as "those principles and practices that argue that conservation goals should be pursued by strategies that emphasize the role of residents in decision-making about natural resources" (Adams and Hulme, 2001, p. 13). Complicating matters, there is also a discussion of compassionate conservation that aims to combine both animal ethics (e.g., animal rights and animal welfare movement as well as disciplines) with conservation of biodiversity based on supporting habitats and species, not necessarily individuals within the species (Wallach *et al.*, 2020). Compassionate conservation about indigenous/local communities in developing countries poses particular challenges and opportunities as well (Kopnina and Baker, 2023). One issue is related to the recognition of "nature rights", which might differ in the context of global West/North or global South and East (Ruales *et al.*, 2024), but also a pragmatic question of "who pays for conservation" (Washington *et al.*, 2021).

Suggestions about expanding traditional forms of interaction with wild nature, as well as more "modern" commercial activities such as eco-tourism (Parker and Khare, 2005; Krüger, 2005; Fennell and de Grosbois, 2023), and conservation education such as the Roots and Shoots program and alternative forms of finance can promote sustainable use of natural resources and conservation by local communities (Kopnina *et al.*, 2022). The need to find the balance between local communities in proximity to protected areas and thriving biodiversity also reveals a gap in understanding how local African beliefs and attitudes toward wildlife reflect on their support of conservation. This chapter will complement a shortage of data on attitudes to wildlife and conservation in Nigeria, seeking to identify barriers and opportunities for inclusive biodiversity conservation.

Indigenous and Local Ontologies of the Environment in the African Context

In contrast to the dominant Western way of thinking, it was argued that indigenous communities, at least in the past, perceived humans as part of nature and not above it (Anderson, 2011; Shoreman-Ouimet and Kopnina, 2015; Taylor *et al.*, 2020). Drawing from these indigenous cosmologies, some legal instruments, and initiatives, such as the "ecocide law", and animal rights and welfare legislation have emerged (Chapron, 2020; Legge and Brooman, 2020). Simultaneously, it was noted that a better understanding of local, indigenous, and/or community beliefs in attitudes regarding biodiversity conservation is crucial for its efficacy (Shoreman-Ouimet and Kopnina, 2015).

While there is literature suggesting that traditional African beliefs were at least partially, or in a broader sense of recognizing the intrinsic value of species, eco-centric (Ojomo, 2011; Conradie, 2019), there is some debate about whether deep ecology, as defined by Naess (1973), or land ethics (Leopold, 1949)are applicable in an African context (Mangena, 2015; Ibanga, 2017; Conradie, 2019). Researchers of traditional African philosophy (and animist religion) concerning animals, plants and other living beings mostly point out a combination of different ethics with the intrinsic value of nature and of particular animals (Chibvongodze, 2016; Horsthemke, 2017; Etieyibo, 2018; Kelbessa, 2018). Increasingly influenced by the views and extractive practices from the Global North or West, but also factors like population growth and expansion of agriculture into previously pristine wilderness areas, some traditional philosophies and practices towards non-human species and biodiversity have become increasingly anthropocentric – utilitarian, functionalist, and economically motivated, with the intrinsic value at times sidelined (Noske, 1997; Etieyibo, 2018; Horsthemke, 2019). Indigenous knowledge and cultural beliefs play significant role in biodiversity conservation (Abdullahi *et al.*, 2013; Bisong and Andrew-Essien, 2010; Louman *et al.*, 2009; Liu, 2006), as African ontology and ethics have been described as "eco-centric, bio-centric and nature-friendly" (Nche and Michael, 2024, Behrens, 2010). In Nigeria for instance, there are traditional, religious and cultural practices tied to indigenous ontologies necessary for biodiversity conservation as different sects of the country attach significance to certain aspects of nature (Ogunade, 2005). Within the Yoruba culture in Nigeria, plants like *Newbouldia laevis* ("*ewe akoko*") are cherished for their medicinal value and highly protected from going into extinction as they are considered very important particularly for coronating traditional rulers, and chiefs (Babalola, 2011, p.127). Some other practices include the protection of monkeys within the "*agbondiekweli*" a sacred forest devoted to the deity *Ndiekweli* in Ozubulu, a town in the eastern part of Nigeria (Nkama *et al.*, 2022). Also, within the Ozubulu community, the fishes and crocodiles in the "*Ogwii*" stream are preserved for the purpose of purity, as well as fishes and alligators in the "*dimugo*" stream are protected being perceived by the people as reincarnators linked to the "*dimugo*" deity (Nkama *et al.*, 2022). Within the indigenous communities, cultural norms and taboos could be utilised as effective wildlife conservation tool. The practice prohibiting the killing of fish in the "*Oso'ro*" river located in the Ogun State metropolis of south-western Nigeria supports the existence of sacred forests built upon cultural beliefs (Babalola *et al.*, 2014). A study of the cultural practices and value system of the communities in the Oban sector of Cross River National Park located in south-eastern Nigeria reveals that hunters are forbidden from killing certain species of animals such as Pythons and Leopards (Jimoh *et al.*, 2012). Our research, among other objectives, queried whether community and Ranger respondents living next to protected areas, recognize the importance of conservation for biodiversity's sake, and question which trade-offs they see, especially in the case of human-wildlife conflicts or co-existence.

An Overview of Biodiversity Conservation in Nigeria

The consequences of habitat destruction and biodiversity loss have led to warnings about a "planetary recession" marked by the collapse of species and ecosystems as well as "human and non-human displacements" caused by an ecological decline (Ripple *et al.*, 2017). To address this ecological degradation, the Half-Earth vision suggests generous protection for terrestrial, freshwater, and marine biodiversity by sharing at least half the planet with other species (Wilson, 2016). Proponents of the Half-Earth rethink "shared" habitat designations ranging from strict protection to sustainable use and management, connecting Protected Areas (PA), supporting communities and wildlife coexistence (Noss *et al.*, 2012; Kopnina, 2016; Wilson, 2016; Dinerstein *et al.*, 2017).

In Nigeria, biodiversity loss is increasing at a "disturbing rate" (Imarhiagbe *et al.*, 2020) with wildlife species such as elephants, hippopotami, and some bird and plant species named among endangered species that are threatened with extinction (IUCN, 2022). Studies from Nigeria further reveal that some aquatic resources some of which are commercially viable fish species such as: "Bagrus filamentosus (silver catfish), Chrysichthys Nigrodigitatus (Bagrid catfish), Heterotis niloticus (African Arowana) and Erpetoichthys calabaricus (Reed fish)" have been identified as threatened and endangered (Emmanuel, Ekpe and Okon, 2016).

Critical social scientists, particularly political ecologists and environmental anthropologists, often explicitly focus on criticism of colonization, racism, and sexism, combining community-level interactions with the involvement of, or on behalf of, marginalized people in the developing world. As the idea of "decolonialism conservation" often refers to the historically marginalised, poor, and vulnerable human ethnic communities, Tarik Bodasing (2019) notes that colonizing entire habitats and discriminating against all other species by humans is not discussed. However, emerging ethical and legal framings of non-human rights and nature rights acknowledge inter-species colonialism, resulting in extinction (Piccolo *et al.*, 2018; Crist, 2019; Taylor *et al.*, 2020). Also, such criticisms overlook the fact that in developing countries or poor segments of society, community members directly rely on natural resources and are much more dependent on their environment (Jones and Murphree, 2013). If the environment is depleted this puts human communities in immediate existential danger as well (Crist *et al.*, 2021; Piccolo *et al.*, 2022; Roka, 2020; Washington *et al.*, 2024).If extinction is seen as unacceptable (Cafaro *et al.*, 2017), the right of other species to exist needs to be considered independent of their utility to humans (Chapron *et al.*, 2019).

Methodology

Study Areas

The study was conducted in eight communities around Yankari Game Reserve (Figure 11.1).

Yankari Game Reserve (YGR)is in the south-eastern part of Bauchi State, Nigeria. The Reserve shares a boundary with Gombe State on its north and eastern parts, and Plateau State on its southern part. It is endowed with many wild animals such as savannah elephants, lions, African buffalo, and several species of non-human primates (Odeniran *et al.*, 2018). The communities visited were Bakin Dutse, Gaji, Gaji Gamu, Gajin Gwaza, Kashara, Kwala, Mai Ari, and Pali. These communities are situated close to the YGR. A total of seventy-five participants consisting of forty (40) men and thirty-five (35) women from these eight indigenous communities were interviewed. The age demography of the participants ranges between sixteen (16) and seventy-five (75) years old, with the female participants much younger than the male participants. These communities were selected because of their proximity to the protected reserve and as indicated in the introduction, local perceptions and participation in conservation efforts is a crucial component of conservation success. We have used convenience sampling in combination with purposive sampling (Etikan *et al.*, 2016).

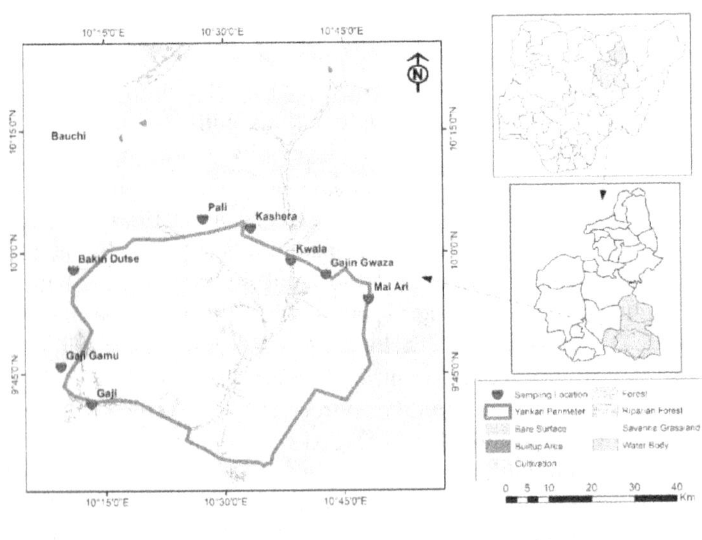

Figure 11.1 Map of study area showing communities visited.

Data Collection

Data on the perception, understanding and activities of residents were obtained through semi-structured interviews conducted by one of the co-authors of this chapter, Naziru, and other assistants. We have chosen semi-structured interviews because they provide both contextually richer data (open questions) and easier to code and quantify data (closed questions) (Schmidt, 2004). The interview questions were translated and conducted in the Hausa language, which is spoken by community residents. In each community, the village heads or their representatives were interviewed first before the other members of the community. Ten people were interviewed in each community, five men and five women except for Gajin Gwaza where the women were not allowed to be interviewed based on their tradition. The village head of Gajin Gwaza indicated that the views of the men represented that of their women folk. The interview questions targeted their understanding of biodiversity conservation and their knowledge of the "Half-Earth" vision which involves the co-existence of humans with different species of animals and plants. Fourteen Rangers (two of whom were volunteers) were also interviewed from the three districts of Alkaleri Local Government Area of Bauchi State where the Yankari Game Rangers live. They were asked questions regarding their motivation for the work they do at the Reserve, their perspective on who they believe benefits from the work they do, their ideas on improving conservation as well as whether their spiritual beliefs have any impact on their work.

Photo Gallery of interview sessions with the communities.

Figure 11.2 Researcher interviewing the village head and elders of Bakin Dutse community.

Figure 11.3 Researcher interviewing a group of women at Bakin Dutse community.

Figure 11.4 Researcher with a research assistant interviewing the men of Gaji Gamu community.

Figure 11.5 Researcher conducting interview at Gajin Gwaza community.

Figure 11.6 Researcher with village head of Kashara community interviewing women of the community.

Figure 11.7 Researcher interviewing the village head and some men in the Pali community.

Figure 11.8 Researcher interviewing some Pali community women.

Figure 11.9 Researcher interviewing the Mai Ari community women.'

Photo Gallery of interview sessions with the Rangers

Figure 11.10 Researcher conducting interview with some Rangers of the Game Reserve.

Data Analysis

The qualitative data results from the interview questionnaires from each participant within the eight selected communities were analysed separately in the spreadsheet to capture the different themes from the varied responses. Afterwards, the responses of the participants were translated into a tabular format for each community with similar themes grouped. Furthermore, the responses of the Rangers, which were

Table 11.1 Overview of Community Interviews

Name of Communities (8) – Bakin Dutse, Gaji, Gaji Gamu, Gajin Gwaza, Kashara, Kwala, Mai Ari, and Pali	
Sample Size	*40 Male Participants and 35 Female Participants*
Knowledge of wildlife and characteristics of species	All 75 participants (100%) possess a basic understanding of the meaning of wildlife, the different types of local animals and plants in the area, wildlife species and their characteristics with varying preferences, and they expressed dislike for pests as harmful to their farms and crops.
Perception of co-existence of humans with animals and plants	69 participants (92%) have a positive view on the idea of peaceful coexistence of all animals and plants with humans, while the remaining six participants (8%) either believed such co-existence would be harmful. One of the female participants specifically stated, "I am positive about co-existing with wildlife. This will work if wildlife professionals are supported by Government". Another participant however noted that: "It is harmful. Except if communities are provided with an alternative form of livelihoods and social amenities".
Possible options required for living with pests	66 participants (88%) suggested payment of compensation to affected persons as an important option for living with animals classified as pests without killing them. While nine participants (12%) further advised that provision of social amenities and employment of indigenes in conservation areas should be included as supplementary options.
Knowledge of Half-Earth vision	Only seven participants (9.33%) indicated they had heard about the concept of Half-Earth vision, while 68 participants (90.67%) did not have any idea about Half-Earth vision before the interviewer explained it.
View on sharing the earth with animals and plants	74 participants (99%) believed sharing the earth with non-humans (Half-Earth) would be beneficial but requires collaboration between the Government, conservationists, and the communities. They suggested that the concept of Half-Earth could help with improved conservation methods, control desert encroachment, and erosion control. In the words of one of the male participants: "This vision will drastically improve wildlife conservation. This will lead to 14 Rangers from 11 select communities in Alkaleri."
Sample size	Local Government Area of Bauchi State, Nigeria (Badara Lungu, Bajama, Fanti, Gale, Galen Duguri, Gar, Maina Maji, Pali, Rimi, Talishot, and Yalo). Two of the 14 Rangers are volunteers who are paid on shift-basis.
Age range	Between 28 and 52 years.

captured in the interview questionnaire, were analysed under six themes representing the focus questions in the survey. A tabular overview of the interview data analysed across the eight communities is shown in Table 11.1, and the data from the Rangers is shown in Table 11.2.

Summary of Findings from the Community Interviews

Table 11.1 above reveals that participants across the selected communities understood wildlife as referring to animals living in the bush, with some adding that it could also include wild plants and animals in the forest as well as animals protected in conservation areas. Participants possessed a good basic knowledge of local animals and plants in the area, wildlife species and their characteristics with varying preferences based on looks and behaviour, whilst expressing their dislikes for pests as being harmful to their farms and crops. From the different responses, it can be concluded that the participants understood the importance of wildlife and

Table 11.2 Overview of Data from the Rangers Interviews

Level of Education	*Twelve participants with secondary school education; one participant with a diploma and one with a bachelor's degree.*
Salary scale	Between ₦33,000 (equivalent to 73.99 USD) and ₦48,000 (equivalent to 107.62 USD).
Length of employment	Between 8 and 23 years.
Accountable authority	The Rangers are accountable to the Bauchi State Government of Nigeria.
Perspective on end beneficiary of services rendered	All 14 participants were of the opinion that their families, community, Government and, in some way, the management of the Game Reserve were end beneficiaries of the services they render as Rangers.
Motivation for work	All the participants noted the motivation for their work revolved around the conservation of nature as well as protection of wildlife resources of the Game Reserve for the benefit of people and the communities.
Ideas on improving conservation	The Rangers were of the view that the ways to improve conservation would involve (1) provision of working equipment to staff of the Game Reserve, (2) providing social amenities for communities around the Reserve, (3) educating local communities around the Game Reserve about the importance of biodiversity conservation, and (4) review of the conservation laws.
Impact of spiritual belief on work	The participants are all of the opinion that their religious/spiritual tenets have a huge impact on the way they carry out their work, as it encourages them to be honest, dedicated, and exhibit integrity.
Free time used for work/hours of work unremunerated	All the participants stated that they utilise their free time to carry out unremunerated work and the hours spent are not particularly specific.

****Note**: The official exchange rate figure used for the purpose of conversion in this study is Naira 446 to 1 USD (as of November 2022).

the conservation of biodiversity. They know the various wildlife species in their area, and they like some species for different reasons, which are personal such as the behaviour of the animal and the potential to attract tourism. They do not like the pests (e.g., snakes and baboons) for reasons including the destruction of crops, risk of physical harm, and disruption of work on their farms. Some of the wildlife species involved in human-animal conflicts identified by the participants include Elephants, Hyenas, Warthogs, Hippopotamus, and some bird species such as Guinea fowl, Owl, and Hawks.

The interview data, however, reveals that 68 out of the 75 participants (representing 90.67%) did not have any prior knowledge about the concept of Half-Earth vision until they were enlightened by the interviewer. Thus, revealing a dearth of understanding about the concept amongst the local communities. However, after the concept was introduced and explained to them by the interviewer, the participants noted that it was viable and would be feasible if conservationists, the Government, and local communities collaborated more effectively. They believe training on alternative livelihoods and the provision of social amenities to people will make the Half-Earth vision successful. 88% of the participants suggested payment of compensation to affected persons as an option to co-exist with animals classified as pests without resorting to killing. Other participants suggested options including collaboration with community leaders to stop encroachment of protected areas, provision of alternative means of livelihood, collaboration of the community with park officials, and provision of social amenities. 99% of the participants believed sharing the earth with non-humans would be beneficial for biodiversity conservation, ecotourism, and community development. They, however, noted it would require collaboration between the Government, conservationists, and the communities.

In essence, most of the participants believe it is feasible to co-exist with wildlife, provided the local communities are provided with social amenities and alternative livelihoods. Providing social services for their basic needs will stop them from being a threat to wildlife survival. They believe the payment of compensation to affected communities for crop raiding will improve the park-community relationship and impact the survival of wildlife positively.

Summary of Findings from the Rangers' Interview

From Table 11.2 above, it can be deduced that the average age of the Rangers that were interviewed was 44.2 years old, the average length of employment was 15.4 years, and the average monthly salary was about ₦39,909.09 (equivalent to about 89.48 USD). They are all accountable to the Bauchi State Government of Nigeria, which is the relevant authority responsible for the management of the Yankari Game Reserve. The data shows that the Rangers are motivated by their passion for protecting the biodiversity of the Game Reserve for the ultimate benefit of the local communities. Most of the Rangers confirmed that when they are not on duty, they offer their time, without compensation, to work in the Reserve or collaborate with other Rangers who are on duty. They offer their services to provide vital information that will lead to the arrest of offenders for the benefit of the Reserve's management. However, the Reserve does not pay Rangers for overtime hours.

All the Rangers interviewed regard their spiritual beliefs as the guiding principle for their honesty and dedication to work. They think that providing adequate working tools and equipment for conservation staff, educating communities about conservation, reviewing conservation laws, and the provision of social amenities to surrounding communities will help improve conservation work in the Game Reserve. Furthermore, they believe that the improvement of their welfare, most especially their remuneration, is key to effective wildlife conservation as their current salaries are quite meagre.

It is important to point out that at the Yankari Game Reserve, all the Rangers (whether staff or volunteers) work in line with the park rules. Volunteer Rangers are officially recognised; they do not work in isolation and their duties are formally organised to prevent cases of connivance with offenders under the guise of volunteering work. The volunteer Rangers are not allowed to organise any informal duty outside the work assigned to them and are only paid allowances for overnight camping as they are not on the payroll of the Bauchi State Government. The overnight allowance is paid by the Wildlife Conservation Society Nigeria, which is the non-governmental organisation that currently supports the Government in wildlife law enforcement in the Yankari Game Reserve.

Discussion

This research has briefly explored the African belief systems and cosmologies about conservation. We have queried whether Community and Ranger respondents living next to protected areas, recognize the importance of conservation for the sake of biodiversity, considered ideas for improving conservation and questioned which trade-offs they see, especially in the case of human-wildlife conflicts and/or co-existence. Based on this research in the context of community beliefs, we have identified the barriers and opportunities for inclusive biodiversity conservation as well as the motivation for staff rendering conservation services. Our findings indicated that our respondents present a range of views that represent a complex cosmology. In line with Ordóñez-Vargas *et al.* (2023), we note that the ecocentric turn, if we can speak of a change of attitudes in the Nigerian context of immediate biodiversity threats, does not exclusively imply positions that are clear-cut anthropocentric or ecocentric, but may be characterised as hybrid and eclectic. Some views align with functionalist, utilitarian, and extractivist logic about biodiversity, while other views reject anthropocentrism and provide alternative forms of interaction with nature– some of which are expressed above in the summary of findings. Furthermore, in line with Frigo *et al.* (2024), we can argue that the data points to a mixture of strong and weak anthropocentric genetics that can synthesize the different positions in environmental ethics and move beyond them toward a more realistic and pragmatic position that many members of the communities interviewed assume. However, it was observed that the previous experiences of the community participants regarding how the Government manage Protected Areas and relates with them as stakeholders played a key role in their responses.

From the interview of the Ranger respondents, we find that the level of commitment and dedication of the Rangers to biodiversity conservation is largely rooted in their religious/spiritual beliefs which are a significant part of the cultural indigenous ontologies required for biodiversity conservation in Nigeria as identified in section 1.3 An Overview of Biodiversity Conservation in Nigeria. above. This manifests in the Ranger's lengthy stay in employment and their willingness to go the extra mile in offering unpaid voluntary services at the Game Reserve when necessary. Thus, buttressing the definition of "meaningful work" in the context of conservation and deep ecology ethics, as explored by Bunderson and Thompson (2009) in their qualitative study on the meaning of work in the zoo-keeping profession in the United States which showed that the zookeepers perceived their work as "a personal calling". Furthermore, the result from the Rangers interview reveals the essence of better remuneration as this could present economic implications for effective biodiversity and wildlife conservation. In this regard, meaningful work has also been inextricably linked to monetary reward being "that which gives essence to what we do and what brings a sense of fulfilment to our lives" (Chalofsky, 2003, p. 74; Pratt and Ashforth, 2003). While work must add to "some greater good" beyond the benefits of the individual (Martela and Pessi, 2018, p. 6), remuneration is significant compensation for work done. According to Arneson (1987, p. 517),

> work for which pay is received is interesting, calling for intelligence and initiative, and in which the worker has considerable freedom to determine how the work is to be done and a genuinely democratic say over the character of the work process and the policies pursued by the employing enterprise.

Expressed in terms comprehensible to sometimes illiterate community members who sometimes have to struggle to meet their basic needs, the questions concerned with environmental ethics are less theoretical. Even more so to members of non-human species that may be critically endangered or driven to extinction. Nonetheless, a few reflections on the "value", especially as expressed in Frigo *et al.* (2024) article and application to the African context, are in order. One could argue that the value arises with the very first living organism, which manifested the agency to "value" those nutrients in its environment that were necessary to maintain its life and proceeded to ingest them – we humans can recognize the existence of that "value," but that "value" exists independently of us.

Some of the proposals voiced by the community members or Rangers in the interviews also pointed in the direction of alternative funding for conservation. Others testified to the need to join/participate in larger projects, some stemming from the Global North or West. Even if the community members know or care little about biology, the findings also point to the possibility of enabling us to "synthesize" biocentric and ecocentric positions into "ecological humanism", which rises "in contrast to the arrogance of controlling nature" (Frigo *et al.*, 2024, p. 12), as illustrated by proposals to bring money to impoverished communities, and more likely disadvantage them in the longer term. Also, returning to the question of the Rights of Nature (e.g., Ruales *et al.*, 2024) and compassionate conservation (Wallach *et al.*, 2020), we note that if we bring in the question of intrinsic values in nature, the mere commercialisation of nature and biodiversity might be seen as simply ethically wrong. Perhaps other proposals by community members, such as the extension of tourism (eco-tourism) may also be seen as feeding into a larger and potentially distractive tourist industry (e.g., Parker and Khare, 2005; Krüger, 2005; Fennell and de Grosbois, 2023). Yet, in each case, the trade-offs between the need to retain livelihoods, protect biodiversity and have realistic alternatives remain paramount.

Conclusion

Generally, the case study indicates the tension between economic and intrinsic motivations. Considering traditional African philosophy, economic motivation is perhaps a product of the Global North or Global West idea of progress and represents shallow, not deep ecology. More broadly, returning to *How Life Works: A User's Guide to the New Biology*, by Philip Ball (2023), and inspired by the views of community members and Rangers from this study, these attitudes emphatically recognize the "agency" of all life forms (although at times prioritising only one – humans). Returning to a more philosophical reflection on ethics, perhaps one could consider that Life (with a capital "L" to encompass the whole "tree of Life," Life on Earth as it has evolved over all of geologic time, including us and everything else that is alive) is the source of ALL "value". Without some sort of life, what would there be? Just rocks, life-less waters, soil bereft even of microorganisms, and geologic material without "interests". What would there be to demand our moral concern? We would not be there either, of course, so even our "weak anthropocentrism" could not get traction if we were determined to continue on a path toward such a "dead world". Consideration of this "dead world" alternative is especially relevant when one considers how quickly diversity of life disappears in Nigeria, the agricultural monocultures used for animal feed, and urban development forming "concrete jungle" take the place of vibrant communities of life.

References

Abdullahi, J., Usman, I., Samaila, G., and Zuni, A. (2013) 'Importance of indigenous knowledge in biodiversity conservation: A case study of communities surrounding Kpashimi forest reserve, Niger Sate, Nigeria', *Journal of Environmental Science, Toxicology and Food Technology*, 5(6), pp. 10–17. http://dx.doi.org/10.9790/2402-0561017

Adams, W. M., and Hulme, D. (2001) 'If community conservation is the answer in Africa, what is the question?', *Oryx*, 35(3), pp. 193–200. https://doi.org/10.1046/j.1365-3008.2001.00183.x

Akindele, E. O., Ekwemuka, M. C., Apeverga, P., Amusa, T. O., Olajuyigbe, S., Coker, O. M., Olaleru, F., Fasona, M., Usen, E. N., Ringim, A. S., and Adedoja, O. A. (2021). 'Assessing awareness on biodiversity conservation among Nigerians: the aichi biodiversity target 1', *Biodiversity and Conservation*, 30, pp. 1947–1970. https://doi.org/10.1007/s10531-021-02175-x

Anderson, E. N. (2011) 'Drawing from traditional and "Indigenous" socioecological theories', in H. Kopnina and E. Shoreman-Ouimet (eds.) *Environmental Anthropology Today*. New York: Routledge, pp. 56–74.

Arneson, R. J. (1987) 'Meaningful work and market socialism', *Ethics*, 97(3), pp. 517–545. https://doi.org/10.1086/292864

Babalola, F. D. (2011) 'Roles of and threats to yoruba traditional beliefs in wilderness conservation in Southwest Nigeria', *USDA Forest Service Proceedings RMRS-P-64*. https://www.fs.usda.gov/rm/pubs/rmrs_p064/rmrs_p064_125_129.pdf

Babalola, F. D., Lawal, I., Opii, E. E., and Oso, A. O. (2014) 'Roles of and threats to indigenous cultural beliefs in protection of sacred forests in southwest Nigeria', *Albanian Journal of Agricultural Sciences*, 13(2), pp. 41–50.

Ball, P. (2023) *How Life Works: A User's Guide to the New Biology*. Chicago: University of Chicago Press.

Behrens, K. (2010) 'Exploring African Holism with Respect to the Environment', *Environmental Values*, 19(4), pp. 465–484. https://doi.org/10.3197/096327110X531561

Bisong, F., and Andrew-Essien, E. (2010) 'Indigenous knowledge systems for promoting community conservation education in a Nigerian protected area', *International Journal of Biology*, 2(2), pp. 149–157. https://doi.org/10.5539/ijb.v2n2p149

Bodasing, T. (2019) 'Looking beyond the past to give African wildlife a future: A critical review of the big conservation lie'. https://www.ecologicalcitizen.net/pdfs/v02n2-16.pdf

Bunderson, J. S., and Thompson, J. A. (2009) 'The call of the wild: Zookeepers, callings, and the double-edged sword of deeply meaningful work', *Administrative Science Quarterly*, 54(1), pp. 32–57. http://www.jstor.org/stable/27749305

Cafaro, P., Butler, T., Crist, E., Cryer, P., Dinerstein, E., Kopnina, H., Noss, R., Piccolo, J., Taylor, B., Vynne, C., and Washington, H. (2017) 'If we want a whole earth, nature needs half: A response to Büscher et al.', *Oryx*, 51(3), pp. 400–400. https://doi.org/10.1017/S0030605317000072

Chalofsky, N. (2003) 'An emerging construct of meaningful work', *Human Resource Development International*, 6(1), pp. 69–83. http://dx.doi.org/10.1080/1367886022000016785

Chapron, G. et al. (2019) 'A rights revolution for nature', *Science*, 363(6434), pp. 1392–1393. https://www.science.org/doi/10.1126/science.aav5601

Chibvongodze, D. T. (2016) 'Ubuntu is not only about the human! an analysis of the role of african philosophy and ethics in environment management', *Journal of Human Ecology*, 53(2), pp. 157–166. https://doi.org/10.1080/09709274.2016.11906968

Conradie, E. M. (2019) 'A (South) African land ethic? The viability of an ecocentric approach to environmental ethics and philosophy', in M. Chemhuru (ed.) *African Environmental Ethics: A Critical Reader*. Switzerland: Springer-Verlag, pp. 127–139.

Crist, E. (2019) *Abundant Earth: Toward an Ecological Civilization*. Chicago: University of Chicago Press.

Crist, E., Kopnina, H., Cafaro, P., Gray, J., Ripple, W.J., Safina, C., Davis, J., DellaSala, D. A., Noss, R. F., Washington, H., Rolston, H., Taylor, B., Orlikowska, E. H., Heister, A., Lynn, W., and Piccolo, J. J. (2021) 'Protecting half the planet and transforming human systems are complementary goals', *Frontiers in Conservation Science*, 2. https://doi.org/10.3389/fcosc.2021.761292

Dinerstein, E., et al. (2017) 'An ecoregion-based approach to protecting half the terrestrial realm', *Bio Science*, 67(6), pp. 534–545. https://doi.org/10.1093/biosci/bix014

Emmanuel, E., Ekpe, A., and Okon, A. (2016) 'Threatened and endangered fish species in Nigeria, a menace to biodiversity – a review', *African Journal of Education, Science and Technology*, 3(1), pp. 12–26. https://ajest.info/index.php/ajest/article/view/665

Etieyibo, E. (2018) 'African philosophy and nonhuman nature', in G. Hull (ed.) *Debating African Philosophy: Perspectives on Identity, Decolonial Ethics and Comparative Philosophy*. London: Routledge, pp. 164–181.

Etikan, I., Musa, S. A., and Alkassim, R. S. (2016) 'Comparison of convenience sampling and purposive sampling', *American Journal of Theoretical and Applied Statistics*, 5(1), pp. 1–4. https://doi.org/10.11648/j.ajtas.20160501.11

Fennell, D. A., and de Grosbois, D. (2023) 'Communicating sustainability and ecotourism principles by ecolodges: A global analysis', *Tourism Recreation Research*, 48(3), pp. 333–351. https://doi.org/10.1080/02508281.2021.1920225

Frigo, G., Ifanger, L. A., Greco, R., Kopnina, H., and Hillerbrand, R. (2024) 'Shallow vs. Deep geoethics: Moving beyond anthropocentric views', *Journal of Agricultural and Environmental Ethics*, 37(2), pp. 1–18. https://doi.org/10.1007/s10806-023-09920-y

Holechek, J. L., Cibils, A. F., Bengaly, K., and Kinyamario, J. I. (2017) 'Human population growth, African pastoralism, and rangelands: A perspective', *Rangeland Ecology & Management*, 70(3), pp. 273–280. https://doi.org/10.1016/j.rama.2016.09.004

Horsthemke, K. (2017) 'Animals and African ethics', *Journal of Animal Ethics*, 7(2), pp.119–144. https://doi.org/10.5406/janimalethics.7.2.0119

Horsthemke, K., (2019) 'Animal rights and environmental ethics in Africa: From anthropocentrism to non-speciesism?', in M. Chemhuru (ed.) *African Environmental Ethics: A Critical Reader*. Springer, pp. 239–253.

Ibanga, D. (2017) 'Is deep ecology inapplicable in African context: A conversation with fainos mangena', *Filosofia Theoretica: Journal of African Philosophy, Culture and Religions*, 6(2), pp. 101–119. https://dx.doi.org/10.4314/ft.v6i2.6

Imarhiagbe, O., Egboduku, W. O., and Nwankwo, B. J. (2020) 'A review of the biodiversity conservation status of Nigeria', *Journal of Wildlife and Biodiversity*, 4(1), pp. 73–83. https://doi.org/10.22120/jwb.2019.115501.1096

IUCN Red List (2020) 'IUCN red list media releases'. Available at: https://www.iucnredlist .org/resources/media

IUCN Red List (2022) 'The IUCN red list of threatened species 2022(2)'. Available at: https://www.iucnredlist.org/

Jimoh, S. O., Ikyaagba, E. T., Alarape, A. A., Obioha, E. E., and Adeyemi, A. A. (2012). 'The role of traditional laws and taboos in wildlife conservation in the Oban Hill sector of Cross River National Park (CRNP)'. *Nigeria. Journal of Human Ecology*, 39(3), pp. 209–219. http://dx.doi.org/10.1080/09709274.2012.11906513

Jones, B., and Murphree, M. W. (2013).) 'Community-based natural resource management as a conservation mechanism: Lessons and directions', in B. Child (ed.) *Parks in Transition: Biodiversity, Rural Development and the Bottom Line*. London: Earthscan, pp. 63–103.

Kelbessa, W. (2018) 'Environmental philosophy in African traditions of thought', *Environmental Ethics*, 40(4), pp. 309–323. https://doi.org/10.5840/enviroethics201 840431

Kopnina, H. (2016) ' Half the earth for people (or more)? Addressing ethical questions in conservation', *Biological Conservation*, 203, pp. 176–185. https://doi.org/10.1016/j .biocon.2016.09.019

Kopnina, H., and Baker, L. (2023) 'Conservation, animal wellbeing, and indigenous participation at an elephant sanctuary in Mondulkiri, Cambodia', *Society and Animals*, pp. 1–22. https://doi.org/10.1163/15685306-bja10126

Kopnina, H., Muhammad, N. Z., and Olaleru, F. (2022) 'Exploring attitudes to biodiversity conservation and Half-Earth vision in Nigeria: A preliminary study of community attitudes to conservation in Yankari game reserve', *Biological Conservation*, 272, p. 109645. https://doi.org/10.1016/j.biocon.2022.109645

Krüger, O. (2005) 'The role of ecotourism in conservation: panacea or Pandora's box?', *Biodiversity & Conservation*, 14, pp. 579–600. https://doi.org/10.1007/s10531-004 -3917-4

Legge, D., and Brooman, S. (2020) 'Reflecting on 25 years of teaching animal law: is it time for an international crime of animal ecocide?', *Liverpool Law Review*, 41, pp. 201–218. https://doi.org/10.1007%2Fs10991-020-09253-0

Leopold, A. (1949) *A Sand County Almanac*. New York: Ballantine.

Louman, B., Fischlin A., Peter, G., and others. (2009) 'Forest ecosystem services: A cornerstone for human well-being' in: S. Risto, B. Alexander, and K. Pia (eds.), *Adaptation of Forests and People to Climate Change - A Global Assessment Report*. International Union of Forest Research Organisations (IUFRO).

Liu, J.L. (2006) 'Traditional knowledge and implication to development from anthropological perspective', *Academic Journal of China's Agriculture University*, 2, pp. 133–141.

Mangena, F. (2015) 'How applicable is the idea of deep ecology in the African context?', *Filosofia Theoretica: Journal of African Philosophy, Culture and Religions*, 4(1), pp. 1–16. https://doi.org/10.4314/ft.v4i1.1

Marine Protection Atlas (2021) 'The marine protection atlas (MPAtlas)'. Available at: https://mpatlas.org/

Martela, F., and Pessi, A. B. (2018) 'Significant work is about self-realization and broader purpose: Defining the key dimensions of meaningful work', *Frontiers in Psychology*, 9. https://doi.org/10.3389/fpsyg.2018.00363

Naess, A. (1973) 'The shallow and the deep, long-range ecology movement. A summary', *Inquiry*, 16(1–4), pp. 95–100. https://doi.org/10.1080/00201747308601682

Nche, G. C., and Michael, B. O. (2024) 'Perspectives on African indigenous religion and the natural environment: beings, interconnectedness, communities and knowledge systems', in *Phronimon*, p. 14621, University of South Africa Press. https://doi.org/10.25159/2413-3086/14621

Nkama, C. L., Okoro, K. N., Egbule, E. (2022) 'Eco-preservation through the lens of igbo beliefs and practices: A re-imagination', *Religions*, 13(11) p. 1066. https://doi.org/10.3390/rel13111066

Noske, B. (1997) 'Speciesism, anthropocentrism, and non-western cultures', *Anthrozoös*, 10(4), pp. 183–190. https://doi.org/10.2752/089279397787000950

Noss, R. F., et al. (2012) 'Bolder thinking for conservation', *Conservation Biology*, 26, pp. 1–4. Available at: https://www.jstor.org/stable/41416122

Odeniran, P. O., Ademola, I. O., and Jegede, H. O. (2018) 'A review of wildlife tourism and meta-analysis of parasitism in Africa's national parks and game reserves', *Parasitology Research*, 117, pp. 2359–2378. Available at: https://link.springer.com/article/10.1007/s00436-018-5958-8

Ogunade, R. (2005) 'Environmental issues in yoruba religion: Implications for leadership and society in Nigeria', A paper prepared for Science and Religion: Global Perspectives, Conference presentation June 4–8, 2005. Available at: https://www.obafemio.com/uploads/5/1/4/2/5142021/environmental_issues_in_yoruba_religion.pdf

Ojomo, P. A. (2011) 'Environmental ethics: An African understanding', *African Journal of Environmental Science and Technology*, 5(8), pp. 572–578.

Oko, P. E., Odey, D. U., and Anake, C. B. (2022) 'Impact of population growth on biodiversity loss in Boki agro-ecological rainforests, cross river state, Nigeria', *Global Journal of Pure and Applied Sciences*, 28, pp. 99–105. https://dx.doi.org/10.4314/gjpas.v28i1.12

Olaoti-Laaro, S. O., Muritala, D. S., Odebunmi, C. A., Alagbe, A. V., Bamgboye, A. M., Agboje, I., Fadoyin, A.S., Bamigboye, O. A., Obakpolor, V., and Haastrup, N. O. (2024) 'Biodiversity conservation in Nigeria: Prospects and challenges', *Journal of Research in Forestry, Wildlife and Environment*, 16(1), pp. 68–75.

Ordóñez-Vargas, L., Gonzalez, L. C. P., and Prieto-Rios, E. (2023) 'An econcentric turn in the transitional restorative justice process in Colombia', *International Journal of Transitional Justice*, 17(1), pp. 107–122. https://doi.org/10.1093/ijtj/ijad003

Parker, S., and Khare, A. (2005) 'Understanding success factors for ensuring sustainability in ecotourism development in Southern Africa', *Journal of Ecotourism*, 4(1), pp. 32–46. https://doi.org/10.1080/14724040508668436

Piccolo, J. J., Taylor, B. Washington, H., Kopnina, H., Gray, J., Alberro, H., and Orlikowska, E. (2022) '"Nature's contributions to people" and peoples' moral obligations to nature', *Biological Conservation*, 270, p. 109572. https://doi.org/10.1016/j.biocon.2022.109572

Piccolo, J. J., Washington, H., Kopnina, H., and Taylor, B. (2018) 'Why conservation scientists should re-embrace their ecocentric roots', *Conservation Biology*, 32(4), pp. 959–961. https://doi.org/10.1111/cobi.13067

Pratt, M. G., and Ashforth, B. E. (2003) 'Fostering meaningfulness in working and at work', in K. S. Cameron, J. E. Dutton, and R. E. Quinn (eds.), *Positive Organizational Scholarship: Foundations of a New Discipline*. San Francisco, CA: Berret-Koehler, pp. 309–327.

Ripple, W. J. et al. (2017) 'World scientists' warning to humanity: A second notice', *Bio Science*, 67(12), pp. 1026–1028. https://doi.org/10.1093/biosci/bix125

Roka, K. (2020) 'Community-based natural resources management', in W. Leal Filho, *et al.* (eds.) *Life on Land: Encyclopaedia of the UN Sustainable Development Goals*. Cham: Springer International Publishing, pp. 161–174.

Ruales, J. G., Hovden, K., Kopnina, H., Robertson, C. D., and Schoukens, H. (eds.) (2024) *Rights of Nature in Europe: Encounters and Visions*. New York: Routledge.

Schmidt, C. (2004) 'The analysis of semi-structured interviews', in U. Flick, E. von Kardoff, and I. Steinke(eds.) *A Companion to Qualitative Research*. London: Sage Publications Ltd, pp. 253–259.

Shoreman-Ouimet, E., and Kopnina, H. (2015) 'Reconciling ecological and social justice to promote biodiversity conservation', *Biological Conservation*, 184, pp. 320–326. https://doi.org/10.1016/j.biocon.2015.01.030

Sobere, N. U., and Ihua-Madunenyi, B. (2019) 'The impact of human population on biodiversity conservation in Nigeria: The need for legal intervention', *Journal of Law Policy and Globalization*, 92, pp. 156–166. https://doi.org/10.7176/JLPG/92-16

Taylor, B., Chapron, G., Kopnina, H., Orlikowska, E., Gray, J., and Piccolo, J. J. (2020) 'The need for ecocentrism in biodiversity conservation', *Conservation Biology*, 34(5), pp. 1089–1096. https://doi.org/10.1111/cobi.13541

Wallach, A., Batavia, C., Bekoff, M., Alexander, S., Baker, L.; Ben-Ami, D., Boronyak, L., Cardilini, A., Carmel, Y., Celermajer, D., Coghlan, S., Dahdal, Y., Gomez, J., Kaplan, G., Keynan, O., Khalilieh, A., Kopnina, H., Lynn, W., Narayanan, Y., Riley, S., Santiago-Ávila, F., Yanco, E., Zemanova, M., and Ramp, D. (2020) 'Recognising animal personhood in compassionate conservation', *Conservation Biology*, 34(5), pp. 1097–1106. https://doi.org/10.1111/cobi.13494

Washington, H., Chapron, G., Kopnina, H., Curry, P., Gray, J., and Piccolo, J. J. (2018) 'Foregrounding ecojustice in conservation', *Biological Conservation*, 228, pp. 367–374. https://doi.org/10.1016/j.biocon.2018.09.011

Washington, H., Piccolo, J., Gomez-Baggethun, E., Kopnina, H., and Alberro, H. (2021) 'The trouble with anthropocentric hubris, with examples from conservation', *Conservation*, 1(4), pp. 285–298. https://doi.org/10.3390/conservation1040022

Washington, H., Piccolo, J. J., Kopnina, H., and Simpson, F. O. L. (2024) 'Ecological and social justice should proceed hand-in-hand in conservation', *Biological Conservation*, 290, p. 110456. https://doi.org/10.1016/j.biocon.2024.110456

Wilson, E. O. (2016) *Half-Earth: Our Planet's Fight for Life*. New York: Liveright, W.W. Norton & Company.

Zungum, I. U., Imam, T. S., Ahmad, A. G., Abubakar, T., Benjamin, B., and Daya, M. G. (2019) 'Impact of typha grass on biodiversity loss of hadejia-nguru wetland located between Jigawa and Yobe states of Nigeria: A review', *Journal of Applied Sciences & Environmental Management*, 23(10), pp. 1847–1853. https://doi.org/10.4314/jasem.v23i10.13

12 Ecophenomenology and Deep Ecology

Toward an African Lived-Ecology of the Human–Non-human Relationship

Kenneth Uyi Abudu

Introduction

The human-non-human relationship is regarded as one of the central subjects of most theories of environmental conservation – deep ecology included. The environment within and across cultures, we can agree, is made up of both human and non-human entities alike. While in some cultures, this relationship is considered as that of a "master-slave" on the basis of exploitation, some cultures consider this relationship as relational. The question of whether or not humans relate to non-humans in a relational, intrinsic sense, or exploitatively, has taken different dimensions. The obvious reason for this is that there have been concerns worldwide regarding cases of deforestation, destruction of rainforests, poaching, endangering species of various plants and animals, to mention a few. All these negative occurrences allude to the position that the human-non-human relationship is declining and continues to decline. In all of this, of particular concern to philosophers, especially philosophers of the environment across the world, is the quest to rethink the relationship between the human and non-human in ways that ensure that this relationship is preserved, and the ecosystem conserved.

The quest to address this declining human-non-human relationship and, by extension, mitigate the lasting consequences of this declining relationship on the natural environment has led philosophers to postulate several theories. Environmental Ethics, Environmental Pragmatism, Deep Ecology, and Eco-Phenomenology, to mention a few, are part of the theories that have been put forward by philosophers, especially Euro-American philosophers. This is not to say that African philosophers have not been engaging in theorising the need to rethink the human-non-human relationship in the ecosystem. For instance, in African traditions, we can talk of Segun Ogungbemi's "Ethics of Nature-relatedness" (1997); Thaddeus Metz's "An African Theory of Moral Status: A Relational Alternative to Individualism and Holism" (2012); and Kevin Behrens' "An African Relational Environmental and Moral Considerability" (2014), to mention a few. From Segun Ogungbemi to Kevin Behrens, a cursory look at their theories of the environment alludes to their emphasis on both ethical and ontological imperatives when it comes to the human-non-human relationship. The ethical imperative of the human and non-human relationship is simply based on the argument that we as humans have a moral obligation to protect and conserve the natural environment, while

DOI: 10.4324/9781003480914-13

the ontological imperative suggests that all beings, humans and non-human, are interconnected, and as such, humans are not expected to exploit the non-human since we are all in one chain of being. The implication of this is that while there has been emphasis on the ethical and ontological imperatives in addressing the declining human-non-human relationship, little or nothing has been done in the aspect of developing an ecosophy that bridges the gap between Ecophenomenology and Deep Ecology, especially in relation to African traditions. Ecophenomenology is seen as the attempt to synthesise ecological thinking with phenomenology, where phenomenology is understood as the study of the structure of human experience. It is an attempt to employ phenomenological methods and insights in the quest to address the decaying ecological situation worldwide. Thus, phenomenology, led by its own momentum, becomes a philosophical ecology, that is, a study of the interrelationship between organism and world in its metaphysical and axiological dimensions (Brown, 2003, xii). Here, ecophenomenology seeks to blend both the ethical and ontological imperatives in ensuring a balanced ecology. Deep ecology, on the other hand, is concerned with three central ideas: The first is that we should ask deeper questions about our relationship with the world around us. Particularly, it is this idea that gives deep ecology its name. The second central idea of deep ecology is the attempt to replace anthropocentric (human-centred) forms of thinking, valuing, and acting with ecocentric (ecology-centred) forms of thinking, valuing, and acting, while the third idea is that of developing a particular kind of approach to our personal relationship with the world around us that Arne Naess refers to as "Self-realization!"(Fox, 2020).

Central to the whole discourse of ecophenomenology and deep ecology, flowing from the above, is that through our ontic experience as humans, we recognise that the ecosystem does not consist of the human species alone, but also other non-human. However, Taodvine noted that there are two salient contradictory positions regarding the relationship between the human and non-human in the ecosystem: they are that humans are part of nature; and at the same time, humans are separated and alienated from nature (2021). This, no doubt, presents a paradox that relies mainly on how we experience the world as humans in our different spaces. Since the way humans experience the natural environment differs across the world, it could be inferred that both ecophenomenology and deep ecology have a tendency to embrace interdisciplinary and multidimensional approaches, which sometimes involve the use of cultural and/or indigenous constructs to address environmental problems, with the African traditions in mind here.

My concern in this chapter is to develop an often-considered underexplored area of African environmental philosophy, *African lived-ecology*, an ecosophy that integrates eco-phenomenology, deep ecology, and some practices of conservation in African traditions for the sake of maintaining an ecological balance between the human and non-human in the ecosystem. The adoption of phenomenology as a method of inquiry in interrogating the environment, though recent, necessitated an *ecological turn* in the human-non-human relationship. Particularly, it ensured a novel approach, beyond the ethical and ontological imperatives, in showing that our lived experiences of nature should dictate how we relate to and preserve it – an idea central to deep ecology. Having this point in mind, I begin first by showing the intersection between eco-phenomenology and deep ecology. Deep ecology presents a kind of environmentalism that prioritizes the flourishing and well-being of human and non-human in the ecosystem. In other words, deep ecology affirms that both humans and non-humans have intrinsic values, and the exploitation of either could lead to ecological disharmony. In what follows, I proceed to examine some practices such as totems and taboos, and how they form the basis for conserving the non-human in African traditions. I then proceed to develop *African lived-ecology* – an ecosophy that integrates eco-phenomenology, deep ecology, and some conservation practices in African traditions. I conclude by showing how the *African lived-ecology* presents a viable ecosophy for exploring and conserving the human-non-human relationship, not just for immediate gains, but also for future generations.

Ecophenomenology and Deep Ecology: Ontic Experience for Lived-Ecology

Ecophenomenology, also known as ecological phenomenology or phenomenological ecology, is considered a recent approach to environmental philosophy and specifically, the human-non-human relationship that seeks to employ the methods of phenomenology in the quest to rethink the current environmental crises being witnessed the world over. As an environmental philosophy that combines ecology and phenomenology, ecophenomenology seeks to ground our conservation of the environment as humans on our ontic experience of the ecosystem. Ecophenomenology, sharing a similar background with phenomenology as a method of philosophy, attempts to challenge the dominance of scientism and/or naturalism in the quest to address environmental challenges. In the context of phenomenology, Edmund Husserl had challenged the dominance of naturalism over rationalism and argued that philosophy should become a rigorous science that concerns itself with the structure of experience. Husserl asserts that "from the beginning, naturalism has resolutely pursued the idea of a rigorous scientific reform of philosophy and even believed at any given time, both in its earlier and modern forms, that it has already realized this idea" (1931, p. 253). Husserl's thought, therefore, extends to the scientific conception of the ecosystem and particularly the human-non-human relationship. Specifically, the implication of the scientific conception of the environment led to a plethora of arguments that science and technology are responsible for the decline in the ecosystem as we have it today. This explains why eco-phenomenology rejects naturalism in the discourse of the environment, and by this, it seeks to ground all our experiences of the ecosystem in reason and truth. For Neil Evernden, "those approaches to nature that strip it of all experienced qualities leave us with an unrecognizable abstraction, and certainly not with any version of nature that could have inspired our initial appreciation" (1993, p. 1).

According to Michael Zimerman (1997), Martin Heidegger agreed that naturalism could be held responsible as one of the factors for the current environmental crisis across the world, which ecophenomenology seeks to address. According to him:

> Is the scientific understanding of nature itself part of the problem? As Heidegger has maintained … modern science may be itself an integral part of the very problem that is defined as "environmental crisis." Heidegger is even partly right in his claims that science is inseparable bound up with the modern project of the technological domination of nature – that science like technology "sets upon" nature aggressively and (as the rhetoric of Bacon and Kant would have it) "tortures" nature to give answers to questions of their own devising – then an exclusively, or even predominantly scientific understanding of nature would prevent us from even comprehending what is at stake, let alone at a salutary solution.
>
> (1997, p. 138)

Ecophenomenology, to Taodvine, is such that it will ensure a revision of what he understands to be a dichotomous relation between the natural and transcendental attitudes. In order to overcome this dichotomy, he proposes that:

> Phenomenology needs environmental thinking … it requires the insights inspired by our current ecological consciousness to revise its own methodological procedures. In fact, a thoroughgoing phenomenology of nature leads inevitably into a phenomenology of phenomenology, since it must provide an account of the relation between transcendental and mundane subjects, between reflective and pre-reflective experience, and between ontological and ontic modalities of the world.
>
> (1999, p. 125)

What can be gleaned from the above excerpt is that phenomenology through an ecological lens presents a novel approach to the human-non-human relationship. In recent times, thinkers such as Charles S. Brown and Ted Taodvine have argued that eco-phenomenology is important in the quest to understand the environment beyond ontological and ethical imperatives. In particular, Brown, for instance, states that the phenomenological method of "bracketing" presents the realistic assumptions of everyday consciousness and by this, "we are in a position to see that the world and the things in it are only given to us through the interpretative and meaning-bestowing function of our intentional acts" (Brown 2003, p. 6). The essence of this bracketing as a phenomenological method in the interrogation of the ecosystem is to project an attitude of respect for our ability to experience the environment as it is.

According to Violeena Deka, the essence of ecophenomenology is to "understand the human-nature dialogue in a new light; to study 'nature' as it appears to a human experience as a 'meant object'. Its attempt is to understand the 'meaning', 'value', 'signification' and 'motivation' that underlies human experiences of nature" (Deka, 2019, 21). Thus, as an ecological philosophy, ecophenomenology is not only concerned about nature, but how we as humans experience it. In the quest to reimagine the ecosystem, ecophenomenology begins with a viewpoint of respect for experience. This is similar to Henry Thoreau and John Muir who characterized the world in such a manner that experiencing it is an essential component of it, so demonstrating the wider potential of experience. Therefore, ecophenomenology critiques earlier environmental theories that function from a naïve point of view and describes the forms and patterns of experience. Returning to the "things themselves" is the way to describe experience. This is important so that humans could gain their lost touch with nature. According to David Abraham, the essence of ecophenomenology is to show the palpable difference between the experienced world, or worlds, of indigenous cultures as distinct from the Euro-American civilizations. Eco-phenomenology, therefore, is seen as the Western philosophical tradition that has most forcefully called into question the modern assumption of a single, wholly determinable, objective reality (1996, p. 31). For Charles S. Brown and Ted Taodvine, the basis of ecophenomenology is that without having an adequate experience of the ecosystem, appreciating it may continue to elude us. Consequently, we must strip ourselves of all the scientific and naturalistic tendencies toward the ecosystem. They sum this argument up by stating that:

> From the beginning, phenomenologists have taken an interest in this process of abstraction by which the world as we experience it is gradually transformed into the naturalistic conception of the world taken for granted by contemporary science. One point of agreement among phenomenologists is their criticism and rejection of the tendency of scientific naturalism to forget its own roots in experience. The consequence of this forgetting is that our experienced reality is supplanted by an abstract model of reality—a model that, for all of its usefulness, cannot claim epistemological or metaphysical priority over the world as experienced. The return to "things themselves" and the critique of scientific naturalism both point in the direction of much contemporary environmental thought.
>
> (2003, ix)

Evidently, phenomenological ecology and/or ecophenomenology is auspicious in its attempt to provide a replacement for the reductionist and scientific understanding of the natural environment. In other words, eco-phenomenology brings to bear the obliteration of our romance with objectivity, our idea that values are strictly anthropocentric, and other legacies that were laid by the famous French thinker René Descartes. Monika Langer (1989) corroborates this by asserting that the Cartesian worldview dichotomises reality, thereby denying reciprocity to the non-human, while also making humans dominate nature. As a result, ecophenomenology seeks to challenge this ontological divide by affirming the value of interrelationships between humans and the non-human world. Eco-phenomenology is also seen as a form of ecological turn that is not only considered a phenomenological approach to the human and nature relationship beyond the ethical and ontological imperatives, but it is considered an ecological turn as it seeks to employ phenomenological concepts and methods in the quest to interrogate an experiential relationship between humans and nature. When it comes to contemporary environmental challenges, ecophenomenology is seen as a novel approach. Ecophenomenology aims to investigate the causes of environmental crises, going beyond the parameters of conventional environmental philosophical inquiries. This entails reevaluating our basic conceptions of nature, the self, and the relationship between humans and non-humans as well as issues of epistemology, ontology, and aesthetics. The approach used in ecophenomenology specifically considers our firsthand experiences to determine what it implies to be an aspect of an ecosystem. According to ecophenomenologists, we must start with our perception of nature in order to respond to environmental challenges. How people perceive the natural world in the pre-reflective subjective phase is the fundamental question, not what the relationship between humans and nature should be. Ecophenomenologists might thereby overcome the surface-level sense of detachment from nature by using phenomenological approaches to uncover the potential of devoting close consideration to the encounter with nature. Ecophenomenologists contend that we are disconnected from the real experience of our surroundings by our anthropocentric conceptions of subjectivity, scientific objectivity, materialism, and minimality. We uncover new aspects of the significance and worth of nature as we rediscover our perceptual, bodily, and intersubjective relationships with the world around us.

According to David Wood, our relationship with nature can be assessed (or rather, re-assessed) by carrying out a phenomenological critique of the structure of that relationship. How we live meaningfully with the natural world is the domain of eco-phenomenology (Wood, 2003: 213). According to Brown and Taodvine, ecophenomenology is defined as the intersection of ecological thinking with phenomenology; the momentum that drives each toward the other begets a new cross-disciplinary inquiry. In furtherance of this, Brown and Taodvine unequivocally affirmed this by stating that:

> Eco-phenomenology is based on a double claim: first is that an adequate account of our ecological situation requires the methods and insights of phenomenology; and, second, that phenomenology, led by its own momentum, becomes a philosophical ecology, that is, a study of the interrelationship between organism and world in its metaphysical and axiological dimensions.
>
> (Brown and Taodvine 2003, xiii)

The essence of eco-phenomenology as advanced by Brown and Taodvine is to ensure that humans' experience of nature is stripped of all forms of naturalistic biases. Hence, their book *Eco-Phenomenology: Back to Earth Itself* examines the intersection between phenomenology as a philosophy of consciousness and experience in relation to the natural environment. Summarily, eco-phenomenology aims to investigate things exactly as they are – that is, without bias or prejudice. Secondly, it aims to get beyond the dualistic notion that there is a binary relationship between humans and the natural world. Put another way, there is a connection between humans and the natural world; the third issue is demonstrating that humans' intentionality is constantly directed not only toward humans but also toward other non-human entities in the environment.

Deep ecology, grounded on ontological relationality, is a monist environmental philosophy that states no dichotomy between the subject (human) and the object (non-human). It argues that the flourishing of all human and non-human entities in the ecosystem is necessary for environmental sustainability. According to Bill Devall:

> What I call deep ecology in this paper is premised on a gestalt of person-in-nature. The person is not above or outside of nature. The person is part of creation on-going. The person cares for and about nature, shows reverence towards and respect for nonhuman nature, loves and lives with nonhuman nature, is a person in the "earth household" and "let being be," let nonhuman nature follow separate evolutionary destinies.
>
> (Devall 1980, p. 303)

Deep ecology was coined by Arne Naess, a Norwegian philosopher and environmentalist in "The Shallow and the Deep, Long-Range Ecology Movement: A Summary" (1973), where he describes deep ecology as an ecological philosophy that recognizes that all living beings in the ecosystem have intrinsic value, irrespective of how humans conceive them. Naess conceptualises deep ecology as a deeper questioning about human life, society, and nature that goes beyond the so-called factual scientific level of self and earth wisdom. The position Naess is emphasising here is that humans will view nature from a vantage position if and only if they avoid seeing nature as something meant to serve their purpose. This explains why he further states that "the essence of deep ecology is to ask deeper questions. Naess goes on further: the adjective "deep" stresses that we ask why and how, where others do not" (Naess 1973, p. 43). Deep ecology as an environmental philosophy questions humans' approaches to economy and value priorities, especially as it relates to the exploitation of the natural environment. In other words, deep ecology has to do with a conscientious rising of ecological awareness on the basis of a philosophical and spiritual attitude that involves the unity of all living things in the natural environment. Deep ecology stresses the need for a paradigm shift from the anthropocentrism that has over time characterized humans' relationship with the non-humans, to a more biocentric one. It has been argued over time that deep ecology maintains a non-anthropocentric standpoint when it comes to the relationship between humans and the non-human. This explains the sharp distinction between deep ecology and shallow ecology. While deep ecology propagates non-anthropocentrism, shallow ecology encourages anthropocentrism as it often protects the interests of the affluent, especially in developed countries.

The core principle of deep ecology as an ecophilosophy states that there ought to be an interdependent relationship among organisms, that is, all organisms, humans included, are interdependent. Deep ecology also suggests that the undue interference of humans in the natural order of things, especially as it has to do with the exploitative tendencies of humans in the natural world, poses a threat. According to Naess, deep ecology may be construed from different perspectives, but there are some inherent principles that unite these perspectives. These principles are eight (8), and they are as follows:

1. The well-being and flourishing of human and non-human life on Earth have a value in themselves. These values are independent of the usefulness of the non-human world for human purposes.
2. The richness and diversity of life forms contribute to the realisation of these values and are also values in themselves.
3. Humans have no right to reduce this richness and diversity except to satisfy vital needs.
4. The flourishing of human life and cultures is compatible with a substantial decrease in the human population. The flourishing of non-human life requires such a decrease.
5. Present human interference with the non-human world is excessive, and the situation is rapidly worsening.
6. Policies must therefore be changed. These policies affect basic economic, technological, and ideological structures; the resulting state of affairs will be deeply different from the present.
7. The ideological change is mainly that of appreciating life quality (dwelling in situations of inherent value) rather than adhering to an increasingly higher standard of living. There will be a profound difference between big and great.
8. Those who subscribe to the foregoing points have an obligation directly or indirectly to try to implement the necessary changes (Naess 2005).

The above principles of deep ecology as outlined by Naess mean that deep ecology offers a perspective to environmental philosophy that is not limited to the general Western ecological knowledge, which places emphasis on the superiority of man over other living things in the environment. Deep ecology maintains a non-anthropocentric view about nature. Furthermore, the advocates of deep ecology argue that environmental philosophy must take into cognisance the idea that nature not only has inherent value (though contestable) but also that it exists independently of human wants and desires. In summary, therefore, the following can be said about deep ecology: First, deep ecology is strongly against anthropocentrism. Secondly, deep ecology calls into question the prevalent worldview in the Western world, which sees man as the fundamental basis of values; and thirdly, deep ecology advocates that ecocentrism should replace anthropocentrism. Ecocentrism suggests that the biosphere is the centre of value, not man. In other words, ecocentrism promotes a holistic idea of value that sees humans and non-human species as relational.

The intersection between eco-phenomenology and deep ecology lies primarily in our ontic experience of the world, that is, experiencing the world as it is for the sake of conservation. For phenomenologists, experience is important as it marks the beginning of all philosophical evidence. By this, its intersection with deep ecology as an environmental theory that deals with human-non-human relationships lies primarily in experiencing the natural environment as it is without any form of prejudice. According to Deka, "patient description of experience and of the means by which abstract scientific knowledge is constructed from experience is an essential prolegomenon to reconnecting our relationship with nature" (2019, p. 50). True to Deka's position, our lived experiences as humans point to the fact that we live in an environment amongst other non-human entities, and any attempt to dominate the non-human in our immediate environment has long-term implications for us as humans. Thus, the intersection between ecophenomenology and deep ecology can be summed up by stating that our ontic experience of our immediate environment is important in recognising the need to cultivate attitudes that promote the flourishing of all, human and non-human.

Christian Diehm interviewed Arne Naess and subsequently published an article entitled "Deep Ecology and Phenomenology" as a follow up to this interview. In his article, Diehm pointed out that Naess agreed that phenomenology was necessary for formulating new ontologies for a reorientation about the environment. Particularly, Naess highlighted the importance of phenomenology and its intersection with deep ecology in *Ecology, Community and Lifestyle*. The significance of a phenomenological understanding of the environment is "to give a description of the immediately apprehended" (1989, p. 51). Such a description, says Naess, brings about "no epistemological ego reaching out to see and understand (1989, p. 66), no subject detached from a world of *Ding-an-sich* into which secondary and tertiary qualities are projected. Instead, phenomenological reflection discovers the subject already at home in the world, immediately involved with a richly textured, meaningful "life-world" (Diehm, 2004, p. 22). Phenomenology as a method in philosophy is common in the works of Maurice Merleau-Ponty, Jean-Paul Sartre, Gabriel Marcel, and Martin Heidegger, to mention a few. It is, however, interesting to note here that Naess found the phenomenology of Heidegger worthy in his attempt to show its intersection with ecophenomenology. In *Four Modern Philosophers*, Naess explains Heidegger's approach to phenomenology as:

> Heidegger describes his own method as phenomenological. His aim is a pure description of state of affairs – with no explanations, theories, or hypothesis. The phenomenologist's motto is "to the facts!" What he describes, therefore, must be something immediately intuitable for the uninhibited and unprejudiced observer.
>
> (1969, pp. 192–193).

The Heideggerian conception of phenomenology, writes Naess, "cuts across the distinction between me and the world" and thus catches sight of something that cognitively precedes the subject-object distinction" (p. 196). Evidently, Naess relied on Heidegger's conception of phenomenology in establishing an intersection between eco-phenomenology and deep ecology. Here, two things could be gleaned: one is that the ecosystem is to be described as it is, and secondly, there exists no dichotomy between the self and nature – being one of the metaphysical groundings for deep ecology.

According to Ecce de Jonge, deep ecology has four approaches: *Deep Questioning, A Metaphysical Ground for an Environmental Ethics, A Political Movement*, and *An Activist Approach to the Environmental Crisis*. Of all these approaches, the one that connects with my aim in this chapter, especially the intersection between ecophenomenology and deep ecology, is the second approach: *A Metaphysical Ground for an Environmental Ethics*. For this approach, it is that there is no distinction between the self and the other; particularly, "a deep ecological approach to metaphysics of nature differs from an environmental ethics" (Jonge, p. 3). Thus, just as ecophenomenology seeks to bridge the ontological dichotomy between the human and the non-human, deep ecology seeks to affirm the metaphysical grounding for all entities in the ecosystem for the sake of flourishing. This is possible through self-realization. In the opinion of Warwick Fox, the deepest idea in deep ecology in its intersection with ecophenomenology is its "esoteric core" – which lies within the doctrine of self-realization (1990, p. 9). Thus, ecophenomenology in relation to deep ecology does not only attempt to challenge the dominance of anthropocentrism, but aims to do so through the doctrine of self-realization. Self-realization is considered a kind of awareness and a structural feature of conscious experience. Particularly, Naess describes self-realization as the basic norm of deep ecology. Through the process of self-realisation, we are able to care for ourselves, and by extension, nature. For Jonge, "Self-realization" leads to a non-anthropocentric attitude of care, obtained through various methods of identification" (2004, p. 55). As a concept that maintains a non-anthropocentric outlook, self-realisation therefore aims to show that entities that make up the ecosystem have intrinsic value. An objection that is tenable here is, to what extent can we say all beings possess self-realisation? Or if all beings possess self-realisation, then what does it imply? On these questions, we can also argue that no sort of co-existence between the human and non-human takes place outside the context of self-realisation as it depends on the existence of other beings to realise its own values. Evidently, grounding deep ecology in eco-phenomenology is important for some reasons: First, it brings the consciousness that as humans, we do not exist in a vacuum; that is, we exist in a factual natural world with other non-humans. Second, we have a duty to protect the non-human for the sake of ecological balance, and third, self-realisation confers intrinsic values on all human and non-human entities, and as such, none should be seen as a means to an end. In all, the importance of integrating eco-phenomenology and deep ecology is to show, amongst other things, that the ontic priority of our perception in establishing that our consciousness as humans recognises the existence of an external world (natural environment) that has intrinsic values.

Conserving the Non-human in African Traditions

In this section, I examine some traditional practices involved in conserving the non-human in African traditions. I rely mainly on non-anthropocentric practices such as totemism, taboos, and the sacredness of the non-human, such as forests, and argue that these practices have spiritual cum ontological import, that is, the basis of conservation in African traditions is the argument that all entities, human and non-human, have what is generally referred to as *life force*, and as such, maintaining ecological equilibrium relies mainly on conserving and showing respect for the non-human. It is instructive to note here that my use of the term "African" here represents the "Bantu-speaking peoples residing in the sub-Saharan part of the continent, thereby excluding the Islamic Arabs in North Africa and white Afrikaners in South Africa, among others" (Metz, 2007, p. 322).

Some thinkers such as J.S. Mbiti (1969), J.B. Callicott (1994), and G. Onah (2009), have argued that African thought is anthropocentric due to its emphasis on man being the fundament of ontology. Particularly, their argument that African thought is anthropocentric is inextricably linked to the African foundational principle of Ubuntu, which states that "a person is a person through other persons". In this context, to them, Ubuntu appears to recognise only humans while occluding the non-human. However, their argument seems untenable as it is obvious that they only pay attention to the ontic experience of Ubuntu and not its ontological import. The ontic experience of Ubuntu focuses on humans alone by recognising that the "person" refers to only humans. However, the ontological import of Ubuntu recognises the existence of the non-human and encourages us as humans to live in harmony with the non-human. Thus, a "person being a person through other persons" does not recognise humans alone as humans. It is believed that humans do not exist in vacuum in African worldview. It is on this basis that Michael Eze stresses that to be human, it is not sufficient to be a living rational or intelligent being; the criteria include a recognition that one's humanity is dependent on harmonious balance and positive relationships with human beings, animals, biological life, non-biological life, spirits, forces, and other inanimate elements that make up our environment (Eze, 2017, 629). True to Eze's assertion, African ontology is seen as a relational ontology that emphasises the co-existence of both humans and non-human entities alike. Thus, in the quest to show that humans recognise the existence of the non-human and owe the non-human a duty to protect, Africans have over time developed some non-anthropocentric practices that emphasise this argument for the sake of conserving the ecosystem. For instance, there are practices such as totemism, taboos, and the protection of forests that are considered sacred. These practices are important to some communities in Africa and they are meant to ensure that the non-humans are respected and preserved. It is instructive to note here that conserving the natural environment in African traditions is not a recent phenomenon. Past generations were aware of environmental degradation and the need for conservation. The practice of conservation in African traditions reflects belief systems such as totems, taboos, rituals, proverbs, norms, myths, and so on. Particularly, the need for conservation in African traditions is necessitated by the belief among Africans that the ecosystem has a spiritual basis which is rooted in the idea that human and non-human in the natural environment were created by the Supreme Being (God) for the purpose of harmonious continuity, and as a result, the practice of conservation is not only seen as both a moral and ontological obligation, but also for the flourishing of the human and non-human in the ecosystem. God is seen as the being that penetrates all things in the environment, living and non-living. As

Leopold Senghor puts it: "the whole universe appears as an infinitely small, and at the same time an infinitely large network of life forces which emanate from God and end in God who is the source of life. It is He who vitalises and devitalizes all other beings, all the other life forces" (1995, p. 49). Ama Mazama resonates with Senghor on the importance of God who manifests as a life force for both human and non-human entities. She asserts that the essence of life force in African ontology is that it permeates and lives within all that is – human beings, animals, plants, minerals, and objects, as well as events. This common energy shared by all confers a common essence to everything in the world, and thus ensures the fundamental unity of all that exists. This energy constitutes the active, dynamic principle that animates creation, and which can be identified as Life itself (2002, p. 219). The consequence of this is that the non-human must be treated with respect, empathy, and reverence, for whoever exploits their being diminishes their life force and does so against God.

Among the Shona people of Zimbabwe, it is believed that spirits operate in the human world through animals, birds, and fish. The implication of this is that an animal that is considered a totem in a clan is treated with reverence. Consequently, each clan has its own taboos and restrictions towards particular animals (Taringa, 2006, p. 205), and as such, any transgression against these animals is generally believed to provoke the ire of the spiritual world (Chemhuru & Masaka, 2010, p. 122). Thus, the practice of conservation as it pertains to Africa is grounded in African spiritual philosophy, a philosophy considered relational. One of the seminal works to have been written by an African philosopher on the practice of conservation which reflects totemism is *Isilwane: The Animal* (1996), a book by Vusamazulu Credo Mutwa. In this book, Mutwa maintains a deep ecological approach to conservation and by extension, the non-human by stating that Africans did not consider themselves to be above other non-humans in the ecosystem. In his words:

> In old Africa we ... believed that we had nature within and beyond ourselves. By making us believe that the highest gods were part animal and part human being, we were taught to look upon animals with great reverence, love and respect (1996, pp. 13, 14). ... The native people of Africa regarded them as a blessing from the gods – as something unbelievably sacred and vital for the continued existence of human beings. Black people believed that animals were the blood of the earth and that as long as there were migrations, criss-crossing the country, human existence on earth was guaranteed.
>
> (1996, p. 15)

In furtherance of the above excerpt, Ikuenobe (2014) writes that "traditional Africans see mountains, trees, rivers, and different animals as representations or embodiment of deities or spirits, and as such, they are divine, sacred, and are given due reverence" (2014, 5). No doubt that "deities", "spirits", "mountains", and "trees" are seen as part of the African ontological hierarchy. Among the Nso people of Cameroun, Godfrey B. Tangwa states that there exists no dichotomy between "plants, animals, and inanimate things; between the sacred and the profane, matter and spirit, the communal and the individual" (2004, p. 389). Thus, traditional Africans maintain a conservationist approach in their dealings with living and non-living entities in the environment. This position is underscored by some scholars' positions that the treatment of some animals and other non-living things in the African space is based primarily on totemism. According to Murugesan Amirthalingam, totemism is defined as "a mystical or ritual relationship among members of a specific social group and specie, of plants or animals" (2014, p. 22). For F. Abiola Irele and Biodun Jeyifo, totemism is "a system of thought and social practice" that is "based on a relationship between a clan and/or a person and an animal or thing" (Irele & Jeyifo, 2010, p. 380). Irele and Jeyifo further note that totemism became important for the sake of humans establishing a mutual relationship with the non-human in the natural world that they inhabit (2010, p. 38). This implies that some animals are regarded as sacred and by extension, totemized. Regarding how certain animals are totemized for the purpose of conservation in African traditions, Burnett and wa Kang'ethe state that in the Bantu thought system, a totemized animal "was not killed except in self-defense, and it was never eaten. Kindness toward the totem was expected, and in turn, it was expected that the totem might assist in times of need or desperation" (1994, p. 156). According to Fainos Mangena, humans are regarded as the fundament of the African ontological hierarchy, but they cannot exist "without association with non-human through totemism and spiritualism" (2013, p. 30). In the traditional Igbo conception of the environment, Chigbo Joseph Ekwealo points out that "animals are believed to have souls and spirits like human beings and they are therefore respected. There is an agreement that they possess certain psychic qualities that are explication of the spirit in them" (2012, p. 266). This explains why in some Igbo communities, they forbid the killing of some animals such as pythons. Regarding this, Ikechukwu Kanu argues that "no one makes the mistake of killing it (python) voluntarily or involuntarily. When it is done accidentally, it is buried with appropriate religious rituals and rites accorded to it. If it is killed knowingly, it is considered an abomination" (2017, p. 131.). In the Yoruba thought system, totemism is often practised with reference to what is generally referred to as *igbo oro* or *igbo ai wo*. *Igbo oro* in the Yoruba thought system is considered a sacred forest that houses other non-humans that are spiritual in their spaces. Here, trees cannot be cut down indiscriminately without carrying out some rituals. Humans are not allowed to carelessly farm or carry out destructive activities in this forest. The implication this sacred forest has for conservation in the Yoruba space is that the non-human entities there cannot be killed or exploited. In some cases, some forests are tagged "sacred" because the people do not want such forests to be tampered with. By merely describing it as a "sacred forest", people who have anthropocentric/destructive motives would be scared to carry out any non-conservative activity within such forests.[1]

The implication of the various conceptions of totemism in African traditions for the non-human is the realisation that they possess intrinsic values and, by extension, moral status, that is, moral status is granted to non-humans on the basis of the intrinsic values they possess. For instance, if a tree is considered a totemized tree, it implies that such a tree has a higher ontological status, especially when compared to other non-humans, such as stones in the natural environment. Kevin Behrens reiterates this position by stating that:

> African thought extends moral considerability to include all things that are part of interconnected web of life, that is, all individual living things, groups of living things such as families, species and ecosystems, as well as inanimate natural objects such as rivers and mountains.
>
> (2014, 66)

Thus, in an instance where an individual whose clan's totem animal is an elephant, he is prohibited from hunting and killing elephants for food. Such transgression would result in illness or in the loss of the transgressor's teeth (Taringa, 2006, p. 206). This goes for not only animals but other living and non-living entities in the environment. Evidently, an African understanding of how animals and other non-human entities in the environment are to be treated suggests the values of conservationism and the flourishing of all. The obvious reason for this is because "there is no such thing as dead matter: every being, everything – be it only a grain of sand – radiates a life force, a sort of wave-particle; and sages, priests, kings, doctors, and artists all use it to help bring the universe to its fulfillment" (Senghor, 1995, 49). The practice of conservation, which sometimes reflects in totemism in African traditions, is essential for the sake of human survival. Mutwa emphasises this position by pointing out that:

> Africans did not hunt animals for fun. They were hunted for food and for religious reasons... The African people know, just as the Native American people knew, that if you would destroy the environment, you will ultimately destroy the human race.
>
> (1996, p. 19)

In another instance, some animals are totemized in African traditions on the basis that ancestors once assisted them in times of crisis using that particular animal. Some scholars have argued that taboos go with totemism in the attempt to conserve the natural environment in African culture. Killing a totemized animal is a taboo. However, taboos are also meant to conserve endangered species or animals that are considered rare. For example, pythons are often protected, not because they are endangered species, but because they are "symbolic in foretelling events of what the future holds for human beings" (Chemhuru & Masaka, 2010, p. 131) and because killing them will invite a drought. The reason Shona people are prohibited from killing pythons is inextricably linked to the indigenous practice of royal python veneration. In Ogun, a state in southwestern Nigeria, pythons are venerated because they serve "as a means of biodiversity conservation ... because the preservation of groves for the python" also means "the preservation of a wide range of other plant and animal species" (Olaoluwa, 2014, pp. 1, 7). For the Shona people, conserving the python is mirrored in what they call the "Royal python veneration", a practice that makes it possible to approach the python "whenever there is drought". Thus, it is believed in the Shona tradition that "if you kill a python, rain will not fall" (Chemhuru & Masaka, p. 131).

Beyond animals, trees in African space are also conserved on the basis that they are useful for health practices. This explains why Julius Gathogo (2013) says that among the people of Kenya, trees are classified according to their usefulness. According to him, some trees are "classed as building trees (refer to *Mukoigo* or *Bridellamicrantha*); medicinal trees (refer to *Bridellamicrantha*, Muiri or Prunus Africana, etc.); poisonous trees (such as *Rucago* also called *Maesopsiseminil* (2010, p. 57). The medicinal trees are sometimes referred to as "God's trees" as they are not to be cut down unnecessarily; they are sacred and endowed with healing powers. Thus, trees are conserved alongside the forests that inhabit them. Forests are seen as hidden places during inter-ethnic or inter-clan feuds. The practice of conservation in African space with reference to the forest, according to Andrew A. Kyomo (2010), has a proverbial expression: "The forest is our skin and if one removes the skin of a human being, the result is death" (2010, p. 60). Here, conservation does not only imply that we revere the natural environment and everything present in it, it also implies that we humans are promoting and sustaining our own lives. Summarily, the practice of conservationism in African culture is based on beliefs such as totemism and taboos. By this, there is a strict adherence to wildlife and forest conservation. The spiritual and cultural values possessed by some forests, no doubt, will discourage deforestation since deforestation might be a threat to a totemized animal or endangered species.

Many objections could be raised against the traditional practices of conserving the non-human in African traditions. For instance, to what extent do Africans still revere these non-humans that make up the ecosystem? This question becomes important as recent environmental crises in Africa no doubt call into question the traditional practices of conserving the non-human in African traditions and, by extension, an African-lived ecology. Ikuenobe poignantly argues that this problem has both external and internal inclinations. For the external inclination, it is the colonial legacy, while the internal inclination is the corruption and ineffectiveness of governments in various African states that have contributed the most to the environmental problems in Africa (2014, p. 15). Another objection that is tenable here is the challenge of modernity via scientific and technological advancements. There has been degradation of the natural environment in African traditions, leading some important species in the biotic community to permanent extinction. A typical example of this can be seen in the Niger Delta region of Nigeria, where the exploitation of crude oil has, over the years, led to negative environmental consequences such as oil spillage, environmental degradation, and deforestation, to mention a few.

The question of totemism as espoused by the aforementioned African philosophers and how it permeates the African worldview in the reverence for the non-human raises some critical outlooks. First, there is the challenge of modernity in terms of religion. True to these scholars' positions, the traditional African maintains a biocentric relationship with the non-human. However, the infiltration of African traditional religion by Christianity and Islam has for a long time posed a challenge to the practice of totemism in African traditions. Chinua Achebe describes this in his *Things Fall Apart* by stating how a recently converted Christian extremist killed a python that was considered a totem. He further describes this Christian extremist as "the outsider who wept louder than the bereaved" (1958, p. 148). This, no doubt, calls into question the attempt to recognise that the non-human in African traditions possess intrinsic values and moral status. This basically alludes to the argument of some scholars that African thought is anthropocentric in nature. In the words of J. Baird Callicott, "African thought, orbits, seemingly, around human interests" (1994, p. 158). However, counter-objections have been raised against the argument that African thought is anthropocentric. According to Behrens, African thought has some elements of "holism", that is, the theory that a harmonious relationship is the basis of African environmental thought. This relationship is based on the recognition that humans do not exist independently of the non-human.

Toward an African Lived-Ecology of the Human-Non-human Relationship

From the discussion thus far, it is important to ask: What intersection, if any, exists between eco-phenomenology, deep ecology, and the traditional practices of conservation in African traditions in the quest to develop an African lived-ecology? This question can be understood if we consider ontic experience to be the most important when it comes to appreciating the ecosystem. Husserl reiterates this position:

> I am aware of a world, spread out in space endlessly and in time becoming and without end. I am aware of it; that means first of all, I discover it immediately, intuitively, I experience it. Through sight, touch, hearing, etc. in the different ways of sensory perception, corporeal things, somehow spatially distributed are for me simply there.
>
> (1931, p. 6)

Phenomenologists introduced an ecological turn by emphasising the importance of ontic experience in our immediate environment, while deep ecologists introduced a monist ontological worldview for the flourishing of all. By this, we realise as humans that the underlying principle that integrates eco-phenomenology and deep ecology is ontic experience, that is, seeing the world as it is. This is because we are out there in the world, coexisting with other non-human entities in the ecosystem. This singular act of consciousness makes it possible for us to appreciate the whole ecosystem. Furthermore, as humans, we are then able to realize that the non-human possess intrinsic values. These values are independent of the usefulness of the non-human world for human purposes (Naess, 2005). This principle implies, in relation to eco-phenomenology, that our being in the world is not based on nothingness, that is, as humans, we do not live in vacuum as there are other non-humans necessary for our flourishing. Relating this to African traditions, it is instructive to note that our knowledge of the environment (ecology) is purely lived, that is, based on ontic experience – we see things, especially the non-human, as they are without prejudice.

Our experience and/or consciousness as it relates to other non-humans in the environment is based primarily on the position that it goes against nature to exploit the non-human, as evident in conservation practices such as totems, sacred forests, and taboos. This forms the basis of an *African lived-ecology*, a type of ecology based on the phenomenological understanding that we as humans owe other non-humans an ecological obligation; this ecological obligation is simply to understand that our relationship with the non-human is somewhat relational, and abiding by this ecological obligation of relationality could go a long way in reassessing our relationship with the non-human from an exploitative one to a more conservative one.

As Africans, we tend to approach the natural environment, non-human included, from an unbiased position. This helps us to grasp events and objects as they are, and by extension, we are able to grasp their essence – essence based on the need for conservation, knowing fully well that they possess life force. The implication of this for African lived-ecology is that this unbiased position makes it possible to see the non-human as embodied species that are necessary for the flourishing of the overall ecosystem – for instance, if I find myself in a community that forbids the killing of a certain animal, sighting the animal means I am seeing an embodied species that is forbidden to be killed. Consequently, I am expected to avoid acts that are likely to be injurious to such an animal. This implies that I am an active species in the natural environment. According to Senghor, how we experience the natural environment as Africans is not in a passive manner. He states that "the African is, of course, sensitive to the external world, to the material aspect of being and things … he is sensitive to the tangible qualities of things – shape, colour, smell, weight, etc… (p. 48)". When it comes to an objectively known reality, the subject is not passive; rather, the subject is an active component of reality, one that interacts with and shapes it to comprehend, experience, and know it. Lest we forget that ecophenomenology was introduced in the quest to bridge the ontological dichotomy between the subject and the object. By bridging this ontological divide, it gives room for deep ecology as a monist environmental philosophy that promotes the flourishing of all. Thus, in African lived-ecology, there exists no such thing as an unexperiential or unknown reality; if anything is not known via experience, it cannot be called a reality. Something doesn't become real until it is experienced. This, no doubt, resonates with ecophenomenology as a type of environmental philosophy which seeks to study/experience the natural environment as it is, with the individual playing an active role in it. By extension, it has implications for deep ecology and African lived-ecology. As Africans, we do not just experience the natural environment as it is, but there is the consciousness to always conserve the environment, knowing that we are not the only entities that make up the environment – an idea which forms the basis of conservation through totemism and taboos. In this case, when we see sacred or totemized animals, we are seeing an embodied species that is capable of relationality. Our perception and attitude, in this case, only present to us species that are not to be killed for any purpose. Perhaps, we can consider Reinhardt Grossman's conception of a scientist's description of a tree from both scientific and phenomenological perspectives. To him:

> The botanist takes for granted that there are trees and then proceeds to tell us fascinating things about them. Now, when we adopt the phenomenological attitude, we do not deny that he is right. But we put his existence assumption in brackets, and try to describe the phenomena precisely as they present themselves to us. For example, we may study the object of a mental act of seeing a tree, the precise 'what' of what we are seeing this moment, irrespective of whether our perception is correct, of whether there are trees, and of whether there are any perceptual object at all. If you have a hallucination and see a tree, the object of your mental act, as an object, may be precisely the same as the object of a true perception. You realize, of course, what notion of an object is here at work: it is the notion of d *noema.*
>
> (Grossman 2013, 139–140).

In African traditions, our observation and/or description of the non-human, especially totemized ones, is such that it does not encourage exploitation and cannot be understood purely from a scientific point of view. Put differently, sighting a totemized animal means that we do not conceive it beyond what our consciousness presents to us. Thus, the basis of an African-lived ecology is the recognition that consciousness is essential in conserving the non-human in the ecosystem, making it essential for ecological equilibrium. This ecological consciousness in African-lived ecology is not peculiar to humans alone. The non-human does as well, and this is based primarily on the nature of the (reciprocal) relationship between humans and the non-human. What this means is that as humans, we are able to experience our environment phenomenologically for the sake of maintaining a reciprocal relationship with the non-human. However, there is ample evidences in African traditions that the non-human also experiences the human. For instance, in communities where an animal, let us say a rabbit, is not considered a totemized animal, upon sighting the human person, the rabbit automatically becomes the subject and takes off to the nearest place of refuge because, essentially, its consciousness at that moment towards the human is that the human would kill it. In the same phenomenological experience of the human towards the rabbit, what the human sees is an animal meant for consumption (meat). In another culture where the same rabbit is considered a totem, you are likely to see the rabbit in lived spaces with humans. This means that phenomenologically, the rabbit also experiences the humans and vice versa, and this forms the basis for a lived-ecology of the non-human in African traditions.

In *Things Fall Apart*, Chinua Achebe gives a vivid description of what I would consider an African-lived ecology by accounting for how the python is revered. According to him, "the royal python was the most revered animal in Mbanta (Things Fall Apart) and all the surrounding clans. It was addressed as "our father" and was allowed to go wherever it chose, even into people's beds" (1958, p. 126). Thus, the basis of eco-phenomenology, deep ecology, and how it helps us think about African-lived ecology is the recognition that there is no ontological dichotomy between the human and the non-human. This is because, in some cases, the human is seen as the object, while the non-human is seen as the subject, depending on who is the subject of the experience. In dealing with the ecosystem in general, we then realise that we could be in a reciprocal relationship with other non-humans in the environment, and as such, the ecosystem should be conserved at all cost, that is, the ecosystem becomes something that has intrinsic value. This, no doubt is necessary for the flourishing of all – a key aspect of deep ecology as an environmental philosophy.

In summary, African lived-ecology focuses on three primary issues: firstly, it aims, on the basis of ontic experience and consciousness, to investigate the environment in its natural state, without any form of bias or prejudice. Secondly, it aims to transcend the dualistic ontology that bifurcates the human and the non-human. The third issue is demonstrating ecological consciousness on the part of humans, towards knowing that the non-human has intrinsic values, thereby making conservation a priority.

Conclusion

This chapter is an intercultural exercise as it shows an integration of eco-phenomenology and deep ecology, environmental theories prevalent in the Euro-American worldview, with some traditional practices of conservation in African traditions in the quest to develop an African lived-ecology, broadly understood as a type of ecology that prioritises our conservation of the non-human based on experience. The chapter begins by integrating eco-phenomenology and deep ecology, demonstrating clearly that ontic experience is important in the attempt to understand the ecosystem for preservation. Next, the chapter examines some traditional practices of conserving the non-human in African spaces. It then proceeded to show an integration of eco-phenomenology, deep ecology, and some practices of conservation in African spaces, using this as a paradigm in developing an African lived-ecology. An African lived-ecology presents a deep ecological approach that emphasises experience as the most important aspect in the ecosystem, recognizing that humans owe the non-human an ontological duty of protection for the flourishing of all. Traditional practices of conservation, as found in totems, taboos, and the protection of sacred forests, resonate with the underlying principles of eco-phenomenology and deep ecology. In other words, our ontic experience as humans forms the basis of our ecological consciousness, which is a foundation for us to appreciate and conserve our ecosystem. This ontic experience also provides us with a deep ecological understanding that humans do not live in a vacuum; instead, we co-exist with other non-humans, and as such, there is no ontological divide between us. Consequently, we have to protect the ecosystem at all costs to avoid ecological disequilibrium.

In recent times, humans have been grappling with some environmental crises and their lasting negative consequences. These consequences, no doubt, speak volumes of humans' inability to conserve the ecosystem. This, according to Behrens, has seen the rise in our need for food and energy. Particularly, as humans, we continue to threaten and deplete natural resources, destroy habitats and ecosystems, and cause pollution and degradation at an unprecedented scale. It is now said that our generation is presiding over the sixth mass extinction the earth has experienced, a cataclysmic risk to biodiversity that threatens to be more serious than that of the mass extinction caused by an asteroid hitting earth that ended the age of the dinosaurs (2017, pp. 191–192). From Behrens' position, it becomes important to reassess our relationship with the non-human and the ecosystem as a whole to avoid further damages. African lived-ecology, an ecosophy grounded in eco-phenomenology and deep ecology, it is crucial to note, has prospects in the quest to mitigate the negative effects of these environmental crises, not just for immediate gains, but for generations to come. Behrens further notes this salient point: "Africans also expect that the current generation should develop an attitude of gratitude towards their predecessors for having preserved the environment on their behalf" (2012, p. 189). If this is anything to go by, it simply implies that African lived-ecology could provide the much-needed ecological framework for the preservation of not only the human-non-human relationship, but also for the flourishing of all in the ecosystem.

References

Abraham, D. *The Spell of the Sensous*. New York: Vintage, 1996.

Achebe, C. *Things Fall Apart*. London: Heinemann, 1958.

Amirthalingam, M. Significance of Totemism among the Tribes of India. *ECO NEWS*, 19(4), 2014.1–5.

Behrens, K. G. An African Relational Environmentalism and Moral Considerability. *Environmental Ethics*, 36(1), 2014: 63.82.

Behrens, K. G. The Imperative of Developing African Eco-philosophy. In I. E. Ukpokolo (ed.) *Themes, Issues and Problems in African Philosophy*. London: Palgrave Macmillan, 2017: 191–204.

Behrens, K. G. Moral Obligations towards Future Generations in African Thought. *Journal of Global Ethics*, 8(2&3), 2012: 179–191.

Brown, C. S. The Real and Good: Phenomenology and the Possibility of an Axiological Rationality. In C. S. Brown and T. Taodvine (eds.) *Eco-Phenomenology: Back to the Earth Itself*. New York: SUNY Press, 2003: 3–18.

Burnett, G. W. and KamuyuwaKang'ethe. Wilderness and the Bantu Mind. *Environmental Ethics*, 16(2), 1994: 145–160.

Callicott, J. B. *Earth Insights: A Survey of Ecological Ethics from the Mediterranean Basin to the Australian Outback*. Berkeley: University of California Press, 1994.

Chemhuru, M. & Masaka, D. Taboos as Sources of Shona's People's Environmental Ethics. *Journal of Sustainable Development in Africa*, 12(7), 2010.: 121–133.

Deka, V. From Ontology to Ethics: A Merleau Pontyan Approach to the Human-Nature Relationship. Being a PhD Thesis submitted to the Indian Institute of Technology, in Partial Fulfillment of the Requirements for the Degree of Doctor of Philosophy, 2019.

Devall, B. The Deep Ecology Movement. *Natural Resources Journal*, 20, 1980: 300–322.

Diehm, C. Deep Ecology and Phenomenology. *Environmental Philosophy*, 1(2), 2004: 20–27.

Ekwealo, C.J. Metaphysical Background to Igbo Environmental Ethics. *Environmental Ethics*, 34(3), 2012: 265–274.

Evernden, N. *The Natural Alien: Humankind and Environment*, 2nd ed. Toronto: University of Toronto Press, 1993.

Eze, M.O. *Humanitatis-Eco* (*Eco Humanism*): An African Environmental Theory. In A. Afolayan and T. Falola (Eds.) *The Palgrave Handbook of African Philosophy*. New York: Palgrave Macmillan. *2017*. 621-632.

Fox, W. Deep Ecology and Virtue Ethics, August 27, 2020: http://www.philosophynow.org/issues/26/Deep_ecology_and_virtue_Ethics.

Fox, W. *Toward a Transpersonal Ecology: Developing New Foundations for Environmentalism*. London: Shambhala, 1990.

Gathogo, J. Environmental Management and African Indigenous Resources: Echoes from Mutira Mission, Kenya 1912–2012. *Studia Historiae Ecclesiasticae*, 39(2), 2013: 1–13.

Grossman, R. *Phenomenology and Existentialism: An Introduction*. London: Routledge, 2013.

Husserl, E. *Ideas: General Introduction to Pure Phenomenology*. The Hague: Martinus Nijhoff Publishers. *1931*.

Ikuenobe, Polycarp. Traditional African Ethics and Colonial Legacy. *International Journal of Philosophy and Theology*, 2(4), 2014: 1–21.

Irele, A. and Jeyifo, B. Totemism. *The Oxford Encyclopedia of African Thought*. Oxford and New York: Oxford University Press, 2010: 380–383.

Jonge, E. *Spinoza and Deep Ecology: Challenging Traditional Approaches to Environmentalism*. London and New York: Routledge, 2004.

Kanu, I.A. Book review: A Hermeneutic Approach to African Traditional Religion, Theology, and Philosophy. IGWEBUIKE: An African Journal of Arts and Humanities. 2017, 3 (17): 129-133.

Kyomo, A.A. The Environmental Crisis as a Pastoral Challenge in Africa. *Theology and Environmental Responsibility. Ethiopian Journal of Environmental Studies and Management*. Nairobi: Action, 3(2), 2010: 54–64.

Langer, M.M. *Merleau-Ponty's Phenomenology of Perception: A Guide and Commentary*. 1989 London: Palgrave Macmillan.

Mangena, F. Discerning Moral Status in the African Environment. *Phronimon*, 14(2), 2013: 24–44.

Mazama, A. 2002. Afrocentricity and African Spirituality. *Journal of Black Studies*, 33(3), 2002: 218–234.

Mbiti, J. *African Religions and Philosophy*. London: Heinemann, 1969.

Metz, T. An African Theory of Moral Status: A Relational Alternative to Individualism and Holism. *Ethical Theory Moral Practice*, 32(1), 2012: 387–402.

Metz, T. Toward an African Moral Theory. *The Journal of Political Philosophy*. 15(3), 2007: 321–341.

Mutwa, C. *Isilwane: The Animal*. Cape Town: Struik, 1996.

Naess, A. Deep Ecology: Some Philosophical Aspects. In A. Drengson and H. Glasser (eds.) *Selected Works of Arne Naess*. Dordrecht, the Netherlands: Springer, 2005.

Naess, A. *Ecology, Community and Lifestyle: Outline of an Ecosophy*. Cambridge: Cambridge University Press, 1989.

Naess, N. *Four Modern Philosophers: Carnap, Wittgenstein, Heidegger, Sartre*. USA: University of Chicago Press. 1969.

Naess, A. The Shallow and the Deep, Long-Range Ecology Movement: A Summary. *Inquiry*, 6, 1973: 95–100.

Ogungbemi, S. An African Perspective on the Environmental Crisis. In Louis J. Pojman (ed.) *Environmental Ethics: Readings in Theory and Application*. Belmont, CA: Wadsworth Publishing Company, 1997.

Olaoluwa, S. When was the Ecological Turn? African Ecocriticism and the Limits of Colonial Claims. Paper presented at the International Conference on Migrations of Knowledge: Potentials and Limits of Knowledge Production and Critique in Europe and Africa, Carl von Ossietzky University Oldenburg, Germany, 2014.

Onah, G. The Metaphysics of Peace in African Traditional Religion and Culture, October, 13, 2009.. http://www.afrikaworld.net/afrel/goddionah.htm

Senghor, L.S. Negritude: A Humanism of the Twentieth Century. In F.L. Hord and J.S. Lee (eds.) *I Am Because We Are: Readings in Black Philosophy*. Amherst, MA: University of Massachusetts Press, 1995: 45–54.

Tangwa, G. Some African Reflections on Biomedical and Environmental Ethics. In K. Wiredu (ed.) *A Companion to African Philosophy*. Oxford: Blackwell Publishers, 2004: 387–395.

Taodvine, T. Ecophenomenology after the end of Nature. In Jerome Melancon (ed.) *Transforming Politics with Merleau-Ponty: Thinking the State*. Lanham, MD: Rowman and Littlefield, 2021. 1-26. –

Taodvine, T. Naturalizing Phenomenology. *Philosophy Today*, 1999, 43 (4): 124–131.

Taringa, N.T. How Environmental is African Traditional Religion? *Exchange* 35(2), 2006: 191–214.

Wood, D. What is Ecophenomenology. In C.S. Brown and T. Taodvine (eds.) *Eco-Phenomenology: Back to the Earth Itself*. New York: SUNY Press, 2003: 211–233.

Zimerman, M. Deep Ecology. In L. Embree, *et. al* (eds.) The Encyclopedia of Phenomenology, Dordretch: Kluwer Academic Publishers, 1997. 138.

13 Deep Ecology, Irreducibility, and African Relational Thinking

Mitchell-Ron Black and Juan Oeschger

Introduction

This chapter will explore the foundations, limits, and strategic potential of African Relational Thinking (ART) in the realm of ecological governance across the continent today. ART refers to a way of reasoning nearly ubiquitous in African thought that presupposes the ontological interrelatedness of all things in the world, conceived as an indivisible complex whole. Ecological governance – or the process of decision-making about environmental issues at the governmental level or higher – represents both a challenge and an opportunity for ART. As such, our focus will be on the relationship between the two.

Engaging ART using Deep Ecology, a framework for developing ecological consciousness, and critical complexity theory (CCT), a post-structural lens on complex systems theory, we will show how they offer both overlapping and complementary perspectives for understanding ART and explaining its notable absence from prominent discussions on ecological issues, such as those surrounding the mining and burning of coal. Arguing for the existence of a kind of gap in African ecological governance, between ART in the worldviews of its people and a lack thereof in the worldviews of its governors, we see a contest of rationalities emerging against the dominant regime of neo-colonial governance. This presents philosophers and activists alike with an urgent point of political reflection and action – an ecopraxis Rubicon of sorts – that divides the present and the future of Africa (and the world) along a contested set of intellectual "border" concepts, such as "the state," "development," and "the environment."

These concepts function epistemically, ontologically, and politically as "instruments of regimentation that constrain freedom [...] entrap reason, keep it under siege, and imprison it" (Chimakonam & Ogbonnaya, 2021, p. 2), making them the preeminent legacy-tools of hierarchy, division, and coloniality that relate concepts to one another in pejorative ways. Our aim in engaging "borders" in relation to ecological governance is thus to offer the platform as a potential tool for mediating borders and reconfiguring limits in a way that leverages those excesses in meaning for which existing borders cannot account to create room for ART to legitimately prefigure the process of ethical decision-making in African environmental policy spaces today.

DOI: 10.4324/9781003480914-14

We will proceed by first offering some substantive clarifications before introducing ART in ART, Ecological Governance, and Contesting Rationalities, and adumbrating the challenge it poses to African ecological governance by detailing the nature of the aforementioned gap. Section three will actualise the complex political reality which the said challenge points to through the example of coal discourse in South Africa to highlight the urgent need for new strategies to overcome it. Thereafter, in Section four, we deploy the conversational method to explore the overlaps and complementarities between CCT and Deep Ecology before provisionally advancing an element of the latter, ecosophy writing, as a means for generating shared worldviews linked to better ecological governance outcomes in Section five before we conclude.

Framing ART, Deep Ecology, and CCT

Before continuing any further, we make the following clarifications to avoid unnecessary confusion. Firstly, we neither understand Deep Ecology as a stand-alone philosophy, nor as a system of thought, nor even an ethical or meta-ethical theory. Rather, following David Rothenberg's analysis of Deep Ecology as a potentially revolutionary movement intent on stimulating the production of ecophilosophy (Rothenberg, 1986, pp. 14–20), we will be engaging Deep Ecology as an evolving/dynamic/living/floating platform – for collaborative decision-making in a context where reality is understood as a relational field of interrelations, or "the totality of our interrelated experience" (Naess, 1989, p. 55).

Secondly, we will not be engaging the basic Deep Ecology platform as set out by Arne Naess in relation to his Ecosophy T itself. Rather, we will be engaging it as "Platform T," and only in relation to the Deep Ecology platform developed by one of this paper's authors, Juan Oeschger, out of his Ecosophy A. We will refer to this as "Platform A." The reasons for departure are: (a) it was intended by Naess that individuals develop their own Ecosophies and Deep Ecology platforms (Naess, 1989, p. 37), such as Bartlomiej Lenart's 2010 *Enlightened Self-Interest: In Search of the Ecological Self (A Synthesis of Stoicism and Ecosophy)* and Chimakonam's *Ohanife* (2018) and (b) the basic platform uses language in some instances that is incompatible with ART, such as a separation between human and non-human entities based on the fact of life. Where Deep Ecology, for instance, argues that non-human life is just as valuable as human life, Platform A argues that in our ontology, life is a concept neutral of value based on the ART understanding of ontological equality (Chimakonam, 2017).

Finally, following Minka Woermann (2016) in Bridging Complexity and Post-Structuralism, we have resolved to use CCT as a reflexive meta-philosophical lens for contemplating the nature of complex phenomena, their organisational character, and their ethical implications. We do this to "challenge and transform our models and practices in a bid to engage responsibly with the complexities of our world," such as competing duties, alterity, and over-determination, to confront the "unavoidable normative dimension of being together with others in the world" (Woermann, 2016, pp. 3–4). As such, it is the possibility of decision-making, or the complexity of actually taking ethical action, that reveals "the very impossibility of calculating or systematising morality" as a necessary condition for the possibility of ethics at all (Woermann, 2016, p. 164). Given this, we will be using the terms "ethics" and "decision-making" interchangeably to allude to the tragic nature of ethical praxis, in that ethics is fundamentally a gamble: a moment of intervention that determines right and wrong against the contextual excesses and unknowns of embedded being (Woermann, 2016, pp. 105–108). In opposition, we refer to the more typical use of the term ethics – meaning attempts at systematising categorical prescripts for right action in regard to how one ought to act – as "meta-ethics" and engage it at the level of organisational character. This helps us to maintain our focus on relationality, or the "interaction between objects (regardless of their nature), rather than the objects themselves" (Woermann, 2016, p. 27).

ART, Ecological Governance, and Contesting Rationalities

At the foundation of African thinking, argue Jonathan O. Chimakonam and Uchenna L. Ogbonnaya (2021, p. 5), lies the basic principles of *nmekọ*, translated as "relationship or communion," and trivalency, a logical orientation that rejects bivalency and the dialectical movement towards synthesis. This *a priori* relationality, according to Lebisa J. Teffo and Abraham P.J. Roux (2003, p. 196), allows African thought to offer "holist" explanations for the nature of reality that touch on a level of fundamentality beyond what science can offer by accounting for the world as a complex totality consisting inextricably of both visible elements, such as people, communities, animals, mountains, and forests, and invisible elements, such as ancestors, spirits, and even concepts. We tentatively call this metaphysical perspective, equally identifiable in the work of Francis Etim (2019), Lesiba Baloyi and Molebogeng Makobe-Rabothata (2014), Crispinous Iteyo (2009), and Magobo B. Ramose (2003), a form of "relational monism."

Owing to its side-stepping of the paradigm of simplicity and the techniques of disjunction ("separating that which is linked") and reduction ("unifying that which is diverse") that come with it (Morin, 2008, p. 39), the relational monist worldview forces us to do the intellectual work of engaging the "enormous complexity of the universe" while "maintaining a strong empirical flavour" without resorting to the types of dualisms and bifurcations that are the "stock-in-trade" of Western thought, at least since the time of Plato (Teffo and Roux, 2003, p. 196). The world is, as such, a whole – an irreducibly complex and self-organising system of "insuperable indivisibility [that] is capable of assuming a plurality and diversity of forms which manifest its character" (Ramose, 2009) – that cannot be reduced to the sum of its parts, or any simple constitutive rule set for their relation to one another as meaningfully isolated elements (Woermann, 2016, pp. 30–31). Of primary concern, then, is the "complex relationship[s]" between humans and the life-world in which they find themselves that makes it difficult, if not impossible, to "distinguish metaphysics, social theory, and morality in African thinking" (Teffo and Roux, 2003, p. 198).

This often means that there are few, if any, "limited" or "restricted" claims that can be made about the world or its constituent elements in meaningful isolation from one another because, at an ontologically significant level, a claim about the nature of something is not just a claim about that thing: it is also a relational claim about its context and complex dual-identity as a complementary intersection between the thing-in-itself and the network of relations into which it is embedded. And so, any claim about the nature of a thing is not solely a claim about its nature: it is also a claim about its ontological relation to other things that are, upon analysis, themselves always constitutive of each other, and of the thing that is. In this way, identities constantly constellate and re-constellate each other. By emphasising this, ART, similarly to CCT, encourages us as conscious beings within the network of Being, to recursively reflect on inter-relation from a holistic perspective, an act which, by the nature of being and knowing, changes us in the process just as a conversation between brothers forms each, and so each goes on to form others through lifetimes of conversations.

Thus, to comprehensively discuss one aspect of being – say ethics – to the exclusion of others – metaphysics, epistemology, ontology, politics, aesthetics, and more – is not entirely possible or honest: no aspect of being exists a-contextually or in a realm of pure abstraction as the thing-in-itself separate from the world to which it must ultimately be brought to bear.

And so, we can see how even concepts exist in a substantive sense and come to influence the world in material ways, though the type of existence they hold is different from, say, a hammer. Chimakonam and Ogbonnaya use the concepts of *ife enu* and *ife ezi* to distinguish the former from the latter as "thing that is up there, meaning abstract reality" and "thing that is out there, meaning concrete reality" as bound subcategories of *ife ụwa*, "things that are in the universe" in their theory of *Nmekọka* metaphysics (Chimakonam and Ogbonnaya, 2021, pp. 30–31). From an African metaphysical point of view, then, the material shapes the immaterial and the immaterial shapes the material: their relationship is not one of dualism requiring negation, compromise, and synthesis, but rather of complementarity, allowing room for "notional solidarity, "creative struggle" and "mutual engagement" towards a higher level of inter-being (Chimakonam and Ogbonnaya, 2021, p. 5). Relationalism such as this is not merely a descriptive feature of entities that exist in spatio-temporal proximity; it is the fundamental nature of reality at base that can best be made sense of when approached using a rationality that is emic to Africa, lest we continue the trend of "carrying the logic of [one] system into a different system and attempting to employ it to assess the assumptions of the new system" (Chimakonam and Ogbonnaya, 2021, pp. 21–22).

As a result of colonialism and neo-colonial globalisation, Africa is today subject to an etic regulative order that determines the organisational character of African society both within countries and across the continent. Some examples of this include Rwanda's post-genocide endeavour to reach Sustainable Development Goals (SDGs), Nigeria's anti-corruption campaigns based on international best practices, and Ghana's electoral processes informed by international and regional bodies. In fact, an example can be made in some form regarding each country on the continent. Whilst these examples seem benign, and it is not the purpose of this chapter to offer a critique of them, these examples serve to show how governmentality in Africa is very rarely authentic in terms of space and place, indicating a divide between governance practices and the lived experiences and cultures of those being governed. To begin modelling the relationship between ART and the disciplinary matrix which imposes limits on it, as will be shown in the next section, we find the Foucauldian concept of governmentality specifically useful for framing the discourse as one of contesting rationalities before bringing it to bear on ecological governance in Africa.

Governmentality and governmentalities are analytic frameworks for analysing how the relationships "between power and subjectivity" and "techniques of power and forms of knowledge" mediate "techniques of rule [...] tied to 'technologies of self' [through] specific types of rationality, regimes of representation, and interpretive models" (Brockling et al., 2011, p. 2.) When looking at ART and the governing of ecological crises across Africa, governmentality analysis allows us to excavate "the systemic ties between forms of rationality [and] technologies of government," which, in our instance, refers to the ties between dualist, disjunctive, or reductive worldviews and the ineffectiveness of ecological governance in Africa. Our focus will be on the political and the means by which "it is produced in the first place" through the establishment of "divisions and distinctions" that prefigure the boundary conditions of a problem and also determine the range of legitimate possible solutions that can be advanced (Brockling et al., 2011, p. 13).

The nation-state represents one such technology. It establishes a specific governmental rationality, among, but always over, the plurality of existent rationalities by coercively ordering society into hierarchical relationships between the governing and the governed, thereby introducing and establishing disjunction as the norm. Once in place, the nation-state is forced to "compete" on the stage of nation-states, contoured and textured along asymmetric North-South power relations that direct the flow of progress from periphery to centre. In doing so, it becomes increasingly subject to a global regime of disciplining technologies that require it to conform to certain criteria of acceptability and legitimacy, such as those reforms and policy trade-offs elicited from the new governing class within emerging states by the World Bank's Structural Adjustment Programmes in exchange for access to global capital markets and an invite to join the "civilised" world around the turn of the millennium.

Subversively, "neoliberal" (Clarke, 2019, p. 157), or more accurately, "Afro-neoliberal" (Climate Justice Charter Movement, 2022, p. 39), strategies enacted through multilateral trade agreements and treaties made African leaders increasingly amenable to the enculturation of "European grids of intelligibility" (Praeg, 2019, p. 3), thereby establishing a globally legitimate, and exploitable, regime of intra-continental inter-national relations. In this way, governmentalities are always relational, operating at the sites of power constituted between actors to functionally and organisationally pre-determine the limited range of permissible individual and collective subjectivities to be created. Lastly, where established, governmentalities function to undermine, discipline, or otherwise erase non-conforming subjectivities, as evidenced by the dominance of, for example, industrial agriculture globally where once stood a diverse plurality of indigenous land relationalities and food systems.

Nature, unlike humanity North and its penchant for enclosure and parochialism, does not trade in borders and limits, and nowhere is this crucial difference clearer than in the consistent ineffectiveness of attempting to govern ecologies and humanity's overall relationship with the environment from within the borders of the nation-state. Marie-Louise Clarke (2019, p. 159) illustrates this well in her consideration of the ocean ecosystem in relation to African environmental ethics when she writes:

> When we talk of the ocean, it is even more obvious that these binaries [local/global, national/transnational, North/South] just do not stack up. Ecological systems span political boundaries. Indeed, Cetaceans and migratory birds do not follow national boundaries, but migrate across huge distances [and] marine ecosystems themselves are fluid, cross boundaries and pay no attention to governed borders.

When it comes to ecological governance, we argue the following: Africa's relational monism more closely coheres with the complex nature of ecosystems and, through ART, provides a way of engaging the world as a whole and attempting to navigate its complexities. However, the nation/state, as just one mechanism for bifurcation among other dualisms ranging from the governed/governing and the developed/developing to the self/other and the subject/object, imposes onto ART a disciplinary limit. To effect ecological governance in Africa, one practising ART may argue, requires that we engage ecosystem governance on the relational terms of the ecosystem and not the political terms of the nation-state. To this, the technologies of the nation-state reply that such is impossible; our borders are the limits of our responsibility in all areas but trade and humanitarian cooperation – what happens on that side of the line is not only beyond our control, but it is not our problem. Herein the ecopraxis gap emerges – that which prevents the rationality underlying ART, stuck in the lecture hall or the initiation ceremony, from graduating into the world of policy and cooperative decision-making in any meaningful way.

Complexities in Ecological Governance, or the Question of Coal in South Africa

To further our analysis and provide a concrete example with which we can engage the meditative potential of Platform A, this section will briefly describe the emerging discourse in South Africa between local coal mining advocates and climate justice forces, as the future of an industry and the world is hotly contested. As will be shown, this discourse tragically encapsulates the intersection between competing rationalities, ecological governance, and the complexities introduced by context.

As an ongoing debate, the issue around coal will be framed between two sides: The first, embodied in the words of the long-standing Minister of the Department of Mineral Resources and Energy (DMRE), Samson Gwede Mantashe, who serves as a representative of the ruling liberation movement *cum* governing party, the African National Congress (ANC). The second, from the forces of Climate Justice across South Africa, such as the Climate Justice Coalition (CJC), a "coalition of trade unions, civil society, grassroots, and community-based organisations" (Climate Justice Coalition, 2024), and the Climate Justice Charter Movement (CJCM), a policy-based movement with over 250 civil society endorsements whose aim is the adoption of the Climate Justice Charter into law by the South African National Legislature in line with Section 234 of the current South African Constitution (Climate Justice Charter Movement, 2024).

Essentially articulating the state's position, Mantashe's argument can be boiled down to three arguments (Evans, 2022). Firstly, Evans (2022) frames the coal sector as a familiar resource necessary for attending to "South Africa's development needs" and thus something that the state has "a duty to protect" and "grow" in its "impact on society." He attributes this status to the sector based on its high return-on-investment, contribution to GDP, and the plentiful availability of coal, arguing that the state should allow it to "benefit as long as it can." Secondly, speaking from an economic redress perspective, Evans argues that coal both supports economic growth through revenue generation and job creation and represents "one of the sectors of mining that has actually transformed" (2022). To abandon coal, then, would be to reverse the gains of the ANC's Broad-Based Black Economic Empowerment policy, which has transformed some of the capital generating potential of coal mining and supporting industries, like logistics, from white hands to black ones. Lastly, he ties the state's ability to fulfil its electricity provisioning mandate to coal in how it has already made substantive investments into both coal-power-generating infrastructure and the coal sector when compared to renewable energy, which he regards as a hopeful fancy yet to be proven in its efficacy – "You build for the future on what you know and what you have" (Evans, 2022).

The CJCM (2022), reflecting on the ANC's relationship with coal, writes of Mantashe's "development needs" argument that he embodies the persona of a "carbon nationalist stuck in the 20th century," hell-bent on "another round of carbon-centred industrial development." In seeming ignorance of the reality of the climate crisis, which demands the rapid transition away from all fossil fuels, they contend that Mantashe and his party are engaging in "disaster capitalism" by making ecological governance decisions that intensify the heating of the planet and increase the potential for "climate catastrophe" in South Africa. Rebutting the carbon nationalist argument as "misleading workers" (the ANC is aligned with the Council of South African Trade Unions, of which the National Union of Mineworkers is a part), the CJCM brands the ANC as "effectively the party of carbon capitalism in South Africa" and frames their tenure in government as "the biggest obstacle to an accelerated and deep just transition" away from fossil fuels. This approach to ecological governance essentially "locks" South Africa into a path paved with "extreme climate impacts" that has "implications for the 1.5°C overshoot on a planetary scale."

Alternatively, Alex Lenferna, General Secretary of the CJC Steering Committee, "standing together [with] communities that are facing the frontline impact" of coal pollution and climate change, argues for a fair and just set of "transitional policies" to move South Africa away from fossil fuels that leave neither "ghost towns" nor a "legacy of unemployment" in its wake (Evans, 2021). He contends that the transition is inevitable, even supported by the economic models produced by Mantashe's own department. Invoking the authority of globally agreed-upon climate science, he remarks that "we [the people of earth] can't afford to dig up and burn new coal, oil or gas" or finance the "harmful, extractive and pollutive projects" they require. Renewables like solar and wind are now the cheapest and most ecologically sound energy options, and so the further pursuit of coal will "cost us higher energy prices, fewer jobs, and more pollution." Concluding his remarks, he notes that on its current trajectory, the state-owned national power utility, Eskom, has a plan for a faster switch to renewables and even wants international help to do it. Between it and a renewable future is "Mantashe, and the DMRE [as] the only ones standing opposed to it."

Greenhouse gas emissions are a complex issue at multiple scales; of this, there can be no doubt. However, like ocean ecosystems, atmospheric emissions transgress political boundaries and require nation-states to overcome state parochialism and cooperate in unprecedented ways that often run counter to hegemonic capitalist logics of accumulation linked to growth economics. And this, at least according to Mantashe, would infringe on the national interests of the South African state, jeopardise its ability to fulfil its mandate, and destroy coal mining as a sector.

For Mantashe, the requirement for local interventions to address global problems stands in opposition to the state's historico-political obligation to (a) redress the economic legacy of Apartheid and the persistent racial inequality it gave way to, b() pursue a "modern" development agenda (including GDP growth), and (c) generate sufficient electricity to meet national demand. Contextually, then, a postcolonial neo-colony like South Africa, at least from the perspective of its Afro-neoliberal governors, finds itself in an intractable situation made all the worse by the recent sudden increase in coal demand on the heels of the war in the Ukraine. Consequently, Europe's supply of natural gas has been compromised, and they have been "forced" to reopen multiple coal-fired power plants to avoid an energy crisis (Hoyle and Wallace, 2022). The South African coal sector, championed by Mantashe, is not one to waste a good crisis, and so has welcomed the reinvigorated investment climate under the banner of competitive advantage, industry protectionism, and redressing unequal development, all narrow economic ends conditioned by a rationality unique to the nation-state's governmentality under global capitalism and so governed to approach the issue of coal on the basis of nation-state delimitation and economic opportunity, as discussed towards the end of the previous section. This is a "limited" or "restricted" model of state responsibility.

Conversely, the climate justice forces contend the view of the state on the basis of *longue durée*, conditioned by a relational rationality linked to an emergent global-South governmentality whereby the ethical duty of the state aligns closer with long-term ecology, transformation, and social justice than with outdated economics, development, and "progress as ideology" (Ani, 1994). This is apparent through the glocal lens with which they lay out their concerns, balancing the scientifically supported "planetary scale" implications should South Africa deepen its coal extractivism and the hyper-local need for "transitional policies" that do not result in "legacies of unemployment." Herein, statements like "we can't afford to dig up and burn new coal, oil and gas" are couched in relational worldviews and pluriversal value matrices, relating them to local and transnational needs, immediate and long-term goals. Herein, the model of state (and individual) responsibility is holistic or more "general" in character.

While the hair-raising slowness of this complexity-laden dialogue hinders effective and timely climate action, we also see it as a tragically necessary part of living together with others in the world. By using the conversational method (Chimakonam, 2017, 2021a, 2021b), we can see how the discourse laid out above is not one beyond resolution, somehow resigned to a state of irreconcilable binaries. Rather, what we hear are two significists – *nwa-nsa*, the proponent, and *nwa-nju*, the opponent who each "convey an idea or the signified" (Chimakonam, 2021a, p. 10) – engaged in heated exchange and the creative struggle between ideas at the level of societal politics. Herein, conversation loses all of its conventional and literal meanings, and gradually a process or method unfolds in their place. Now, meanings, and not ideas, oscillate between the *benoke* point and the tension of incommensurables in the conversational mode, trying to remain within the realm of "mutual intelligibility" (Chimakonam, 2022, p. 164) wherein meaning-making remains a possibility. However, the value-binding question of context cannot be avoided, and so the conversation must eventually be decided – this is the decision-making moment of ethics, the switch from the complementary mode to the contextual mode of logical evaluation with the hope that both significists will emerge in a state of binary complementarity, reformed, instead of synthesised. In a way, it is as Afrikan poet Breyten Breytenbach had written about the relationship of Afrikaans people and otherness: "*Ons moet kompos wees, ontbindend om wéér in ander vorme te kan bind,*" translated as "We must be as compost, decomposable to be able to bind again in other forms" (Breytenbach, 2015, p. 12). Similarly, the process of conversation could be said to be a kind of compost politics, of inter-relation, divergence, convening, formation, and reformation of one significist by another and who together must rise to a new, shared level of greater meaning.

Given the consequential nature of this discourse for the future, the extreme time pressure for its urgent resolution cannot be discounted in favour of simply "letting things play out as they will." And so, some strategic intervention is required, with the hope that African societies will overcome the gap between ART and colonial technologies of governance. In the next section, we offer ecosophy writing as one possible meditative tool.

Mediating Complexity through the Deep Ecology Platform

The principles of relationality, contextuality, and complementarity are basic moral principles in African philosophy and can be found in various expressions across many traditions in Africa south of the Sahara (Chimakonam, 2022). The principle of relationality states that actions are always relational instances. In conversational philosophy, this principle finds expression in the law of *Njikoka,* which states that truth is relationally determined on the condition of the presence of a false opposite. The principle of contextuality states that the moral value of actions is dependent on context, expressed in the law of *Nmekoka* in conversational philosophy, which states that if something is true in a specific context, then it cannot be false in the same context. Finally, the principle of complementarity states that opposing moral values are complementary and not contradictory, as seen in the law of *Onona-etiti*, which states that something can be both true and false in a complementary mode of thought. Chimakonam argues that the first two laws, *Njikoka* and *Nmekoka*, allow for the pursuit of the good for the self and the other, whilst the third law, *Onona-etiti*, grounds the pursuit of the good for all (Chimakonam, 2022). We refer here to conversational philosophy in order to draw attention to an important facet of ethics in African relational thought, that all actions are relational. The implication of this understanding is that stakeholder involvement is necessary in decision-making to the largest possible extent, at all levels, and across contexts.

In a relational understanding of the nature of the universe, it is understood that there are at a minimum as many relations as there are entities squared, in addition to the reflexive relation each entity has with itself. Deep ecology and Complexity theory emphasise proximity, locality, and stakeholder involvement when it comes to decision-making. Identifying all relevant stakeholders as stakeholders requires an awareness and consideration of as many relations as possible, within a relational context, made possible by growing an ecological conscience. Deep ecology refers to ecological consciousness, firstly as an ever-widening understanding of the self as embedded in a relational field of being, and secondly as understanding all relations, even those with the environment, as constituent of the self (Naess, 1989). We take as given that ecological consciousness is an aspect of ART. Deep ecology, as a movement, invites individuals to engage in the act of articulating their personal Ecosophies.

An ecosophy is a coinage by Naess, combining the Greek *oikos*, meaning household, and *sophia*, meaning wisdom. The English "eco" is used to indicate that, as in the study of ecology, "eco" (oikos) is extended to include all the world, not just the immediate household. When deep ecology speaks of ecology, it refers to "the interdisciplinary scientific study of the living conditions of organisms in *interaction with each other* and with the surroundings, organic as well as inorganic" (Naess, 1989, our emphasis added). As a scientific field, ecology emphasises the importance of diversity, complexity, and symbiosis of life forms within a given ecosystem and so is primarily concerned with relationships between beings as essential to the nature of the beings themselves (Naess, 1989). From the ecological idea that all things are interconnected in an ecosystem, we can formulate the idea of ecophilosophy to signify the study of problems common to ecology and philosophy, or the study of philosophical problems from an eco-relational disposition.

Philosophy, in the deep ecology movement, can refer to one of two things: either a field of study (as in academic philosophy) or an individual's worldview and set of values. Ecosophy refers only to the latter conception of philosophy, which is the worldview formulated by the individual based on ecological principles in response to questions involving the self and the environment (Naess, 1989). Naess' definition of an ecosophy reads as a "philosophical world-view or system inspired by the conditions of life in the ecosphere" (Naess, 1989, p. 37). A growing body of ecosophies would prove the need for ecological governance that is sensitive to the irreducibility of systems and open to revisability.

One example of an ecosophy based on ART is *Ohanife*, an ecological worldview founded on Conversational Thinking articulated by Chimakonam (2019). Another example is Ecosophy A, an ecosophy in development by Juan Oeschger, one of the present authors.

Ecosophy A is an African relational philosophy from South Africa. It is an articulation of the present author's worldview. Ecosophy A does not benefit from a specific cultural heritage, including that of the present author. Ecosophy A is an ecosophical exercise, which is an active conversation between ART and Deep Ecology as movements. Ecosophy A is based on a relational understanding of reality found in ecology, physics, and *Nmekọka* metaphysics (Chimakonam and Ogbonnaya, 2021, p. 23). Ecosophy A conversationally engages the Deep Ecology platform earlier referred to as "Platform T" and identifies areas of congruency, as well as the limits of congruence, between Deep Ecology and the ART of Ecosophy A. Deep Ecology relies on an intentional vagueness in its formulation of Platform T in order to establish a prima facie agreement between most, if not all, relational and ecological worldviews. Ecosophy A recasts the platform as Platform A, which retains the essence of Platform T but makes certain changes to the original formulation in order to emphasise ontological variance and equality in line with Conversational Thinking and which forms part of the author's lived experience and intuitive beliefs:

1. Entities in the universe exist equally, independent of notions of usefulness and value.
2. Life sustains life, and more diverse life sustains more life.
3. Humans can use their capacity for reason to determine the point where the requirements of sustaining life have been met, and to progress with exploitation and killing beyond this point would not be sustaining life.
4. The flourishing of human life and cultures is compatible with a substantial decrease of the human population. The flourishing of non-human life requires such a decrease (Naess, 2008.).
5. Present human interaction with non-humans and each other is largely exploitative and damaging and needs to become predominantly symbiotic and mutually beneficial instead.
6. In view of the foregoing points, policies must be changed. The changes in policies affect basic economic, technological, and ideological structures. The resulting state of affairs will be deeply different from the present and will make possible a more joyful experience of the connectedness of all things (Naess, 2008).

7. The ideological change is mainly that of appreciating life quality (dwelling in situations of inherent value) rather than adhering to an increasingly higher standard of living. There will be a profound awareness of the difference between big and great (Naess, 2008).
8. Those who subscribe to the foregoing points have an obligation directly or indirectly to participate in the attempt to implement the necessary changes (Naess, 2008).

It may be noted that in the above eight points, the points referencing Naess indicate that the point has remained unchanged from Platform T. As with the formulation of Platform T by Naess, it must be emphasised that the worldview espoused by Ecosophy A is a personal worldview, not intended to be universally subscribed to, but rather to clearly state the grounds from which the author would embark on a decision-making process. Deep Ecology and Ecosophy A invite the reader and the entire reading public to engage in ecosophy writing and to engage the platform in order to clearly define their own base for decision-making. Reflexively engaging our own views in this manner will establish clearly where we agree and highlight where we do not, so that we can identify instances where our views can act as complements in order to establish points of complementarity where a decision can be made that best resembles the interests of the stakeholders involved, whilst remaining conscious that these moments of complementarity are transient and ephemeral. In the current absence of a proven method to engage in truly collaborative and representative decision-making, ecosophy writing offers itself as a viable way to add one's voice and beliefs to a growing data set which forwards ecological consciousness in decision-making.

This movement requires participants to generally agree with the tenets of the platform, which is formulated based on a relational approach towards metaphysics, ontology, and ethics. Any overwhelming disagreement between an ecosophy and the platform T or platform A would indicate a metaphysical/ontological clash, making Deep Ecology as a movement of platforms attractive only to those who subscribe to a relational/ecological understanding of the nature of reality, or at the very least a mode by which those who do not can be involved in the process of developing their own ecological worldview. In this sense, Deep Ecology becomes a "middle ground" of sorts where many diverse systems – all of which agree, at least in principle, with Deep Ecology – can be represented and engage in decision-making. Disagreement with the platform indicates that stakeholders are not in a position to move forward with ecological governance decision-making yet as a moment of complementarity between opposing worldviews has not been reached, whilst agreement in general would highlight a moment of shared ontological understanding, establishing a clear base from which decision-making can depart.

The conversational process of decision-making to which Ecosophy A subscribes brings to the table an emphasis on the tentativeness or provisionality of action. We cannot at first consider all possible outcomes of an action; however, we can say that as long as we make our decision in the complementary mode, i.e., at a moment of shared understanding where our differences have been creatively consolidated into a stronger position, we are making the decision that is best for all entities involved, of which we are aware. In the complementary mode, we as the decision-makers can identify ourselves with the object of the decision and, based on the assumption that we choose what is best for ourselves, we can say that these decisions are acceptable in the given context at the given time.

The rejection of synthesis by the conversational method prevents us from universalising moral actions, as we are aware that we are seeing only a fraction of the picture when we deliberate on our actions. By engaging in creative struggle for shared meaning, we become aware of the shortcomings of our solutions to problems, as new relationships are constantly being discovered. However, we are able to make better decisions after every turn towards complementarity. The context-dependence of value in Conversational Thinking serves to remind us in our normative deliberations that what is right in one context might not be right in another; hence, the conversational process is continuous and does not end when a decision is made. We can, however, say that we have made the best decision for the relationships involved with the information accessible to us at the time.

By admitting the certainty that new information will prove choices wrong in retrospect, we avoid the hubris of pretending to understand the intricacies of the universe perfectly. In doing so, we also inherit a sensitivity to the aneconomic dimension of our decision-making model (Woermann, 2016, p. 181). This allows us to perform our actions with the knowledge that we are at least acting in good faith and in clear conscience, and that we have considered more than we have previously by involving previously excluded ecological stakeholders in our thinking process. We remain steadfast in our intention to do good by admitting a lack of perfect knowledge and a revisability to our norms, for it also leaves us with the opportunity to do even better in the future at a higher level of deliberation.

CCT also allows us to understand decision-making as a liminal space where complex systems interact. Each individual entity's worldview is described as a complex system constellated amongst a plurality of complex systems, whilst each context in which an entity exists is itself a complex system equally constellated by other complex systems. Deep Ecology furnishes us with the concept of Gestalts, borrowed from psychology, as a tool to identify and demarcate systems and how they interact. Gestalt, in Naess' philosophy, is succinctly defined as "an entity that cannot be arbitrarily separated into subject, object, and medium" (Naess, 2002, p. 175).

This understanding of gestalt gives more philosophical weight to the idea that the whole is more than the sum of its parts, that what we are dealing with is a complex system, as opposed to a complicated system. As stated before, a complex system, unlike a complicated system, has no central organising principle, meaning that no conceptual scheme can exhaustively explain such a system (Cilliers, 1998, pp. VIII-IX; Poli, 2013). Gestalt thinking suggests that reality can be conceptualised as a gestalt, which implies the constant possibility of a gestalt switch. A gestalt understanding of complex systems allows for what in Deep Ecology is known as the gestalt switch, where new information about one part of a system changes our understanding of the system as a whole. Such a gestalt switch can be accounted for using CCT's notion of non-linear causality (Woermann, 2016, pp. 30–31).

Just so, being confronted with the ecosophy of another entity, or having to confront one's own ecosophy through ecosophy writing, would impact our understanding of the system we find ourselves in, thus shifting how we view the system as a whole. In the language of Deep Ecology, a shared gestalt understanding between entities regarding a system could indicate a moment of complementarity. Notions of complicated and complex systems, the former referring to closed systems which in theory can be fully understood, though not necessarily modelled (Woermann, 2016, p. 41), complement ART as it furnishes us with a conceptual tool to understand the nature of the problems and questions that complicated colonial technologies of governance give way to. The efficient distinction between complex and complicated systems could greatly benefit decision-making as a first point of departure, which would inform our methodological response to specific problems or questions, where the problems of complicated systems are approached from the logic within the system, whilst the problems of complex systems are to be engaged in from the paradigm of complexity (Morin, 2008, pp. 37–38), a creative struggle for meaning. The gestalt conception becomes, then, a tool with which to identify such systems and begin to identify one's relationship with and within them.

Linking back to our example, we find a rich interplay between complicated and complex systems. Mantashe operates from within the closed conceptual schemas of Afro-neoliberal capitalism and the nation-state. But, a growing body of ecosophies could possibly make a gestalt shift in the perspectives of leading decision-makers a reality capable of coterminously shifting the identity of both the decision-maker and the mode of ecological governance as a matter of organisational recursion within African ecological governance understood as a complex system, or gestalt, premised on a complex notion of non-linear causality (Woermann, 2016. pp. 30–31). By highlighting an economic dimension of modelling, where restricted models cannot account for the excesses in their fixed meanings which general models give rise to, a relational understanding of complex systems (such as the self, the community, and the environment, and their mutual interlinkages, interdependencies, and relationalities) suggests that the complicated technology of the nation-state (a restricted economy) may be displaced by a more complex organisational relatedness (a general economy), oriented towards continuous, cooperative, and provisional meaning-making between self and other.

Ecosophical Reflections on the Complex Other in African Ecological Governance

Our understanding in CCT and Deep Ecology, an understanding compatible with ART, that entities exist as constitutive of all possible relations with all other entities precludes a definitive subject-object distinction in ethical deliberation. This implies a co-constructivist understanding of the world, where we create and are created by others coterminously (Woermann, 2016, p. 180). In a similar vein, the gestalt understanding of Deep Ecology suggests that entities cannot be arbitrarily separated into subject, object, and medium (Naess, 2002, p. 175). The role of ecosophy writing in our analysis is to establish this awareness, that ethics is a collaborative process geared toward collective meaning-making as collaborative decision-making within a relational universe. As any decision made in a relational or systems understanding will necessarily impact all relations, and all relations are constitutive of the decision-maker, the decision-maker thus becomes all stakeholders. Decision-making becomes "a structural condition for any engagement with complex phenomena" (Woermann, 2016, p. 180).

When modelling a decision as a general economy, an open or complex system that includes the modeller, like *the* environment, one develops sensitivity to the aneconomic dimension of modelling *an* environment (to the exclusion of the modeller) that enculturates into the model the excesses in meaning for which a restricted economy cannot account. Inversely, when modelling something as a restricted economy, a closed or complicated system separate from the modeller, the aneconomic dimension is foregone, rendered external as that which is lost in the modelling process to the detriment of the general system. Although a general economy can only be engaged by the restricted economies generated from our limited conceptual schemas, senses, and intuitions, the interplay between these economies indicates to us the artificial and only ever provisional nature of closed-system models for decision-making. Thus, "true" or "complete" decision-making models in regard to complex phenomena are unachievable, partly due to our contextually bound rationalities, but also due to the inherently *aporetic* relationship between general and restricted economies of meaning, represented as the aneconomic dimensions of *différance* that render an accurate and categorically true account (im)possible (Woermann, 2016, pp. 64–67, 187).

This reading of CCT and Deep Ecology allows the concept of ecopraxis to emerge from ecosophy. In the preceding section, we claim that ecosophy writing adds to a body of literature, or a data set, which includes other ecosophies as well as world-views loosely grouped under the term Southern Environmentalism (Manzini, 2019). It is our hope that this growing data set becomes ever more representative of the various and diverse relational world-views in existence and may aid decision-makers to possibly consider such views in the decision-making process where otherwise the authors, or proponents, of said ecosophies and world-views might not be directly involved in, or in a position to act as, the explicit decision-makers regarding issues of ecological governance at transnational levels. While this is no mean feat, it remains that the invitation extended by Deep Ecology for us to engage in ecosophy writing is an invitation for all to partake in a process of ecological self-constitution.

By writing an Ecosophy, we actively declare ourselves as existent and we actively insert ourselves in the decision-making process. It is choosing to be heard, choosing to claim the world as ours, our creation, and our home, whilst admitting at the same time that we are at the mercy of the Other, the Other's creation, and the Other's home. We involve ourselves lovingly in the world of the Other when we articulate our ecosophies and, in turn, invite the Other to enter into conversation with us on the basis of our ontological egalitarianism. We admit that living in the world is a collaborative project by doing this, and as such, we are moved to admit that the future of the world is a collaborative project. Ecosophy writing is the act of collaborative meaning-making.

As a movement intended to revolutionise ecological governance, Deep Ecology extends the call to write ecosophies to all who can answer; however, in the current state of ecological governance in Africa, this call is especially directed towards those who are in positions, or who have the ear of those in positions to explicitly choose on behalf of others, for instance, lawyers and policymakers. We forward the articulation of ecosophies as a provisional imperative (Woermann, 2016, pp. 173–179) for engaging in the decision-making process on a transnational governance level.

An articulated ecosophy, informing a decision, holds actions accountable to the worldviews supporting them. It aims at minimising incongruency in our normative logic, or how we reason from our beliefs to our actions. This promotes transparency in decision-making and facilitates collaboration as the premises for our normative arguments are visible and accessible to all. Furthermore, ecosophy writing in Africa is an act of decoloniality, as any ecosophy is necessarily founded on a relational understanding of being and the irreducibility of systems, which provisionally precludes any notions intended to separate, universalise, essentialise, or dominate.

(In) Conclusion

Over the course of this chapter, we introduced ART and framed its apparent limits through the concept of governmentalities, relating rationalities to technologies of government, in consideration of the specific example of coal discourse in South Africa and the governmentality-rationality gap it gave way to. From there, we introduced ecosophy writing as a means for developing ecological perspectives linked to deep ecology platforms, functionally serving as political middle-grounds for contemplating one's shared ontological base, comparative complexity, and relatedness to the other. Such a process, we argued, may aid in forming a unified, but not uniform, front for resisting neo-colonial technologies of governance and forging new paths for ecological governance in Africa.

Pointing at this site of liminality and generativity, where relational rationalities find themselves stranded, we position deep ecology as both a movement intended to revolutionise the decision-making process and a mediating tool for doing so. This unfolding, wherein closed, or restricted, systems of meaning, through ecosophy writing, are exposed to their aneconomic, or general, dimension and made to engage with it, brings into focus their overlapping ontological grounds and complementary ethical grounds for rejecting the regulative order and disciplinary matrices of the nation-state technology. Without concluding, we reflexively advance the potential of calling for ecological decision-makers to engage in ecosophy writing as a provisionally appropriate strategy for tackling the governmentality-rationality gap in the face of ecological collapse.

Bibliography

Ani, M. "Donna R." (1994) *Yurugu: An African-centered Critique of European Cultural Thought and Behavior*. Trenton: Africa World Press, Inc.

Baloyi, L. and Makobe-Rabothata, M. (2014) 'The African Conception of Death: A Cultural Implication', in L.T.B. Jackson et al. (eds.), *Toward Sustainable Development Through Nurturing Diversity: Selected papers from the Twenty-First Congress of the International Association for Cross-Cultural Psychology*. Melbourne: International Association for Cross-Cultural Psychology, pp. 260–272.

Breytenbach, B. (2015) *Parool/ Parole: Versamelde Toesprake/Collected Speeches*. Cape Town: Penguin Books.

Bröckling, U., Krasmann, S. and Lemke, T. (eds.) (2011) *Governmentality: Current Issues and Future Challenges*. First issued in paperback. London: Taylor and Francis Group.

Chimakonam, J.O. (2017) 'What Is Conversational Philosophy? A Prescription of a New Theory and Method of Philosophising, in and Beyond African Philosophy', *Phronimon*, 18(1), pp. 114–130.

Chimakonam, J.O. (2019) 'Ohanife: An Account of the Ecosystem Based on the African Notion of Relationship', in J.O. Chimakonam (ed.) *African Philosophy and Environmental Conservation*. New York: Routledge, pp. 120–134

Chimakonam, J.O. (2021a) 'On the System of Conversational Thinking: An Overview', *Arụmarụka: Journal of Conversational Thinking*, 1(1), pp. 1–45.

Chimakonam, J.O. (2021b) 'What Is This Thing Called the System of Conversational Thinking (SCT)? a Reply to Critics', *Arụmarụka: Journal of Conversational Thinking*, 1(2), pp. 137–168.

Chimakonam, J.O. and Ogbonnaya, L.U. (2021) *African Metaphysics, Epistemology and a New Logic*. Cham: Springer International Publishing.

.Chimakonam, J.O. (2022) *A Journey Through the Contemporary Philosophy Of The African Continent, Country By Country* (2022). SOAS University of London. Available at: https://www.youtube.com/watch?v=_dm9dJf8UzM (Accessed: 27 June 2022).

Cilliers, P. (1998) *Complexity and Postmodernism: Understanding Complex Systems*. London: Routledge.

Clarke, M.L. (2019) 'New Waves: African Environmental Ethics and Ocean Ecosystems', in M. Chemhuru (ed.) *African Environmental Ethics: A Critical Reader*. New York: Springer International Publishing (The International Library of Environmental, Agricultural and Food Ethics, 29), pp. 153–173.

Climate Justice Charter Movement (2022) 'CJCM Strategic Perspectives: Accelerate the Deep Just Transition Now!'. Available at: https://cjcm.org.za/download/8161c42e-f6d1-424e-a792-e43f48f2c5d0 (Accessed: 8 April 2024).

Climate Justice Charter Movement (2024) 'About the Climate Justice Charter'. Available at: https://cjcm.org.za/about (Accessed: 8 April 2024).

Climate Justice Coalition (2024) '"About Us", Climate Justice Coalition'. Available at: https://climatejusticecoalition.org/about-us/ (Accessed: 8 April 2024).

Etim, F. (2019) 'Ontology of African Ritual', *Advances in Applied Sociology*, 09(01), pp. 1–14.

Evans, J. (2021) 'Activists Demand Renewable Energy and Community Consent in Mining', *Daily Maverick*. Available at: https://www.dailymaverick.co.za/article/2021-09-22-gwede-must-go-activists-voice-demands-on-renewable-energy-and-community-consent-in-mining/ (Accessed: 8 April 2024).

Evans, J. (2022) 'Energy Minister Gwede Mantashe Explains His Reasons for Sticking with Coal', *Daily Maverick*, 13 January. Available at: https://www.dailymaverick.co.za/article/2022-01-14-the-just-transition-energy-minister-gwede-mantashe-explains-his-reasons-for-sticking-with-coal/ (Accessed: 8 April 2024).

Hoyle, R. and Wallace, J. (2022) 'Coal Runs Hot as Russia-Ukraine War Risks Europe's Climate Agenda', *Wall Street Journal*, 4 March. Available at: https://www.wsj

.com/articles/coal-runs-hot-as-russia-ukraine-war-risks-europes-climate-agenda-11646402902 (Accessed: 8 April 2024).

Iteyo, C. (2009) 'Belief in the Spirits of the Dead in Africa: A Philosophical Interpretation', *Though and Practice: A Journal of the Philosophical Association of Kenya*, 1(1), pp. 147–159.

Lenart, B. (2010) 'Enlightened Self-Interest: In Search of the Ecological Self: A Synthesis of Stoicism and Ecosophy', *Praxis*, 2(2), pp. 26–44.

Manzini, N.Z. (2019) 'African Environmental Ethics as Southern Environmental Ethics', in M. Chemhuru (ed.) *African Environmental Ethics: A Critical Reader*. New York: Springer International Publishing, pp. 111–126.

Morin, E. (2008) *On Complexity*. Cresskill: Hampton Press, Inc.

Naess, A. (1989) *Ecology, Community and Lifestyle: Outline of an Ecosophy*. Print. Translated by David Rothenberg. Cambridge: Cambridge University Press.

Naess, A. (2002) *Life's Philosophy: Reason & Feeling in a Deeper World*. Interview by Haukeland, P.I. Print. Huntford, R. (trans.). Athens: University of Georgia Press.

Naess, A. (2008) *Ecology of Wisdom: Writings by Arne Naess*. Drengson, A. and Devall, B. (eds.). Berkeley: Counterpoint.

Poli, R. (2013) 'A Note on the Difference Between Complicated and Complex Social Systems', *Cadmus*, 2(1), pp. 142–147.

Praeg, L. (2019) 'Philosophy on the Border', in L. Praeg (ed.) *Philosophy on the Border: Decoloniality and the Shudder of the Origin*. Pietermaritzburg: University of KwaZulu-Natal Press, pp. 3–25.

Ramose, M.B. (2003) 'The Philosophy of Ubuntu and Ubuntu as Philosophy', in P.H. Coetzee and A.P.J. Roux (eds.) *The African Philosophy Reader: A Text with Readings*. London: Routledge, pp. 270–280.

Ramose, M.B. (2009) 'Ecology Through Ubuntu', in M.F. Murove (ed.) *African Ethics: An Anthology for Comparative and Applied Ethics*. Pietermaritzburg: University of Kwazulu-Natal Press, pp. 308–314.

Rothenberg, D. (1986) 'Introduction: Ecosophy T - from Intuition to System', in Naess, A. (ed.) *Ecology, Community and Lifestyle: Outline of an Ecosophy*. Print. Rothenberg, R. (trans.) Cambridge: Cambridge University Press, 1989.

Teffo, L.J. and Roux, A.P.J. (2003) 'Metaphysical Thinking in Africa', in P.H. Coetzee and A.P.J. Roux (eds.) *The African Philosophy Reader: A Text with Readings*. 2nd ed. New York: Routledge, pp. 192–208.

Woermann, M. (2016) *Bridging Complexity and Post-Structuralism*. New York, NY: Springer.

14 Integrating Deep Ecology, Degrowth, and Ubuntu

Foundations for an Ecosophy

Aïda C. Terblanché-Greeff

Introduction

It can be argued that the current dominant human-nature relationship is fundamentally anthropocentric, where human interests and well-being are prioritised. That is, humans have intrinsic value (i.e., value in and of themselves), whereas nature has instrumental value (i.e., value based on utility for humankind). Humans are seen as superior to, and removed from, nature. In general, humans – as *homo economics* – are focused on how the natural environment can be used to facilitate economic development in the dominant Western capitalistic paradigm. This dominant economic orientation, with its focus on unbridled growth, has led to the global ecological crisis faced by humankind in the 21st century. However, "responses to the global ecological crisis cannot be based on modes of thinking that created the problems in the first instance and would therefore require responses that are counter to such thinking" (Le Grange, 2015, p. 307).

One such response is made by the Norwegian philosopher Arne Næss, who faults the Global North for the arrogance of its human-centred instrumentalisation of nature. In 1973, he made a distinction between shallow ecology and deep ecology (see Næss, 1973). Shallow ecology focuses on the symptoms of the ecological crisis affecting humankind. On the other hand, deep ecology focuses on more fundamental issues, like the underlying cause of natural resource depletion, viz. the dominant, anthropocentric Western capitalistic paradigm.

Næss is considered the founder of the Deep Ecology Movement. The movement gained traction in the West in the late 1970s. It was later underpinned by its eight principles that formed the broadly construed Deep Ecology Platform (see Næss and Sessions, 1984). Ultimately, the Deep Ecology Movement called for a renewed emphasis on the interconnectedness of living beings and the natural environment, and a prominent change in human behaviour and values to facilitate environmental sustainability and harmony with nature. This marks a movement from anthropocentrism (based on shallow ecology) to ecocentrism (based on deep ecology). Ecocentrism holds that all of nature (including humans as part of ecosystems) has intrinsic value, and moral consideration should be extended to the environment as a whole regardless of its utility to humankind (Le Grange, 2015, p. 305).

DOI: 10.4324/9781003480914-15

Ideally, this move from anthropocentrism to ecocentrism would change the hierarchy in the dominant human-nature relationship such that humans are considered a part of, and therefore equal to, nature. Nonetheless, the main focus of developed countries in the Global North remains anthropocentric. This became prevalent with the focus on sustainable development: the notion that current human societies meet their needs without compromising the ability of future generations to meet their needs (see World Commission on Environment and Development [WCED], 1987).

The sustainable development paradigm has been adapted according to capitalistic ideals, where economic growth is prevalent in sustainable development strategies (Asara *et al*., 2015, p. 380).The capitalistic traits of sustainable development paired with the current socio-economic-environmental crisis motivated the formulation of an alternative reaction to sustainable development and growth, namely *degrowth* (Zozuˇlakova, 2016, p. 187). "Degrowth concerns appeared from a grassroots social movement that arose as a critique of growth, and that has tried to raise awareness about alternative lifestyles that can be more sustainable" (Cosme, Santos and O'Neill, 2017, p. 327).

The proponents of the Degrowth movement still do not agree on a set definition of their central concept. Nonetheless, as a starting point, Degrowth can be described as:

> a voluntary transition towards a just, participatory, and ecologically sustainable society... The objectives of degrowth are to meet basic human needs and ensure a high quality of life, while reducing the ecological impact of the global economy to a sustainable level, equitably distributed between nations.
> (Research and Degrowth, 2010, p. 524)

A third theory to consider in debates on "ecological and social limits to economic growth, and new ways to deal with resource scarcity without compromising human well-being" (Cosme, Santos and O'Neill, 2017, p. 321) can be found in African philosophy; more specifically, the aspects of Ubuntu as it pertains to environmental ethics (see Terblanché-Greeff, 2019). The aspects under consideration are Ubuntu communitarianism; Ukama as a concept that refers to relatedness between humans, non-humans, and nature; and the principle of sufficiency, which prescribes a lifestyle based on "enoughness ".

Deep Ecology, Degrowth, and Ubuntu call for a human-nature relationship that differs from the one prevalent under the Western capitalistic paradigm. Various scholars have investigated the links between Ubuntu and Deep Ecology (see Le Grange, 2015; Mangena, 2015; Ibanga, 2017), as well as Ubuntu and Degrowth (see Hoeft, 2018). However, something that has not yet been done is the simultaneous comparison between Deep Ecology, Degrowth, and Ubuntu. In this chapter, I explore the potential overlap between the principles of the Deep Ecology Platform, the tenets of Degrowth, and the aspects of Ubuntu prevalent in environmental ethics. I argue that Ubuntu Degrowth – as a novel ecosophy[1] – can be posed as a viable, nuanced framework for promoting a harmonious human-nature relationship that is sustainable. Attention will be given to the theoretical landscape concerning the three theories under investigation. I start by discussing the aspects of Ubuntu prevalent in environmental ethics (Terblanché-Greeff, 2019), followed by the tenets of Degrowth (Demaria *et al.*, 2013). To avoid repetition, I unpack the principles of the Deep Ecology Platform (Næss and Sessions, 1984), while simultaneously offering a comparative analysis between the three theories. The analysis sets the stage for the final section, in which I present Ubuntu Degrowth as an ecosophy.

Aspects of *Ubuntu* Pertaining to Environmental Ethics[2]

Ubuntu (as a value system) is difficult to define, and scholars have yet to agree on a common definition. Nonetheless, Ubuntu (humanness) is a cultural concept from sub-Saharan Africa embodied by the maxim "umuntu ngumuntu ngabantu" (in Nguni languages), which translates to the pervasive maxim "a person is a person through other persons". When relating the maxim to African environmental ethics, it can be rewritten more distinctly as "a person is a Person through others". More specifically, "the word 'person' (small letter *p*) refers to individual humans, whereas 'Person' (capital letter *P*) refers to the personhood, selfhood, and humanness[3] a person should strive for by interacting with 'others' (humans, non-humans, and nature) to *become* fully human" (Terblanché-Greeff, 2019, p. 97).

According to Thaddeus Metz (2011, p. 537), "[o]ne can be more or less of a person, self or human being, where the more one is, the better". From this, the main goal in life should be to *become* a Person – to attain Ubuntu through moral interactions with others. This gives Ubuntu ethics its relational rationality.

To unpack Ubuntu as it pertains to environmental ethics, attention will be allocated to three prominent aspects: (1) Ubuntu communitarianism, (2) Ukama as relatedness, and (3) the principle of sufficiency. These aspects are derived from the critique against the Western capitalistic paradigm prevalent in the Global North (see Terblanché-Greeff, 2019).

Ubuntu Communitarianism

The relational rationality of Ubuntu ethics suggests that a person is fundamentally group-orientated. Ubuntu (humanness) is marked by communitarianism, whereby the person does not exist in isolation, as is prevalent in the individualistic-orientated capitalistic paradigm of the Global North. Instead, the person is inherently social and communal, focusing on meaningful social relationships and interdependency between various persons.

Ubuntu communitarianism is underpinned by harmonious relationships, which is a "combination of two logically distinct ways of interacting, namely, identifying with others and exhibiting solidarity with them" (Metz, 2022, p. 91; see Metz, 2011, 2016, 2017, 2019, 2022). Metz (2022, p. 93) elucidates that *identity* entails "considering oneself part of the whole, being close, participating, sharing a way of life, belonging, and thinking of oneself as bound to others…". Moreover, *solidarity* can be described as "achieving the good of all, being sympathetic, sharing, promoting the common good, engaging in service, and being committed to others' good…" (Metz, 2022, p. 93).

Although Metz (2011) did not originally intend for the aspects of identity and solidarity to be applied to humans, non-humans, and nature (hereafter referred to as *others*) as identified in the maxim "a person is a Person through others", I argue that his formulation of harmonious relationships can be employed to this end. This is based on the concept of *Ukama*. Melanie Carstens and Rita Presser (2024, p. 1) state that:

> "[U]buntu (humanness) and [U]kama (relatedness), as predominantly observed in southern Africa, are essentially immersed in human-nature relationality. This type of relationality considers everything as interconnected, and therefore that nothing happens in isolation, and that the well-being of humans is inextricable from the well-being of nature.

Ukama as Relatedness

It can be argued that a focus on harmonious relationships – interaction based on identity and solidarity – extends to human-nature relationality. This becomes pertinent when considering the concept of Ukama (as a broader concept than Ubuntu). Ukama, prevalent in Zimbabwean Shona traditional thought, refers to relatedness; that is, relatedness between *others*.[4]This focus on relationality becomes apparent when we consider that "no person is complete in him/herself; s/he is fully human in as far as s/he remains a part of the web of life, including creation and the earth" (LenkaBula, 2008, p. 378). Lesley Le Grange (2012, p. 338, emphasis added) states that:

> [U]buntu needs to be understood as a microcosm of (or concrete expression of) a broader concept [U]kama, which means relatedness – relatedness to the entire cosmos. Understood in this way, [U]buntu is not only concerned about the development of the [moral] self in an individualistic way, but that the development of the true self is inextricably bound up in relationships with other humans *and non-human nature*.

It can thus be said that harmonious human-nature relatedness plays a fundamental role in *becoming* a Person. This *becoming* facilitated through relatedness does not merely entail caring for the self and other humans; instead, there is a duty to care for *non-humans and nature* because they have intrinsic value.

The focus on harmonious relationships and relatedness is significant when discussing Ubuntu and, by extension, Ukama as concerns the human-nature relationship in African environmental ethics. Akin to Ubuntu (where communitarianism is prevalent), Ukama exemplifies a sense of interdependence (identity) and collective responsibility (solidarity), thus promoting an ethic that values community over individualism (as prevalent in the capitalistic paradigm). Concurrently, Joseph P. Hungwe and Amasa P. Ndofirepi (2021, p. 114) opine that Ukama finds resonance in communitarianism as prevalent in Ubuntu.

Ukama is based on holism: a person's well-being has ontological meaning in relationality with others (Murove, 1999, p. 84). Ukama further encompasses *beneficence*, such that:

> [a]ll societies accept this universal principle [of beneficence] as essential for true and enduring well-being, though Africans go a step further [with Ukama] and stress that all existence, including being human, can best be understood and interpreted in the context of one's symbiosis with [*others*].
>
> (Ndofirepi and Shanvanana, 2015, p. 432)

According to Ukama, and akin to solidarity in Ubuntu, "[t]his unity [between *others*] translates into a belief in mutual sharing and co-operative participation" (Ndofirepi and Shanvanana, 2015, p. 433).

Ukama has a macro cosmovision. Its emphasis on relationality entails a bond between past, present, and future generations as part of the web of life. That said, Ubuntu, as a concrete form of Ukama, prescribes moral responsibilities to intergenerational others. Specifically, "[r]esponsibility towards others is founded on Ukama, and this dual moral responsibility that extends to past and future generations is based on respect and gratitude" (Terblanché-Greeff, 2019, p. 99). Ubuntu thus prescribes a "way of life that sustains human and non-human life in all shapes and forms, and across generations" (Carstens and Preiser, 2024, p. 4).

The dual moral obligation placed on the present generation is based on identity and solidarity in Ubuntu. That is,

> [t]he identity of a person [based on a we-orientation] includes both past and future generations. The current generation acknowledges that predecessors lived in solidarity which extended to intergenerational relatedness, and thus the current generation must act altruistically towards future generations.
>
> (Terblanché-Greeff, 2019, p. 99)

Accordingly, moral responsibility should be based on harmonious relationships with *others* through the principle of sufficiency.

The Principle of Sufficiency per Ubuntu

When we consider contemporary statements regarding how the current generation must meet their needs without compromising the ability of future generations to meet their own needs, the distinction between differing kinds of needs is often not clearly stipulated. It is useful to distinguish between basic and false needs. Basic needs are those that are required for survival and flourishing. False needs refer to "economic and material 'wants' that are not considered essential for survival" (Terblanché-Greeff, 2019, p. 99). *Homo economicus* is under the delusion that false needs are equivalent to basic needs. This delusion is fuelled by the individualistic capitalistic Western paradigm, where "Homo œconomicus – economic human – pursues false needs in the name of economic growth and development with limited attention being allocated to the differentiation between basic and false needs" (Terblanché-Greeff, 2019, p. 100). This lack of differentiation and focus on non-specific needs (whether it be basic or false) contributes to the contemporary human-nature relationship marked by the domination and commodification of nature.

Ubuntu prescribes a nuanced human-nature relationship that stands in sharp contrast to the needs-fulfilment behaviour of *Homo economicus*. Relationality is not purely human-centred and includes *others* based on Ukama (relatedness). In *becoming* a Person through meaningful relationships with *others*, it would be counter-intuitive to allocate instrumental value to nature to satisfy individualistic self-interest and the unnecessary accumulation of material wealth. Instead, there should be a focus on *sufficiency* whereby a person's needs are met in relation to those of *others* (Murove, 2014, p. 40); "By striving for a lifestyle based on sufficiency, the person and the community only utilize resources in nature to satisfy their basic needs" (Terblanché-Greeff, 2019, p. 102). To ensure all persons have access to the resources needed to live a dignified life, goods should not be wasted or hoarded. Rather, sufficiency calls for a lifestyle of "enoughness" wherein resources are distributed justly for all to flourish.[5] This need for flourishing and dignity extends not only to humans but to *others* as well. This is prescribed by the principle of sufficiency as an aspect of Ubuntu as it pertains to environmental ethics.

Ubuntu takes an antagonistic stance to *Homo economicus's* "market-orientated economic logic of maximalisation [sic]" (Van Binsbergen, 2001, p. 58). Ubuntu is "anti-egoistic as it discourages people from seeking their own good without regard for, or to the detriment of, others and the community" (Munyaka and Motlhabi, 2009, pp. 71–72). This is based on harmonious interactions based on identity. When considering interactions based on solidarity, Ubuntu prescribes the principle of sufficiency "whereby the present and future generations are provided with resources to meet their basic needs through acts of altruism and the achievement of shared goals" (Terblanché-Greeff, 2019, p. 99).

Tenets of Degrowth

This section outlines Degrowth's most prominent tenets (or sources)[6] as proposed by Federico Demaria, François Schneider, Filka Sekulova, and Joan Martinez-Alier (2013, pp. 195–201). Although the given tenets are not exhaustive, they are a good start to articulate the main themes in the Degrowth movement. The entangled degrowth tenets are: 1) Ecology, 2) Critiques of development, 3) Meaning of life and well-being, 4) Bioeconomics, 5) Democracy, and 6) Justice.

Tenet 1: Ecology

The focus on ecology is one of the most prominent Degrowth tenets. It argues that ecosystems have intrinsic value regardless of their instrumental value to humankind (Demaria *et al.*, 2013, p. 195). The attention to ecology highlights the tension between ecosystems and industrialised systems of (over)production and (over)consumption, with ecological destruction brought about by the (industrialised) development predominant in Western capitalism. Degrowth thus establishes itself as a possible solution to the tension created by this problematic contemporary human-nature relationship. Degrowth is characteristic of the *res communis* ("common property" or "common things") approach, which states that the ecosystem, and its parts, should be cared for and shared in a communal manner. This calls for a move away from the individualism of *Homo economicus*. Significantly, then, "...*res communis* implies an integration of humans in nature [as part of the ecosystem]..." (Demaria *et al*. 2013, p. 196).

Tenet 2: Critiques of Development

The second tenet of Degrowth is concerned with the critique of development, which is a social construct imposed by the Global North on the rest of the world. The critique is levelled against the homogenisation of diverse cultures due to globalisation and the concomitant use of various contemporary technologies paired with production and consumption modes prevalent in the Global North.[7]

Sustainable development focuses on alternative, greener, and improved development. One might argue that sustainable development is one model that might address some of the challenges caused by unbridled development and growth. Yet, this is not the case in the context of Degrowth, where the construct of sustainable development is considered an oxymoron (Demaria *et al*., 2013, p. 196): how can development be sustainable if it is often at the expense of nature owing to this need to continuously maximise utility as *Homo economicus*? Degrowth not only critiques development – whether sustainable or unrestricted development – but calls for a paradigmatic shift away from this socially constructed concept of sustainable development toward a new vision of a global society where production and consumption are significantly reduced. Degrowth therefore emphasises a state of affairs that is "*different,* not only *less*. Degrowth signifies a society with a smaller metabolism... a society with a metabolism which has a different structure and serves new functions" (D'Alisa, Demaria and Kallis, 2014, p. 4).

Moreover, the proponents of Degrowth advocate for displacing the dominant growth imperative and focusing on *sufficiency* (Hoeft, 2018, p. 16). Concurrently, importance should be allocated to "economic relations based on sharing, gifts and reciprocity, where social relations and conviviality are central. The focus here is on the change in the structure of values and the change in value-articulating institutions" (Demaria *et al*., 2013, p. 197). Degrowth proponents argue against measuring prosperity (or the "good life") based on economic indicators (like gross domestic product [GDP]); instead, social and environmental factors should be incorporated in assessing well-being and quality of life.

Tenet 3: Meaning of Life and Well-Being

In contemporary society, *Homo economicus* often lives a lifestyle that is marked by excessive working, earning, buying, and selling. The focus is on material accumulation, which contributes to one's meaning of life and subjective well-being. The third tenet of Degrowth critiques this approach to meaning and well-being based on economic indicators. Degrowth proponents call for a different approach to finding meaning and well-being; one which involves the "movement for voluntary simplicity, reducing individual consumption while seeing simple life as liberating and profound rather than restraining and limiting" (Demaria *et al.*, 2013, p. 197). The aim is to promote lifestyles that prioritise meaningful relationships and shared experiences over well-being founded on material accumulation.

Tenet 4: Bioeconomics

In the fourth tenet of Degrowth, Bioeconomics, principles of biology and ecological economics are integrated to steer economic systems to levels of sustainability and resilience. Simultaneously, it features an advocacy for reducing overall economic activity to achieve biophysical balance, well-being, and social justice. Bioeconomics emphasises the need to utilise economic systems that function within ecological limits and prioritise well-being over interminable growth. There is also a call to question and redefine concepts of progress and prosperity in ways that are compatible with the long-term health of humans and nature. "Degrowth has a discriminating approach to reducing economic activity. It seeks to scale down ecologically destructive and socially less necessary production …, while expanding socially important sectors like healthcare, education, care and conviviality" (Hickel, 2021, p. 1108).

Additionally, in accordance with tenet 2, critique is also levelled against techno-optimism, the "belief in ecological modernisation which claims that new technologies and efficiency improvements are key solutions to the ecological crisis" (Demaria *et al.*, 2013, p. 198). Technological innovation is a highly debated subject in the Degrowth movement. Degrowth proponents question the ability of such innovations to surpass environmental limits while simultaneously sustaining unchecked economic growth.

Tenet 5: Democracy

The fifth tenet is based on the need for a deeper or different democracy. Degrowth proponents argue that there is a clear lack of debate and democratic dialogue regarding growth, economic development, and technological innovation. Degrowth therefore emphasises the need for more participatory democratic decision-making when formulating alternative political and economic systems. This further entails the re-evaluation and decentralisation of power relations to facilitate the creation of policies that prioritise the well-being of communities and ecosystems over the interests of *Homo economicus*. According to Demaria *et al.* (2013, p. 199), opposing viewpoints exist between "those who defend present democratic institutions considering the risks of losing what we have achieved (a more reformist strand), and those who demand completely new institutions based on direct and participatory democracy (more alternative, or post-capitalist vision)" (see Demaria *et al.*, 2013 for more detail).

Tenet 6: Justice

The sixth tenet of degrowth discussed by Demaria *et al.* (2013) focuses on justice and a "degrowth" of inequality. Firstly, there is a common misconception among economists that economic growth is the answer to improving the living conditions of the poor. This problematic assumption is further based on the improbability of *Homo economicus'* voluntary income reduction and redistribution of wealth. Some proponents of Degrowth argue that it is exactly this trickling-down of resources that is needed to address some global injustices. Following this trickle-down hypothesis (see Snowdon, 2006), "degrowth opts for less competition, large scale redistribution, sharing and reduction of excessive incomes and wealth" (Demaria *et al.*, 2013, p. 199).

Justice then necessitates such redistribution within and between the economies of the Global North and Global South insofar as global justice demands a "concern for a fair distribution of economic, social and environmental goods and bads [sic] at all time-lines" (Demaria *et al.*, 2013, p. 200). This further demonstrates the need for an intra-generational and intergenerational dimension to justice.

This tenet thus advocates for the fair distribution of resources and opportunities to address social inequalities, such as equitable access to education, employment, and healthcare, to ensure that basic needs are met. Accordingly, Jason Hickel (2021, p. 1108) states that "Degrowth seeks to expand universal public goods and services, such as health, education, transportation and housing, in order to decommodify the foundational goods that people need in order to lead flourishing lives".

Principles of Deep Ecology Compared with Degrowth and Ubuntu

While an *ultimate* ecological philosophy seems somewhat untenable, we have recourse to common ecological themes and concerns to formulate an ecosophy with specific values. Here, I investigate the principles of the Deep Ecology Platform, general tenets of Degrowth, and aspects of Ubuntu to facilitate a comparative analysis highlighting commonalities and differences. I aim to demonstrate how these three theories can complement a context-specific inspired ecosophy (viz., Ubuntu Degrowth). Accordingly, the core values of Deep Ecology (derived from the principles) will serve as a foundation on which Ubuntu Degrowth is formulated as an ecosophy that promotes an alternative human-nature relationship that is sustainable and harmonious.

In order to move towards such a formulation, the eight principles underpinning the Deep Ecology Platform (Næss and Sessions, 1984, p. 5) will be unpacked and compared with aspects of Ubuntu and tenets of Degrowth.

Principle 1: The wellbeing and flourishing of human and non-human life on Earth have value in themselves. These values are independent of the usefulness of the non-human world for human purposes.

Firstly, with the term *life,* Næss refers to the biosphere *in toto*, which includes "individuals, species, populations, habitat, as well as human and non-human cultures ...[that are based on] *all-pervasive intimate relationships*" (Næss, 1995, p. 68, emphasis added). This focus on the relationality of all life prompts Diana-Abasi Ibanga (2017, p. 105) to describe Deep Ecology as an "ecocentric communitarian system of ethics". Furthermore, the emphasis on the intrinsic value of all life, instead of the instrumental value of the environment for humankind, moves us away from anthropocentrism towards ecocentrism.

Turning now to Ubuntu, its two prominent aspects are communitarianism and Ukama (relatedness). These aspects resonate with Deep Ecology as they focus on the interrelationship between the person and *others*. Lesley Le Grange (2012, p. 333) states that the "idea of the oneness of humans with nature expressed in Arne Næss's (1973) deep ecology… might have resided among African people for many centuries through notions such as [U]kama". Furthermore, both Ukama and the Degrowth tenet of ecology (tenet 1) affirm that nature, as part of a community of *others*, has intrinsic, and not merely instrumental, value.

Principle 2: Richness and diversity of life forms contribute to the realization of these values and are also values in themselves.

This second principle features a holistic perspective emphasising the interdependence and interconnectedness between all life forms. A variety of life forms – as part of a multitude of ecosystems – can contribute to the overall well-being and functioning of the ecosystem. The ecosystem's complexity diminishes in conjunction with any reduction in the richness and diversity of its life forms. This richness and diversity enhance the ecosystem's resilience and adaptive capacity when faced with stressors, as the system is self-regulatory and self-maintaining. Consequently, a decrease in diversity can negatively influence functioning within an ecosystem. Furthermore, relationality in ecosystems contributes to richness and diversity, which can be considered values in and of themselves.

The above focus on holism and relationality accords with Ukama as relatedness. Interdependence and interconnectedness are supported by the notion that what happens to the individual affects the whole community, which is prevalent in communitarianism. Furthermore, the richness and diversity of the community (which includes *others*) facilitate flourishing: "[t]hrough the interaction with unique others, a person can subjectively grow and attain Ubuntu – to become truly human (Person)" (Terblanché-Greeff, 2019, p. 98).

In regard to the above, Degrowth recognises the need for diversity in symbiotic relationships between *others*. This interdependence means that the well-being of a person is dependent on the well-being of *others*. Degrowth thus identifies the need for diversity and richness of life forms and calls for various systems that are not confined to the Western capitalistic paradigm. By critiquing the homogenising effects of globalisation, it allows for greater cultural and economic diversity to engender a more communitarian human-nature relationship.

Principle 3: Humans have no right to reduce this richness and diversity except to satisfy vital needs.

The third principle is restrictive regarding the contemporary prevalence of human interference in ecosystems. Interference should be restricted to minimal levels based on the satisfaction of vital needs. Humans should prioritise vital needs over false needs (where the latter results in unnecessary exploitation of nature) to ensure the long-term prosperity of the ecosystem.

In accordance with restrictive interference, Ubuntu's principle of sufficiency highlights basic needs, akin to the vital needs proposed above: "[b]y striving for a lifestyle based on sufficiency, the person and the community only utilize resources in nature to satisfy their basic needs" (Terblanché-Greeff, 2019, p. 102). The principle of sufficiency thus advocates for materialistic minimalism, whereby well-being is based on meaningful relationships with *others* instead of the satisfaction of false needs.

Similarly, it can be argued that Degrowth also proposes a minimalistic lifestyle whereby the basic needs of all are met. This should be achieved through the just (re)distribution of resources (tenet 6). Additionally, in critiquing development and techno-optimism, Degrowth calls for less interference with nature instead of an overreliance on technological innovation that might further exacerbate the negative ecological impact of *Homo economicus* (tenet 4). By promoting a different human-nature relationship and a different societal metabolism, Degrowth emphasises practices that do not compromise the ability of ecosystems to regenerate and support life. This is on par with Ukama as relationality when considering that human well-being is related to the well-being of *others*.

Principle 4: The flourishing of human life and culture is compatible with a substantial decrease of the human population. The flourishing of non-human life requires such a decrease.

The fourth principle is by far the most controversial and heavily critiqued. It states that a smaller global human population will lead to higher overall well-being for all life forms. It is argued that fewer humans will mean less interference and overconsumption and, ultimately, a more sustainable ecological footprint for humankind. This will relieve some of the pressure imposed on ecosystems (viz., pollution, habitat destruction, resource extraction, etc.) to allow them to recover and flourish.

I argue that this is the first Deep Ecology principle that does not apply to Ubuntu. Attaining Ubuntu cannot happen in isolation, and well-being is based on harmonious relationships through interactions marked by identity and solidarity. That said, Ubuntu cannot be categorised as ecocentric in the same manner as Deep Ecology (see Le Grange, 2015). To be sure, Ubuntu emphasises human relationality (albeit not in a predominantly anthropocentric sense); this is not to say that it precludes the possibility of a more intimate human-nature relationality.

Degrowth is relatively silent on the claim that a decrease in the human population is needed to facilitate alternative lifestyles. In fact, tenet 3 focuses on well-being based on meaningful human relationships, which might indicate a more anthropocentric stance in Degrowth.

Principle 5: Present human interference with the non-human world is excessive, and the situation is rapidly worsening.

The fifth principle evaluates the excessive and unsustainable anthropogenic behaviour towards the natural environment. The rapidly worsening effects of extreme human interference (see Intergovernmental Panel on Climate Change [IPCC], 2023) indicate the urgent need to address the problematic human-nature relationship by investigating and changing social norms and values such that this relationship can become more sustainable and harmonious.

Ubuntu prescribes a nuanced human-nature relationship that can be contrasted to the domination over, and commodification of, nature. The principle of sufficiency advocates for a sustainable lifestyle based on "enoughness", which aims to sustain the life of *others* even across generations.

The principle of sufficiency is comparable to sustainable development (as a way of addressing the worsening situation) as regards the intergenerational aspect. However, it is:

> not equated to the sustainable development paradigm, which focuses heavily on development [and technological innovations] as a means for the current generation to meet their basic and false needs while also ensuring that future generations will be able to meet their needs.
>
> (Terblanché-Greeff, 2019, p. 99)

In this regard, Degrowth calls for a move away from sustainable development (tenets 1, 2, and 4) by advocating for a human-nature relationship that is fundamentally different from sustainable development under the Western capitalistic paradigm.

Principle 6: Policies must therefore be changed. These policies affect basic economic, technological and ideological structures. The resulting state of affairs will be deeply different from the present.

The sixth principle builds on the fifth, proposing how the problematic human-nature relationship might be addressed via fundamental and holistic changes in societal systems. It argues that the problematic human-nature relationship is rooted in various structures: economic (capitalism, prioritising growth over sustainability), ideological (anthropocentrism), and technological (development and progress through technological advancements that contribute to environmental degradation). Existing structures and values should be re-evaluated in accordance with the long-term health of the ecosystem and societal values that are more ecocentric, sustainable, and harmonious with nature. Notably, then, "[d]ecisive improvement requires considerable change: social, economic, technological and ideological" (see Næss and Haukeland, 2002, pp. 108–109).

Notably, this focus on policy change is not entirely consistent with the more practical aspects of Ubuntu. Nonetheless, the required change might be achievable on a social, ideological, and economic level. Change on the social level is supported by Praeg (in an interview with Hoeft, 2018, p. 16), who states that "Ubuntu is not learnt through theory, but through practical socialization". That is, through meaningful, practical relationships with *others*, we can bring about societal change. Additionally, Ubuntu can facilitate change on the ideological level insofar as it challenges anthropocentrism by means of the element of Ukama. Finally, economic change becomes possible through the implementation of the principle of sufficiency and livelihoods based on "enoughness".

Akin to Deep Ecology, Degrowth demands policy changes (tenet 5). Moreover, *all* of the Degrowth tenets lend themselves to such changes as they focus on the problematic social, economic, technological, and ideological aspects of Western capitalism. The emphasis on "*different,* not only *less*" highlights the need for policy change that can usher in a future that is radically different from the present.

Principle 7: The ideological change will be mainly that of appreciating life quality (dwelling in situations of inherent value) rather than adhering to an increasing standard of living. There will be a profound awareness of the difference between bigness and greatness.

Principle 7 pertains to the ideological change required for a more harmonious human-nature relationship. Foremost, it highlights the importance of the qualitative and subjective aspects of life (such as healthy communal relationships with others) over the quantitative and objective measures (such as materialistic wealth accumulation) when determining well-being. Accordingly, attention should be allocated to the quality of life (i.e., *greatness*) instead of focusing on attaining a higher standard of living to increase well-being (i.e., *bigness*). Humans should lead a simplified, minimalistic lifestyle, where satisfying vital needs alone can bring about well-being.

The principle of sufficiency clearly prescribes a focus on greatness (through harmonious relationships with *others*) and a move away from bigness by instead focusing on "enoughness". Per Ubuntu (and the aspects of communitarianism and Ukama), it can be said that well-being is determined through meaningful interactions based on identity and solidarity as per the (reformulated) maxim "A person is a Person through others".

This focus on greatness is echoed in tenet 3 of degrowth. The notion that increased consumption and material wealth will lead to increased well-being is challenged. Instead, degrowth promotes alternative measures of well-being by prioritising community, quality of life, and personal fulfilment over material accumulation.

Principle 8: Those who subscribe to the foregoing points have an obligation directly or indirectly to try to implement the necessary changes.

The eighth principle is somewhat prescriptive of human behaviour. Humans who subscribe to the Deep Ecology Platform have an ethical obligation and responsibility to take action to facilitate the societal changes needed to develop an ecocentric, harmonious human-nature relationship. This calls for active engagement with others to encourage recognition of the intrinsic value of all life forms, which can be achieved through direct or indirect action.

Similarly, Ubuntu is also prescriptive as a normative environmental ethic. Ubuntu offers a tenable alternative human-nature relationship that must be considered when faced with the present climate crisis. By adopting aspects of Ubuntu, the needed change can be achieved (on a local or global scale).

Contrariwise, Degrowth is descriptive rather than prescriptive, inasmuch as it is primarily a movement rather than an ethic. Despite this, some Degrowth activists find resonance with (particular) Deep Ecology Platform principles, thus engendering the proposed changes through their direct or indirect actions.

In conclusion, the Deep Ecology Platform is a set of principles that can facilitate the theorisation of various ecosophies based on context and culture. Accordingly, Le Grange (2015, p. 304) states that:

> [i]t is important, however, to understand that 'deep ecology' is a movement not an ecophilosophy. It is a movement whose principles develop from the bottom up and is therefore a grass-roots movement. As a movement, it is not delinked from ecophilosophy/ecosophy, though.

As has been demonstrated, the Deep Ecology Platform can inform the development of diverse ecosophies, such as Ubuntu Degrowth.

Ubuntu Degrowth: An Ecosophy

The core values of Deep Ecology are twofold, based on the principles of the Deep Ecology Platform. The values can be reduced to (1) acknowledging that all life forms have intrinsic value and (2) the need for a human-nature relationship based on harmonious relationships between *others*. These core values are mostly on par with the tenets of Degrowth and the aspects of Ubuntu, thus forming the basis for Ubuntu Degrowth as an ecosophy.

Regardless of the shared core values mentioned above, one prominent difference is the place of the three approaches on the anthropocentric-ecocentric spectrum. Deep Ecology is predominantly ecocentric; Degrowth tenets can be perceived as ecocentric to a lesser degree with some elements of anthropocentrism (albeit not in the most extreme form of anthropocentrism as prevalent in the Western capitalistic paradigm). Ubuntu arguably lies somewhere between Deep Ecology and Degrowth on this spectrum, especially considering that Ubuntu "has very strong ecocentric leanings" (Le Grange, 2015, p. 305).

It has even been argued that Ubuntu overcomes the anthropocentric-ecocentric divide: "[n]urturing the self or caring for other human beings is not antagonistic towards caring for non-human nature (the more-than-human-world) – [U]buntu cannot simply be reduced to a category of anthropocentric or ecocentric" (Le Grange, 2015, p. 307). When considering the reformulated maxim "A person is a Person through others",others," the following rings true: "To *become* more fully human [viz., a Person] does not mean caring only for the self and other human beings but also for the entire biophysical world [viz., *others*]" (Le Grange, 2012,

p. 329, emphasis added). This relatedness (as Ukama) acknowledges that there is a *self* in *others* and vice versa (Ndofirepi and Shanvanana, 2015, p. 434).

This human-nature relatedness is prevalent in aspects of Ubuntu and tenets of Degrowth. Thies Hoeft (2018, p. 19) states that Ubuntu and Degrowth embody forms of communitarianism as they "promote conviviality, living in solidarity with others, and value the interest of other people because their well-being defines one's own well-being. These values demand a mindset which does not prioritise individual interests over the community's interest". This is exemplified by the harmonious communal relationship based on interactions marked by identity and solidarity as prevalent in Ubuntu.

Metz (in an interview with Hoeft, 2018, p. 20) states the following:

> Ubuntu is a natural way to support a Degrowth program… [That is,] [i]f the focus is to reduce production and consumption, or to at least steer the goal of production and consumption towards enhancing relationships instead of keeping us isolated[,] [a]nd if we furthermore say we want to decrease the grossly unequal distribution of wealth[,] that animals and nature deserve respect[,] and that we care for future generations. All those things grow out of any plausible interpretation of Ubuntu.

By drawing on the tenets of Degrowth and the aspects of Ubuntu, a new ecosophy can be posited, viz., Ubuntu Degrowth. As a result, communitarianism, Ukama as relatedness between *others*, and the prescribed principle of sufficiency can facilitate a prominent shift in practices and values to promote a holistic and harmonious human-nature relationship that stands in contrast to the dominant paradigm that enforces the behaviour of *Homo economicus*. Accordingly, Ubuntu Degrowth prescribes harmonious relationships between *others* (human, non-human, and nature) as well as alternative activities, whereby basic needs are met while ensuring a high quality of life, paired with reducing the economy's ecological impact to a sustainable level, marked by social justice.

To elaborate, harmonious relationships between *others* refer to a communitarian orientation and acknowledgement that all life forms have intrinsic value based on Ukama (relatedness). Alternative activities refer to behaviour inspired by the principle of sufficiency. Basic needs – required for survival and flourishing – can be met through lifestyles based on "enoughness", focusing on *greatness* rather than *bigness*. Here, quality of life (and well-being) is measured by meaningful relationships with others instead of material accumulation. Therefore, challenging the dominant growth-orientated development model (under the Western capitalistic paradigm) and implementing the principle of sufficiency may well lead to economic shrinkage, whereby the ecological impact of anthropogenic behaviour can be decreased to a sustainable level within biophysical boundaries. Based on communitarianism and relatedness, the aforementioned effects should then be distributed equitably between communities (and nations) to facilitate large-scale social justice.

From the discussions in this chapter, I propose that r*elational rationality* is the basic norm of Ubuntu Degrowth as an ecosoph*y*. Ubuntu (humanness) prescribes harmonious relationships to become a Person. This requisite moral interaction cannot happen in isolation, and there is a strong element of communitarianism via Ubuntu. This focus on relationality extends to the human-nature relationship based on Ukama as relatedness between *others*. This relatedness further calls for an intergenerational responsibility based on the principle of sufficiency.

Conclusion

Currently, humans dominate and (ab)use nature to such a degree that we will exceed planetary boundaries if we continue as *Homo economicus* under Western capitalism. When considering the problematic human-nature relationship and the climate crisis currently faced by all (humans, non-humans, and nature), it is no surprise that some grassroots movements – such as the Deep Ecology Movement and the Degrowth Movement – call for a renewed approach to human-nature relationality. Investigating the theories underlying these movements insofar as common overlaps can be utilised to develop novel ecosophies is useful. That is, nuanced and particular philosophies or worldviews (based on the broader field of ecophilosophy, such as African environmental ethics) that investigate and address various ecological concerns to advance a harmonious human-nature relationship that differs from the Western capitalistic paradigm.

In this chapter, attention was directed towards aspects of Ubuntu as they pertain to environmental ethics, tenets of Degrowth, and the principles of the Deep Ecology Platform in order to facilitate a comparative analysis, highlighting commonalities and differences between the underlying theories. Ubuntu Degrowth as an ecosophy was then presented to facilitate an alternative human-nature relationship based on the basic norm of *relational rationality*.

To conclude, Ubuntu Degrowth is an amalgamated ecosophy, drawing from Ubuntu, which is prevalent in the Global South, and Deep Ecology and Degrowth, prevalent in the Global North. This amalgamation is motivated by Ibanga (2017, p. 117), who states:

> …[T]he interest of the environmental ethicists should not be limited to some specific territories but should extend to include the entire global ecosystem in a cosmic gestalt sense. It is important to note that one cannot be an African environmental ethicist without simultaneously being concerned with ecological matters and interests in other parts of the globe… Her interest should extend and significantly include the ecological and social issues in parts of the world hither and thither. African environmental ethics should synchronize [sic] with environmental ethics in other parts of thc world. All people in all parts of the world are relationally trapped in the same ecological turmoil.

Notes

1 I argue that an ecosophy is a nuanced philosophy or particular worldview (based on the broader field of ecophilosophy) that investigates and addresses various ecological con-

cerns in an attempt to postulate a harmonious human-nature relationship. Furthermore, an ecosophy is based on a *basic norm*.

2 The content of this section is discussed in Terblanché-Greeff, 2019.

3 The term humanness refers to the formation of "one's (moral) personhood, a prescription to acquire Ubuntu" (Metz, 2011, p. 537).

4 It is important to state that Ukama (in traditional Shona thought) includes relationality with spiritual ancestors or predecessors. In this chapter, reference will only be made to predecessors without a specific spiritual dimension.

5 See Terblanché-Greeff, 2019, for a practical example of the embodiment of the principle of sufficiency.

6 The degrowth sources are based on the original work of Flipo (2007) and have been adapted and expanded on by Demaria *et al.* (2013) who also added the sixth tenet: Justice.

7 See Terblanché-Greeff (2019) for a relevant critique of the use of genetically modified crops in Africa.

References

Asara, V., Otero, I., Demaria, F. and Corbera, E. (2015) 'Socially sustainable degrowth as a social–ecological transformation: Repoliticizing sustainability', *Socially Sustainable Degrowth as a Social-Ecological Transformation*, 10, pp. 375–384. https://doi.org/10.1007/s11625-015-0321-9

Carstens, M. and Preiser, R. (2024) 'Exploring relationality in African knowledge systems as a contribution to decoloniality in sustainability science', *Ecosystems and People*, 20(1). https://doi.org/10.1080/26395916.2024.2315995

Cosme, I., Santos, R. and O'Neill, D.W. (2017) 'Assessing the degrowth discourse: A review and analysis of academic degrowth policy proposals', *Journal of Cleaner Production*, 149, pp. 321–334. https://doi.org/10.1016/j.jclepro.2017.02.016

D'Alisa, G., Demaria, F., and Kallis, G. (2014) *Degrowth: A Vocabulary for a New Era*. New York: Routledge, Taylor and Francis.

Demaria, F., Schneider, F., Sekulova, F. and Martinez-Alier, J., (2013) 'What is degrowth? From an activist slogan to a social movement', *Environmental values*, 22(2), pp. 191–215. https://doi.org/10.3197/096327113X13581561725194

Flipo, F. (2007) 'Voyage dans la galaxie décroissante', *Mouvements* 50(2), pp. 143–151.

Hickel, J. (2021) 'What does degrowth mean? A few points of clarification', *Globalizations*, 18(7), pp. 1105–1111. https://doi.org/10.1080/14747731.2020.1812222

Hoeft, T. (2018) *Towards the Pluriverse: How to connect Degrowth and Ubuntu*. Unpublished MSc Mini-dissertation. Lund University, Sweden. Available at: https://lup.lub.lu.se/luur/download?func=downloadFile&recordOId=8945696&fileOId=8945701 (Accessed: 27 April 2024).

Hungwe, J.P. and Ndofirepi, A. (2021) 'Centering "Ukama" ethic (relatedness) in the COVID-19 pandemic: New normal in African higher education', *International Journal of Higher Education*, 10(5), pp. 112–120. https://doi.org/10.5430/ijhe.v10n5p112

Ibanga, D.A. (2017) 'Is deep ecology inapplicable in African context: A conversation with Fainos Mangena', *Filosofia Theoretica: Journal of African Philosophy, Culture and Religions*, 6(2), pp. 101–119. https://doi.org/10.4314/ft.v6i2.6

Intergovernmental Panel on Climate Change [IPCC] (2023) 'Summary for policymakers.', in H. Lee and J. Romero (eds.) *Climate Change 2023: Synthesis report. contribution of working groups I, II and III to the sixth assessment report of the intergovernmental panel on climate change*. Geneva, Switzerland: IPCC, pp. 1–34, https://doi.org/10.59327/IPCC/AR6-9789291691647.001.

Le Grange, L. (2012) 'Ubuntu, ukama, environment and moral education', *Journal of Moral Education*, 41(3), pp. 329–340. https://doi.org/10.1080/03057240.2012.691631
Le Grange, L. (2015) 'Ubuntu/Botho as ecophilosophy and ecosophy', *Journal of Human Ecology*, 49(3), pp. 301–308. https://doi.org/10.1080/09709274.2015.11906849
LenkaBula, P. (2008) 'Beyond anthropocentricity—Botho/Ubuntu and the quest for economic and_ecological justice in Africa', *Religion and Theology*, 15, pp. 375–394. https://doi.org/10.1163/157430108X376591
Mangena, F. (2015) 'How applicable is the idea of deep ecology in the African context?', *Filosofia Theoretica: Journal of African Philosophy, Culture and Religions*, 4(1), pp. 1–16. https://doi.org/10.4314/ft.v4i1.1
Metz, T. (2011) 'Ubuntu as a moral theory and human rights in South Africa', *African Human Rights Law Journal*, 11(2), pp. 532–559. Available at: https://scielo.org.za/pdf/ahrlj/v11n2/11.pdf (Accessed: 27 April 2024).
Metz, T. (2016) 'An African theory of social justice', in C. Boisen and M.C. Murray (eds.) *Distributive justice debates in political and social thought*. New York: Routledge, pp. 171–190.
Metz, T. (2017) 'Values in China as compared to Africa: Two conceptions of harmony', *Philosophy East and West*, 67(2), pp. 441–465. https://doi.org/10.1353/pew.2017.0034
Metz, T. (2019) 'African communitarianism and difference', in E. Imafidon (ed.) *Handbook of African philosophy of difference*. Cham: Springer Nature, pp. 1–22. https://doi.org/10.1007/978-3-030-04941-6_2-1
Metz, T. (2022) *A relational moral theory*. Oxford: Oxford University Press.
Munyaka, M. and Motlhabi, M. (2009) 'Ubuntu and its socio-moral significance', in M.F. Murove (ed.), *African ethics: An anthology for comparative and applied ethics*. South Africa: University of Kwazulu-Natal Press. pp. 63–84.
Murove, M. F. (1999) *The Shona concept of ukama and the process philosophical concept of relatedness, with special reference to the ethical implications of contemporary neo-liberal economic practices*. Unpublished MA dissertation. University of Natal, Pietermaritzburg.
Murove, M.F. (2014) 'Ubuntu', *Diogenes*, 59(3–4), pp. 36–47. https://doi.org/10.1177/0392192113493737
Næss, A. (1973) 'The shallow and the deep, long-range ecology movement', *Inquiry*, 16, pp. 95–100. https://doi.org/10.1080/00201747308601682
Næss, A. (1995) 'The deep ecological movement: Some philosophical aspects', in G. Sessions (ed.) *Deep ecology for the twenty-first century*. Boston and London: Shambhala, pp. 64–84.
Naess, A. and Haukeland, P. (2002) *Life's philosophy: Reason and feeling in a deeper world*. Athens, GA: University of Georgia Press.
Næss, A. and Sessions, G. (1984) 'Basic principles of deep ecology', *Ecophilosophy* VI, pp. 5–7.
Ndofirepi, A.and Shanyanana, R. (2015). 'Rethinking Ukama philosophy in the context of "Philosophy for Children" in Africa', *Research Papers in Education*, pp. 1–14. https://doi.org/10.1080/02671522.2015.1073773
Research & Degrowth (2010) 'Degrowth declaration of the Paris 2008 conference', *Journal of Cleaner Production*, 18(6), pp. 523–524. https://doi.org/10.1016/j.jclepro.2010.01.012
Snowdon, B. (2006) 'The enduring elixir of economic growth: Xavier Sala-i-Martin on the wealth and poverty of nations', *World Economics*, 1(7), pp. 73–130.
Terblanché-Greeff, A.C. (2019) 'Ubuntu and environmental ethics: The West can learn from Africa when faced with climate change', in M. Chemhuru (ed.) *African environmental ethics: A critical reader*. Switzerland: Springer, pp. 93–109. https://doi.org/10.1007/978-3-030-18807-8_7

Van Binsbergen, W. (2001) 'Ubuntu and the globalization of Southern African thought and society', *Quest,* 15(1–2), pp. 53–89.

World Commission on Environment and Development [WCED] (1987) *Our common future*. Oxford University Press.

Zozu˘lakova, V. (2016) 'Degrowth - A way of social transformation', *ZeszytyNaukowe. OrganizacjaiZarz˛adzanie/Politechnika 'Sl˛aska*, 94, pp. 185–195.

15 The Non-Human in the Context of Colonisation, Modernity and Trauma

Austine E. Iyare

Introduction

Coloniality in Africa remains a central aspect of the continent's historical past and its postcolonial present. Before the onset of colonization, Africa had a long history of complex interactions with Europeans, shaped by regional, cultural, and temporal differences. These relationships encompassed trade, exploration, diplomacy, and cultural exchange. A defining feature of this interaction was the Atlantic slave trade, during which millions of Africans were forcibly transported to Europe and the Americas as part of an institutionalised system of human commodification lasting nearly 350 years. According to Tores Agena:

> Africa suffered a forcible possession of its people by Europe and America. That is the Atlantic slave trade. For nearly 350 years, slave trade was an institutionalized trade in human cargo. It was practiced to supply cheap labour to plantation owners, About 20 million able bodied Africans were taken across the Atlantic between the 15th and 19th centuries. While the trade contributed immensely to the growth and development of Europe, Africa saw its young productive able bodied population depleted.
>
> (2007, p. 33)

This practice, which profoundly disrupted African societies, declined during the industrialisation era, as mechanised farming and industrial production became more profitable than reliance on slave labour. The decline of the transatlantic slave trade, coupled with missionary activities and shifting European economic interests, paved the way for the Berlin Conference of 1884, where European powers partitioned Africa into colonies to secure raw materials for their burgeoning industries. Colonisation marked the beginning of a new phase of domination, involving the exploitation of African resources, labour, and markets for the economic benefit of European powers. Colonisation took various forms, including political domination through control of governance, economic exploitation of resources and labour, cultural imposition of Western languages and values, and settler colonization, which displaced indigenous populations. As Nkrumah (1957) observes, the primary goal of colonialism was economic advancement for the colonising powers, achieved at the expense of African nations. While some argue that colonialism introduced development and modernity, such as infrastructure, governance, and

DOI: 10.4324/9781003480914-16

technological progress (Parashar & Schulz, 2001, p. 864), this narrative often disguises the exploitative foundations of the colonial project. Biodun Jeyifo (2000, p. 45) highlights the intrinsic link between colonialism and modernity, noting that the colonial endeavour was central to the global economic and cultural systems that define modernity. These systems often marginalise African societies, representing them as inferior while elevating European achievements.

The colonial and modernity projects profoundly destabilised African social and economic structures, degrading both the environment and human dignity. The cultural narrative imposed by colonialism depicted African traditions and societies as inferior, reinscribing Africa into the unequal orbit of Western discourse (Nkomo, 2011, p. 374). This erasure of indigenous systems created profound material and psychological deficits, leading to widespread disenchantment, which Max Weber associates with the rationalisation of the world under modernity. In this chapter, we examine how the resulting trauma is beyond the usual human-centric understanding and is also significantly ecological in nature, and it continues to resonate in postcolonial African societies. Addressing this legacy requires a holistic approach to trauma by exploring the principles of deep ecology, which emphasise intrinsic value, diversity, and sustainable human–environment interactions. By adopting these values, it may be possible to mitigate the enduring trauma of colonisation and modernity for both the human and the non-human, fostering a more equitable and ecologically harmonious future for Africa.

Colonisation as Trauma-Producing Institution

Colonization, when viewed through a Western lens, is characterised by the acquisition and exertion of political dominance over another nation, primarily driven by the economic interests of the colonising power. This Western ideological framework was purportedly intended to facilitate the success of modernity in many colonised regions, particularly Nigeria. However, the colonial enterprise has engendered profound adverse effects on both human societies and the environment. Beyond asserting direct control over governmental and political structures within colonised territories, colonisation has led to the systematic exploitation of resources, labor, and markets. Furthermore, it imposed the culture, language, and values of the colonisers upon the colonized, often displacing indigenous populations and undermining their cultural heritage. These detrimental practices have fostered widespread disenchantment, contributing to significant human suffering and widespread trauma affecting both human and non-human entities across Africa.

In both colonial and post-colonial Africa, the consequences of colonialism have been profoundly negative. Contrary to claims that the colonial or modernity project would inspire societal development, the experience across much of Africa demonstrates otherwise. Rather than reaping the anticipated benefits of modernity – such as industrialisation, technological advancement, urbanization, social restructuring, and the rise of democracy – Africa's encounter with modernity has largely been a narrative of misfortune and exploitation. The fundamental objective of colonial empires was the enrichment of the metropole through the extraction of trade goods and natural resources, often pursued with little regard for, or outright neglect of, the local environment (Fairhead, 2000, p. 37).

The environmental degradation caused by colonial activities is particularly evident in countries like Nigeria and South Africa. In Nigeria, multinational corporations engaged in crude oil extraction adopted exploitative practices, while in South Africa, platinum mining operations exhibited similar irresponsibility. These activities caused substantial environmental challenges, demonstrating the damaging ecological impacts of colonial economic ventures. Even though ecological justice emphasises the right to equitable access to a secure and healthy environment conducive to the well-being of all individuals, irrespective of ethnicity, class, or religion (Malangwa & Agbiboa, 2013, p. 77), postcolonial Nigeria has continued to experience the adverse effects of multinational oil corporations. These corporations exploit weak legal frameworks and ineffective tax enforcement to maximise their profits at the expense of environmental and human health.

Rather than fulfilling the purported civilising mission of modernism – characterised by industrial growth, technological innovation, and social transformation aimed at fostering the general well-being of populations – colonialism became an institution that perpetuated trauma. Its destructive legacy includes air and water pollution, loss of biodiversity, degradation of farmland, destruction of wildlife habitats, and severe damage to aquatic ecosystems, all of which have resulted in profound health crises for affected communities (Agena, 2007, p. 38). The cumulative impact of these adverse effects underscores the enduring environmental and socio-economic harm caused by colonial practices in Africa. The Niger-Delta region of Nigeria is one of the best examples of this damage. According to Babatunde Abosede in "Environment Insecurity and Poverty in the Niger Delta", the quest for crude oil and natural resources in the Niger Delta in Nigeria by multinational companies continues to result in the destruction of,

> the land and renewable resources that constitute the mainstay of the local economy of the people of Niger Delta. This has negatively affected traditional economic activities and exacerbated the propensity for conflict and violence in the region resulting in the maiming and killing of innocent people, including women and children.
>
> (2017, p. 40)

The exploitative activities of multinational corporations have inflicted profound trauma and hardship on the lives of colonised peoples. Oil-induced pollution in affected regions has generated both direct and indirect consequences for the communities residing in oil-producing areas. Directly, oil spills have caused the loss of thousands of jobs and led to the extensive, often uncompensated, destruction of personal property. Indirectly, the spills have reduced protein intake due to the decline in fish availability, further compounding nutritional and economic hardships. In addition, exposure to polluted water for drinking, bathing, and cooking has resulted in a myriad of adverse health effects, exacerbating the suffering of these communities.

Life in the Niger Delta, particularly for communities such as Ilaje in Nigeria, has become largely unbearable due to massive oil-induced pollution. As Abosede observes, the depletion of critical livelihoods – most notably fishing – has rendered many residents jobless and destitute, significantly increasing poverty levels. The destruction of fishing waters has not only erased income sources but also deeply disrupted cultural and social identities tied to these traditional livelihoods. Moreover, oil spills and gas flaring have caused severe air and water pollution, damaging trees, crops, and other natural resources essential to survival. The obliteration of these foundational livelihoods has precipitated widespread unemployment, intensifying the poverty and despair experienced by affected populations. Beyond economic devastation, there is a growing epidemic of carcinogenic and respiratory diseases in these communities, linked to prolonged exposure to the toxic byproducts of gas flaring. The radioactive elements released into the air have resulted in symptoms of bronchial and respiratory diseases, with vulnerable populations, including children, disproportionately affected. The Niger Delta region now grapples with compounding layers of trauma: physical, economic, and psychological, as environmental degradation continues to erode the stability and well-being of its inhabitants.

A parallel crisis unfolds in South Africa, where coal and platinum mining activities have caused extensive environmental damage and severe health challenges. According to Andy Bike of the University of Pretoria, respiratory infections among children under five are notably higher during winter months, peaking in March and August (Munnik, 2010, p. 34). The environmental impacts of platinum mining are similarly dire: water pollution from chemical and heavy metal contamination seeps into groundwater and nearby water sources, while air pollution releases harmful particulates into the atmosphere, endangering the health of nearby residents. Additionally, the high thermal gradients resulting from mining disrupt geological stability and complicate mine ventilation, while land degradation from mining activities destroys vegetation and erodes soil integrity. These environmental catastrophes have destabilised ecosystems, exacerbated health disparities, and perpetuated cycles of intergenerational trauma within mining communities.

The cumulative effect of these environmental problems has caused severe health crises, destabilised local ecosystems, and created enduring social trauma. Colonization's legacy of exploitation, cultural erasure, and environmental destruction has inflicted significant psychological and emotional harm, fostering unresolved grief and intergenerational trauma among affected populations. This trauma manifests in systemic inequities such as economic disparities, cultural disintegration, identity fragmentation, and human rights violations. The history of environmental degradation and its resulting health and socio-economic impacts underscore how the colonial project has systematically inflicted suffering and trauma on human and non-human entities alike, leaving scars that persist across generations.

The Anthropocentrism of Trauma

One of the most important issues that deep ecology as a philosophy seeks to challenge is the dominance of anthropocentrism. Anthropocentrism, as it has been argued by some scholars, has over time led to a form of trauma called ecological trauma, with its resultant effect being some of the critical issues that face humankind till date. In this section, I attempt to show the intersection, if any, between anthropocentrism and trauma, that is, the extent to which trauma in African places has over time showcased the dominance of humans above other non-humans. Perhaps it is crucial to explore the meaning of the term anthropocentrism. The term anthropocentrism is prevalent in both ethical studies and environmental philosophy. Within both contexts, the term refers to the position which states that humans have a higher worth or status when put against the non-human. On this note, anthropocentrism amounts to "human chauvinism" (Routler, 1980) or "speciecism" (Singer, 1975). With humans considered the centre of an anthropocentric worldview, it simply implies that it has an ontological basis, especially when pitched against non-humans. Immanuel Kant describes this in his *Critique of the Power of Judgement* as:

> As the sole being on earth who has reason, and thus a capacity to set voluntary ends for himself, [man] is certainly the titular lord of nature, and, if nature is regarded as a teleological system, then it is his vocation to be the ultimate end of nature.
>
> (2000, p. 298)

Evidently, Kant's excerpt captures the anthropocentric worldview that we are emphasising here. The implication of the anthropocentric worldview is that its adherents see nothing wrong in the exploitative activities of man in the natural environment. J.B. Callicott resonates this position as he argues that "only human beings are able to give values to the eco-system and this means that all intrinsic value is grounded in human beings and projected onto the natural object that excites the value" (1980, p. 325). Corroborating Callicott, Devall and Sessions state that: "for thousands of years Western culture has been too intoxicated with the idea of dominance: with dominance of humans over nature, masculine over feminine, wealthy and powerful over the poor, with the dominance of the West over non-Western cultures" (1980, p. 358).

One of the rationales behind the anthropocentric worldview is the argument that humans are the only living beings on earth that possess rationality. Consequently, other beings in the natural environment are seen as subservient to humans, and humans are therefore justified to exploit them. The anthropocentric paradigm has, in recent times, seen an increase in the use of scientific and technological advancements for the sole purpose of exploiting the natural environment. Bill Devall argues in line with this thought as he states that:

> Science is wedded to technology, the development of *techniques* for control of natural processes (such as weather modification). Change ("planned

> obsolescence") is an end in itself. The new is valued over the old and the present over future generations. The goal of persons is personal satisfaction of wants and a higher standard of living as measured by possession of commodities (houses, autos, recreation vehicles, etc.
>
> (Devall, 1980, p. 301)

For some scholars, the issue of anthropocentrism today cannot be discussed in isolation from the Judeo-Christian values as expressed in the Bible. These scholars argue that the Bible contains some verses which, over the years, have served as injunctions in the regulation of the human-non-human relationship, and these injunctions are being used to justify the dominance of humans over other non-human species in the natural environment. For instance, in the book of Genesis, it is written that:

> Then God said, "Let us make man in our image, after our likeness. And let them have dominion over the fish of the sea and over the birds of the heavens and over the livestock and over all the earth and over every creeping thing that creeps on the earth" (Gen. 1:26). So God created man in his own image, in the image of God he created him; male and female he created them. (Gen. 1:27). And God blessed them. And God said to them, "Be fruitful and multiply and fill the earth and subdue it, and have dominion over the fish of the sea and over the birds of the heavens and over every living thing that moves on the earth" (Gen. 1:28). And God said, "Behold, I have given you every plant yielding seed that is on the face of all the earth, and every tree with seed in its fruit. You shall have them for food."
>
> (Gen. 1:29)

The above Bible passages, no doubt, have been interpreted within the context of fundamentalism in the wake of man's relationship with the natural environment. The implication of this is that non-human species are seen as not having a moral status. Lynn White, Jr., in "The Historical Root of Our Ecological Crisis" (1967), argued that the ecological crises being witnessed today are epiphenomenal of the anthropocentric paradigm as outlined by the Judeo-Christian tradition. Thus, the Judeo-Christian has given rise to a man versus nature "war".

The intersection between anthropocentrism as a dominant theory in environmental philosophy, and trauma raises compelling questions, particularly regarding the extent to which ecological harm can be conceptualised through the lens of trauma. While the term trauma is often contested and can evoke varied interpretations, its application is predominantly anthropocentric, focusing on human psychological or physical suffering. However, in the context of this analysis, the focus is on eco-based trauma, a form of harm inflicted upon the environment and non-human entities due to anthropocentric dominance. Anthropocentrism prioritises human well-being at the expense of the non-human and the ecosystem, perpetuating ecological crises such as climate change, deforestation, oil spills, mangrove destruction, and biodiversity loss. These crises contribute to what can be termed ecological trauma – a disruption and degradation of ecosystems with far-reaching consequences. Despite the seemingly physical nature of such harm, it is inseparable from psychological dimensions. Trauma theorists, such as Cathy Caruth, define modern trauma not as the event itself but as the aftermath and lingering effects of the incident. Applying this perspective to the environment reveals that the trauma of ecological harm is manifested in its consequences, such as polluted water, dead aquatic life, and degraded ecosystems, rather than solely in the original act of destruction.

For instance, agricultural runoff polluting a creek results in eutrophic waters depleted of oxygen and lifeless fish, symbolizing the ongoing trauma experienced by the ecosystem. While the environment lacks repressed memory as humans do, the signs of environmental trauma are physically evident and continuously interpretable. Unlike human trauma, which often necessitates accessing suppressed memories to identify its cause, ecological trauma presents a visible narrative of harm and its consequences. Yet, the anthropocentric framework often marginalizes such narratives, perpetuating the perception that trauma is an exclusively human phenomenon.

The entanglement between ecological and human trauma is undeniable. Human exploitation of the environment leads to a cyclical relationship of harm and counter-harm, as ecological degradation rebounds to affect human communities through phenomena such as climate anxiety, displacement, and loss of cultural and economic stability. Trauma theorists have underscored that trauma cannot be entirely understood, even by those who experience it firsthand, making it even more challenging to comprehend the trauma inflicted on the non-human natural world. Nevertheless, the enduring effects of environmental harm – on both humans and non-humans – highlight the interconnectedness of these traumas.

Psychological repercussions also arise from anthropocentrism, further reinforcing the trauma nexus between humans and the environment. Scholars have linked anthropocentric attitudes to fear, denial, and apathy toward the environment. For instance, those less inclined toward environmental consciousness exhibit indifference and destructive behaviors, while fear of nature or its uncontrollable elements fosters anxiety and disengagement. This fear not only disenchants humans from the vitality of the natural world but also perpetuates harmful cycles of neglect and destruction. Anthropocentric denial, as psychoanalysis suggests, acts as a defence mechanism, allowing individuals and societies to avoid confronting the guilt or anxiety associated with environmental degradation. This denial is particularly evident in the refusal to acknowledge catastrophic phenomena like climate change, leading to inaction and exacerbating ecological trauma. The consequences of this denial are catastrophic, both ecologically and socially. Ignoring environmental crises accelerates ecological collapse, intensifies conflicts, and threatens the survival of human and non-human entities alike. As sociologist (Zerubavel2006) argues, silence and denial about environmental justice issues reflect a broader anthropocentric hubris that underpins industrial civilization. This "denial of nature" has become a pervasive disease of modernity, shutting off human imagination, ethics, and creativity in addressing ecological challenges. We can therefore say that there is a clear intersection between anthropocentrism and trauma, manifesting in the ecological crises that affect both human and non-human entities. Addressing this intersection requires transcending anthropocentric worldviews and acknowledging the trauma inflicted upon ecosystems. Doing so necessitates a reimagining of human-nature relationships to promote ecological balance and mitigate the profound ecological trauma that anthropocentrism continues to perpetuate.

In sum, trauma is predominantly understood and discussed through a human-centric lens, often emphasising how individuals, communities, and societies endure and respond to harm. This perspective reflects our anthropocentric worldview, where humans are positioned as the primary agents of value and meaning. Consequently, discussions about trauma often focus on human experiences – such as war, abuse, and natural disasters – while neglecting the profound impacts these events can have on non-human animals and ecosystems. Animals, for instance, can experience psychological and physiological trauma from abuse, captivity, displacement, or witnessing violence. These manifestations of trauma are evident in behaviours like hypervigilance, social withdrawal, or disrupted reproductive patterns. Despite this, animal trauma is rarely treated with the same depth of empathy or scientific rigour as human trauma, highlighting a disparity in our recognition of suffering across species. Ecosystems, too, can be conceptualised as experiencing trauma when subjected to catastrophic disturbances such as deforestation, oil spills, or climate-induced disasters, as has been experienced in Africa in both the colonial and postcolonial periods. These events disrupt the balance and resilience of ecosystems, leading to loss of biodiversity, soil degradation, and altered climate patterns, which can persist for decades or centuries. Such ecological trauma not only affects non-human life but also feeds back into human suffering, as we are deeply dependent on the health of the natural world. However, the language of trauma is seldom applied to these contexts, reinforcing a separation between human experiences of harm and the interconnected suffering of the broader web of life. Expanding the framework of trauma to include non-human entities, as we attempt in the forthcoming section with reference to Africa, would foster a more holistic understanding of harm and recovery, emphasising our shared vulnerabilities and responsibilities within a fragile, interconnected biosphere.

Biospheric (Non-human) Trauma in Africa

Biospheric trauma, with particular reference to the non-human in this chapter, refers to the long-term ecological and environmental distress experienced by ecosystems as a result of human activities that disturb the natural balance. It encompasses a wide range of destructive processes, including deforestation, soil degradation, pollution, loss of biodiversity, and disruptions to climate systems, all of which contribute to the deterioration of the Earth's biosphere. In the context of Africa, biospheric trauma is a product of both historical and contemporary forces, with colonialism serving as a pivotal moment in the destabilisation of Africa's ecosystems. During the colonial era, African land and resources were systematically extracted and exploited to benefit European colonial powers. This led to the destabilisation of indigenous ecosystems, as agricultural practices and resource extraction methods were introduced that were not aligned with local environmental knowledge or sustainability. The widespread changes made to Africa's natural landscapes, including large-scale monoculture farming, deforestation, and the expansion of mining operations, disrupted ecosystems and caused imbalances that continue to affect the continent's environmental health.

The ecological consequences of colonialism are still visible today in many African countries. A key example of biospheric trauma during this period is the extensive deforestation caused by the colonial demand for timber and the need for land to grow cash crops such as cotton, coffee, and cocoa. In countries like Ghana, Uganda, and Kenya, vast areas of forests were cleared to create plantations, dramatically altering ecosystems and contributing to soil erosion and habitat loss. The legacy of these practices is still evident, as these countries struggle with diminishing forest cover, reduced agricultural productivity, and increased vulnerability to floods and droughts. The loss of biodiversity that accompanied colonial land use changes is particularly notable in Central Africa, where extensive logging and mining have reduced the populations of various species and led to the degradation of once-thriving rainforests. The over-exploitation of land under colonial rule not only caused environmental harm but also severed indigenous communities from their ancestral relationship with the land, further compounding the trauma inflicted on Africa's ecosystems.

Another example of colonial exploitation contributing to biospheric trauma can be seen in the environmental devastation caused by the extraction of minerals and fossil fuels. In countries like South Africa, Zambia, and the Democratic Republic of Congo, colonial powers set up mining operations that extracted gold, copper, and diamonds with little regard for the environmental costs. The legacy of these extractive industries has had lasting effects on the landscape, leaving behind polluted waterways, toxic waste dumps, and abandoned mines that continue to pose environmental risks today. The environmental degradation caused by mining activities has resulted in the contamination of rivers, soil, and groundwater, and has disrupted local ecosystems, leading to the loss of agricultural land and the collapse of traditional livelihoods. The mining industry's impact on Africa's biosphere, initiated under colonialism, continues to affect millions of people who live near these sites, exacerbating poverty and increasing health risks. We have also witnessed the exploitation of natural resources and indigenous culture, and intergenerational trauma caused by historic events, such as forced relocation, genocide, and slavery. From the above human activities, it is possible to imagine the wounding, harming, and violently hurting of the non-human natural world that has occurred, affecting different cultures and their environment. This means the environment has also experienced trauma. According to Cathy (1996) Caruth, the consequences of the events and not the event itself constitute modern trauma. Similarly, applying this reasoning to the ecosphere, we might observe the effects of environmental damage rather than the damaging event. For example, the trauma of polluted waterways consists of dead fish and eutrophic water rather than in the source of the pollution, perhaps runoff from a farm (1996)

Even after the formal end of colonialism, neo-colonial forces have continued to exploit Africa's natural resources, exacerbating biospheric trauma across the continent. Multinational corporations, often with the backing of foreign governments, continue to extract oil, gas, and minerals at a scale that exceeds the environmental carrying capacity of many African ecosystems. In countries like Nigeria, Angola, and Equatorial Guinea, the oil industry has been responsible for widespread environmental damage, including oil spills, gas flaring, and the destruction of wetlands. These activities not only disrupt local ecosystems but also contribute to climate change, further intensifying the ecological crisis. In Nigeria, for example, the Niger Delta has been severely impacted by oil extraction, leading to the loss of vital mangrove forests and contamination of the region's fishing waters, which millions of people rely on for their livelihoods. The environmental degradation caused by these extractive industries continues to exacerbate social and economic inequalities in these regions, leading to displacement, loss of food security, and health crises.

The agricultural sector, which was heavily influenced by colonial powers, remains another key source of biospheric trauma in contemporary Africa. The imposition of monoculture farming, particularly in countries like Kenya, Côte d'Ivoire, and Malawi, has led to the overuse of chemical fertilisers and pesticides, which have degraded soil health and reduced agricultural biodiversity. The shift from traditional farming practices, which were often more attuned to the local ecology, to large-scale industrial agriculture has caused a loss of soil fertility and an increase in water scarcity. In Malawi, for example, the continued cultivation of tobacco, a cash crop introduced under colonial rule, has led to soil exhaustion and water depletion, which in turn threatens food security for local populations. The emphasis on export-oriented agriculture, often at the expense of local needs, continues to exacerbate environmental degradation and deepen the vulnerabilities of African ecosystems to the changing climate.

The biospheric trauma Africa faces today is also linked to the global economic and political system that perpetuates neo-imperialist practices. The ongoing influence of international financial institutions, multinational corporations, and global trade agreements continues to shape Africa's economic policies in ways that prioritise resource extraction over environmental protection. Neoliberal economic policies, which emerged after the end of colonial rule, have encouraged the privatisation of land and the promotion of industries that prioritise short-term profits over long-term ecological sustainability. The expansion of agribusinesses and large-scale mining operations, often funded by foreign investors, has led to continued deforestation, water scarcity, and soil degradation. This exploitation, rooted in both historical and contemporary forces, continues to disrupt African ecosystems, pushing many species toward extinction, and leaving large swaths of the continent's land barren and unproductive. The compounding effects of colonial and neo-colonial exploitation illustrate how biospheric trauma in Africa is not just a historical artefact, but an ongoing crisis that requires urgent attention and systemic change.

Concluding Thought: Towards a Holistic Approach to Trauma Studies

African and African diasporic peoples have experienced trauma from colonisation and the accompanying modernity. Moreover, African ecosystems have experienced trauma, harm, and distress that is often not centred in the trauma discourse. The economic interest of Europe and the rest of the West generally have brought hardship to African human and non-human entities, especially in the colonies, as a result of their wanton exploitation of resources and the environment, with complete disregard for the value of life, both the human and non-human in the environment, and this continues today, with Global North actors taking advantage of the weak policies, governance, and institutions in postcolonial Africa to continue exploitation, as we find in the Niger Delta, Nigeria. Polycarp Ikuenobe provides a vivid description of this point by asserting that:

> Many of the environmental problems in Africa are the results of exploitative colonial activities and their legacies. For instance, the exploitative and destructive economic activities of multinational corporations (which are a legacy of colonialism), aided by corrupt and ineffective governments of various African states, have contributed to most of the environmental problems in Africa.
>
> (2014, p. 15)

From the examples of the activities of the multinational corporation cited in both Nigeria, South Africa, and other countries, the suffering and trauma experienced by the individual, the community, the society, and the ecosystem generally is palpable. For example, the negative effect of oil exploration activities on wildlife also precipitated the forced migration of a wide range of apes, especially monkeys that were highly visible within Niger Delta Coastal towns before the advent of oil exploration (Jike, 2004, p. 674) creating an ecological imbalance. For some communities, ancestral homelands were desecrated and converted, as was the case of the Ogoni/Shell crises in the late 1990s, thus severing the link between the living and the pantheon of forebears (Jike, 2004) In the destruction of wildlife, some animals and trees considered as totems were also destroyed. Apart from productive lands lost to these oil productions, there were also more fundamental environmental problems of permanent damage to flora and fauna. The foregoing thus underscores the need for a holistic approach to trauma studies rather than an essentially anthropocentric focus.

A holistic approach to trauma studies provides a profound intersection with the principles of deep ecology, urging a reimagining of trauma that extends beyond human suffering to include the non-human world. Traditional trauma studies have predominantly focused on human experiences of violence, displacement, and suffering, yet such a narrow perspective fails to capture the full scope of the ecological disruptions caused by historical and contemporary exploitation. A more comprehensive trauma framework, rooted in deep ecology, acknowledges the intricate and interdependent relationships between humans, non-human beings, and the ecosystems they inhabit. This recognition of mutual reliance within the biosphere – where the health of humans is inextricably linked to the well-being of the natural world – underscores the necessity of addressing trauma not only as a human phenomenon but as a systemic issue affecting the entire ecological network. In the context of Africa, the traumas wrought by colonisation and modernity have been deeply ecological, manifesting in deforestation, soil degradation, and the displacement of both human and non-human species. By broadening the scope of trauma studies to encompass the suffering of the non-human, we can begin to understand the full extent of environmental damage, creating pathways for more holistic responses to ecological crises. Deep ecology offers a philosophical framework that can mitigate the biospheric trauma caused by colonisation and modernity. Central to deep ecology is the assertion that all entities within the biosphere – human and non-human alike – are interconnected and possess intrinsic value. This principle challenges the anthropocentric worldview that has driven much of the exploitation of natural resources, including in Africa, where colonisation led to widespread ecological degradation. By recognising the intrinsic worth of all beings, deep ecology emphasises the need for humans to adopt a more ethical relationship with the environment, one that transcends utilitarian views of nature as a mere resource for human consumption. The philosophy's call for a rethinking of human interference with the natural world aligns with a holistic understanding of trauma, where the suffering of non-human entities is not secondary to human trauma but is instead an integral part of the larger ecological crisis. Deep ecology advocates for a shift in both individual and collective consciousness, promoting policies and actions that honour the diversity of life and prioritise the flourishing of all species, human and non-human. The application of deep ecology's principles, such as intrinsic value, diversity, and human interference, can provide vital solutions to the ecological traumas that persist in post-colonial Africa. For instance, the principle of intrinsic value urges a re-evaluation of environmental policies that prioritise economic growth over ecological health, as seen in the destructive mining and agricultural practices that continue to harm African ecosystems. By recognising the inherent worth of forests, rivers, and wildlife, and by considering the needs of all species, deep ecology offers an ethical foundation for policy changes that seek to restore balance to ecosystems. Similarly, the principle of diversity emphasises the importance of preserving biodiversity, which is often threatened by the mono-cultural agricultural systems imposed during colonial rule and exacerbated by modern industrial practices. Through the application of these principles, deep ecology can promote a paradigm shift that not only addresses the trauma inflicted on human communities but also heals the scars borne by the land, water, and non-human life forms. This integrative approach offers a pathway to both ecological restoration and social justice, aiming for a world where all beings can flourish in a balanced, interconnected ecosystem.

References

Agena, T. 2007. The relationship between economic activities and environmental degradation in Africa. *Journal of the Historical Society of Nigeria*, 17, 28–40.

Babatunde, A.O. 2017. Environmental insecurity and poverty in the Niger Delta: a case study of Ilaje. *African Conflict and Peace Building Review*, 7(2), 35–59.

Callicot, J.B. 1980, Winter. Animal liberation: A triangular affair.' *Environmental Ethics*, 2(4), 311–338. E-Book.

Caruth, C. 1996. *Unclaimed Experience: Trauma, Narrative, and History*. Baltimore: Johns Hopkins University Press.

Devall, Bill. 1980. The Deep Ecology Movement. *Natural Resources Journal*, 20, 300–322.

Fairhead, J., & Leach, M. 2000. Desiccation and domination: Science and struggle over environment and development in colonial Guinea. *Journal of African History*, 41(1), 35–54.

Ikuenobe, P. 2014. Traditional African environmental ethics and colonial legacy. *International Journal of Philosophy and Theology*, 2(4), 1–21.

Jeyibo, B. In the Wake of Colonialism and Modernity. *Anglophonia/Calliban* 2000, 7, 71-84.

Jike, V.T. 2004. Environmental degradation, social disequilibrium, and the dilemma of sustainable development in the Niger Delta of Nigeria. *Journal of Black Studies*, 34(5), 686–701.

Kant, I. 2000. *Critique of the Power of Judgement*. New York: Cambridge University Press.

Malangwa, B. & Agbiboa, D. 2013. Oil multinational corporations, environmental irresponsibility and turbulence in the Niger Delta. *African Spectrum*, 48(2), 71–83.

Munnik, V. 2010. *The social and environmental consequences of coal mining in South Africa: a case study*. A Joint initiative of Environmental Monitoring Group, Cape Town, South Africa and Both ENDS, Amsterdam, The Netherlands.

Nkomo, S.N. 2011. A postcolonial and anti-colonial reading of African leadership and management in organization studies: Tensions, contradictions and possibilities. *Organization*, 18(3), 365–386.

Nkrumah, K. 1957. *Towards Colonial Freedom*. London: Heinemann.

Parashar, S., & Schulz, M. 2021. Colonial legacies, postcolonial selfhood and the un(doing) of Africa. *Third World Quarterly*, 42(5), 867–881.

Routler, R. 1980. Human chauvinism and environmental ethics. In D.S. Mannison, M.A. McRobbie, & R. Sylvan (Eds.) *Environmental Philosophy*. Dept. of Philosophy, School of Social Sciences, Australian National University.

Singer, P. 1975. *Animal Liberation: A New Ethics for our Treatment of Animals*. New York:Harper Collins.

The Book of Genesis, from Chapter 1;26 – 1;29, in The New International Version. Colorado Springs

White, Lynn Jr. 1967. . The historic roots of our ecologic crisis. In D. Van De Veer & C. Pierce (Eds.), *The Environmental Ethics and Policy Book: Philosophy, Ecology, Economics* (3rd ed.). Belmont, CA: Wadsworth Publishing.

Zerubavel, E. 2006. *The Elephant in the Room: Silence and Denial in Everyday Life*. London, UK: Oxford University Press.

16 General Conclusion

Toward an Inclusive Environmental Epistemology

Kenneth Uyi Abudu, Kevin Behrens, and Elvis Imafidon

The essays in this volume have affirmed that globally, the struggle to improve the fractured human–non-human relationship, due to the promotion of an anthropocentric worldview over an ecocentric one continues, and there is much to learn from the diverse traditions of the world, and this volume has zoomed in on African traditions. Governments around the world continue to map out ways to ensure that environmental crises are properly addressed and managed through policies and legislation. For example, the "Climate Change Act of 2021", an all-encompassing legislation that most countries all over the world adopted to address the issue of climate change, provides insight into one of the many efforts by governments worldwide in the quest to ensure that the ecosystem is well managed to avert further degeneration. Also, academic disciplines, including the humanities, debate and theorise models and frameworks that would improve our relationship with our ecosystems and result in the actualisation of deep ecological commitments. Thus, as different stakeholders such as academics, activists, and governmental and non-governmental actors pursue ways to deal with this global challenge with a global impact, it is fundamental that this is done within the framework of an inclusive episteme, and it is this idea of an inclusive environmental epistemology that has led to the emergence of this volume. This volume emerged from our understanding that a global challenge of this kind and magnitude would require global and diverse voices on a hermeneutics of the environment that acknowledges and draws solutions from diverse and inclusive perspectives. Such perspectives would be inaccessible if there are no deliberate efforts in academia to explore non-Western environmental philosophies.

An inclusive epistemology, on the one hand, is an approach to knowledge production and validation that seeks to include diverse perspectives, experiences, and methods of understanding, rather than privileging dominant or traditional frameworks. It challenges exclusionary practices in knowledge production by recognising the value of marginalised viewpoints, cultural contexts, and interdisciplinary approaches. This framework promotes equity in who is regarded as a knower and how knowledge is validated and guarantees that all stakeholders have the chance to contribute their understanding, irrespective of their background or identity. It is about creating an atmosphere where everyone's knowledge is accepted as part of the repository that deserves to be attended to in terms of validation, justification, or otherwise, thereby fostering a more thorough understanding of a subject. Put

DOI: 10.4324/9781003480914-17

differently, incorporating the varied viewpoints of peers is the basis of inclusive epistemology. Compared to integration, which demands that everyone conform to the prevailing worldview, inclusive epistemology disrupts the centre-margin narrative and promotes a level playing field for knowledge production. This implies that previously marginalised epistemologies do not give up their viewpoints in an inclusive encounter. Inclusive epistemology, then, is an evolving method that enables the collective knowledge to surpass the mere sum of the individual positions rather than only being an additive process. A crucial element in optimising the advantages of cognitive diversity is epistemic inclusivity. An inclusive setting evaluates a variety of scientific theories, enabling the development of other theories based on them. Here, cognitive diversity is important. Cognitive diversity represents the recognition of diverse voices in the process of knowledge production. Put differently, it seeks to include diverse views in the attainment of knowledge. Cognitive diversity is, therefore, central to inclusive epistemology as one must recognise the importance of "difference" in order to appreciate diversity and inclusion. As David Asai puts it, "diversity without inclusion is an empty gesture" (2019: 537). The importance of cognitive diversity for inclusive epistemology is emphasised by Vlasta Sikimic, who asserts that cognitive diversity is capable of crowdsourcing information or bringing up unconventional ideas in knowledge discourse. This implies that cognitive diversity has instrumental value from which inclusive epistemology benefits. As such, cognitive diversity is intertwined with social diversity since, for example, people with different backgrounds are likelier to have different experiences and ideas (2023, p. 756).

Environmental epistemology, on the other hand, evaluates the epistemic beliefs that we hold about our immediate environment. A relatively new field of research, environmental epistemology examines how humans learn about environmental concerns, how knowledge about the environment is produced and sustained, and how our experiences, beliefs, and values influence how we interact with our immediate ecosystem. Environmental epistemology looks at how we learn and produce and validate knowledge about ecosystems, climate change, sustainability, and the environment in general, as well as the ethical, social, and cultural aspects that affect that learning. Some of the questions that environmental epistemology seeks to ask are: How can we determine what is actually destructive or ecologically sustainable? What are the boundaries of human comprehension of intricate ecological systems? Environmental epistemology thus examines the relationship between human cognition and ecological systems, exploring how cultural, scientific, and indigenous knowledge systems shape our understanding of environmental issues and inform decision-making. The concept of ecological knowledge and the ways in which various cultures and societies comprehend their surroundings are important facets of environmental epistemology. How indigenous knowledge, or traditional ecological knowledge (TEK), may enhance scientific understanding of ecosystems is a crucial subject here. For instance, the complex interactions among plants, animals, and the environment have long been known by many indigenous communities and have been transmitted from one generation to the next. However, the narrative of environmental epistemology is dominated globally by the scientific perspective deeply rooted in the

Western Enlightenment tradition rather than drawn from diverse traditions around the world. In contrast to the more reductionist ways typical of Western scientific methodologies, environmental epistemology ought to be more holistic and based on global and diverse lived experience. Western science frequently prioritises observation, experimentation, and empirical data in order to generate knowledge. Critics contend that this approach occasionally ignores or undervalues the subjective, local, and experienced components of environmental knowledge, despite the fact that it has been quite effective in comprehending many facets of the natural world.

Bearing these in mind and aiming to promote the epistemic commitment to learning about the environment from diverse perspectives, we advocate here for an inclusive environmental epistemology as the foundation for developing a practical knowledge base to actualise deep ecological commitments and improve our relationship with our ecosystems. But what makes an environmental epistemology inclusive? Epistemic inclusion and/or inclusive epistemology, as we noted earlier, recognise diverse and marginalised traditions in the production and dissemination of knowledge. An epistemology is inclusive if and only if it encourages traditions that are often marginalised to be part of the global conversation. It promotes cognitive diversity, epistemic justice, and an increase in group knowledge. Extending this to the environment, then, it implies the recognition and assimilation of indigenous knowledge about the environment, ecosystems, and sustainable eco-relationships in the quest to address environmental challenges. It is instructive to note here that knowledge systems differ across the globe, and as such, the environment and how we understand it also differ and it becomes enriched when we collaborate and bring our diverse perspectives together. The implication of this is that the different physical environments can be a rich source of cognitive diversity. Christopher J. Preston captures this thus:

> Different physical environments, both natural and artificial, supply radically different experiences for those who move about them. Those differences are likely to have cognitive significance … Life in a village on the steep slopes of a Himalayan mountain affords different physical challenges from life in Manhattan, and both these environments in turn provide radically different experiences from those afforded to fishermen off the Aleutian Islands of southwestern Alaska.
>
> (2003, p. 71)

Preston further adds that:

> They supply some of the greatest ranges of textures, shapes, sights, and sounds of any kind of environment. Natural environments are inherently unpredictable, autonomous, and structurally dynamic … Rainforests and ice caps (for example) should both be highly valued for the resources they offer the mind. The types of diversity that are characteristic of environments under less control will always make them into particularly valuable sources of cognitive value. Their increasing rarity only elevates that value.
>
> (2005, pp. 133–134)

It is clear that a variety of natural settings may provide us with a range of experiences, obstacles to overcome, and consequently, a range of cognitive resources to draw from. Therefore, an inclusive environmental epistemology means the reliance on diverse cognitive communities and diverse cognitive agents, both individual and collective, in drawing and forming norms, knowledge, and understanding about the environment, and in evaluating already existing knowledge about the environment in relation to inclusion. It portrays environmental knowledge that integrates diverse perspectives, including scientific, indigenous, local, and experiential ways of knowing. It emphasises the importance of valuing and incorporating marginalised voices and cultural contexts to create a more holistic and equitable understanding of environmental issues and solutions.

It is our conviction that for the aims of deep ecology to be achieved, we must explore diverse systems of indigenous knowledge and traditions to make the epistemology of the environment inclusive. The intersection between deep ecology and an inclusive environmental epistemology is crucial to diagnose, understand, and address environmental problems across the world. Thus, incorporating an inclusive environmental epistemology in deep ecology discourse would ensure that all voices – Western, African, Eastern, Middle Eastern, and the like – are heard and taken seriously. What constitutes the environment, the knowledge of the environment, and processes of validation and justification in these aforementioned places share dimensions of similarities and dissimilarities, and as such, adopting only the conventional Western or global north understanding of the environment in addressing specific environmental challenges in different locales around the world, such as in an indigenous African community, may prove futile. While environmental problems are broadly similar all over the world, the solution to specific experiences of such problems may differ and could be gleaned from paying attention to diverse knowledge and attitudes toward the environment and ecosystems in different places. The essays in this volume draw specifically from African epistemic communities on how best to improve human relationships with the non-human. It thus explores an often marginalised perspective in the discourse of the environment, but it is as well limited as there is so much more to explore across the globe for a truly inclusive epistemology of the environment to emerge.

Consequently, for a just, inclusive, and effective deep ecological commitment to emerge, we will need to pay attention to diverse epistemologies in and beyond the African experience. This would yield a decolonial discourse that stays true to the very spirit of deep ecology – the promotion of intrinsic value and worth of all living beings, and by extension, ecosystems, communities, and traditions.

References

Asai, D. 2019. To Learn Inclusion Skills, Make It Personal. *Nature* 565 (7737): 537–538. https://doi.org/10/1038/d41586-019-00282-y

Preston, C. J. 2005. *Grounding Knowledge: Environmental Philosophy, Epistemology, and Place*. Athens, GA: University of Georgia Press.

Sikimic, V. 2023. Epistemic Inclusion as the Key to Benefiting from Cognitive Diversity in Science. *Social Epistemology*,37, 753–765.

Index